BE THE SAND.

BETHESAND.

TobakkoNacht
The Antismoking Endgame

Michael J. McFadden

AEthna Press

Pennsylvania, USA

TobakkoNacht – The Antismoking Endgame

Copyright 2013 by Michael J. McFadden

All Rights Reserved. No part of the material protected by this copyright notice may be reproduced or utilized in any form or by any means, electronic or mechanical, including photocopying, recording, or by any information storage and retrieval system, without written permission from the copyright owner.

The author is neither a medical doctor nor a lawyer. Any opinions contained herein should be taken and accepted on that basis. Any erroneous material is inadvertent and will be corrected in subsequent editions upon notification.

AEthna Press

ISBN 978-0-9744979-1-4

TobakkoNacht – The Antismoking Endgame is dedicated to all those who understand and value the true meaning of freedom, not just for themselves, their friends, their families, or for any lone individual or group at all, but for everyone. We're all in this game together.

Acknowledgements

I can only specifically acknowledge a very few of the many who helped me make *TobakkoNacht* a reality in this space, but my sincere thanks go out to all. *TobakkoNacht* is your accomplishment as well as mine.

I'll start of course with my family: Joseph, Eileen, Karen, and Maureen. Without their love and support, this book wouldn't exist. Next are those who helped me with their gifts and hospitality in helping me spread *Dissecting Antismokers' Brains* throughout the US, the wider world. Larry and Ann Hershey, Harry O'Brien, and Robert Gehrmann helped me when I most needed it, and Audrey, Carol, Sally, and others have spontaneously sent me unexpected gifts that meant more than they realized. Special thanks for the warmth and beds offered in my travels by Belinda, Ellie and her folks, John and Laura, Lol and Bob, Peter, Robert, Samantha, Stephanie, Wiel and Renee.

Then there are those who loaned Word and editing skills: Allen, Bernd, Catherine, Chris, Dave, Debbie, Iro, Jerry, Kevin, Rich, Linda, Tom, and my two mainstay editors, Belinda Cunnison and Susan Uttendorfsky. Beyond simple words, I need to recognize the talented artists of World War Two whose images I have freely adapted from the public domain, as well as the creative skills of Bill Brown, Sam Ryskind, and Susan Wenger. Just as surely, I'm grateful to those who have simply offered friendship, support, and inspiration in my personal life: Bob, Cynthia, Irma, Jacqueline, Jack, Jeanine, Joe, Kenny, Loic, Lynette, Mara, Mike, Mourad, Owen, Phil, and Sister Michael.

I can't overlook my dens of creativity, Philly's wonderful Free Choice pubs: The Pen & Pencil, The Locust, Brownie's, Paddy's, and a few quiet defiers; all happily tolerating the hippie who'd sit at the end of the bar scribbling notes, smoking, and mumbling quietly to himself.

And, of course, there's the practical, intellectual and emotional support I've received through so many emails, blogs and books. I hope someday I can repay all the goodness that's come my way.

Finally, I'd like to remember some our past fighters and their supporters. I believe they continue to get joy out of what we do in their memories: Darlene Brennan, Frank Zaniol, Frazier Webb, Garnet Dawn, Gian Turci, Gordon Finlay, Hans Tegankampf, Laura Gray, Lauren Colby, Lillian Laprade, Linda Hubbard, Marty Ronhovdee, Mary Herron, Otto Muensch, Robert Sanden, Sally Hull, Sam Nettles, Sara Mahler, Sharyn Kuneman, Stan Forrester, Steve Handman, Virgil Kleinhelter, and Warren Klass. To all of you, sincere thanks from those who fight on.

Table of Contents

As nightfall does not come at once, neither does oppression. In both instances there is a twilight where everything remains seemingly unchanged. And it is in such twilight that we all must be aware of change in the air – however slight – lest we become unwitting victims of darkness.

– Supreme Court Justice William Douglas

Author's Preface

In the nasty world of the Internet, bulletin board debaters hiding behind anonymous handles are quick to use extreme imagery and words that they'd be far more hesitant to employ in face-to-face confrontations. In the early days of Internet battles, it was quite common for people disagreeing with those in authority to start jumping up and down while pointing and screaming "Hitler! Nazi!" at the top of their lungs. An Internet pioneer, Mike Godwin, observed that, "As a Usenet discussion grows longer, the probability of a comparison involving Nazis or Hitler approaches One."[1]

In recent years, users of the Internet who propose that a certain person, group, action, or style of action bears some resemblance to what happened in the years of the Third Reich will often find themselves accused of breaking "Godwin's Law," and are usually told that they have left the realm of rational discussion and have *ipso facto* lost the argument being engaged in.

In reality, Mike Godwin never meant for his proposition to be used that way, but such a reaction is, more often than not, correct. The Nazis and the Holocaust do not stand alone in the horrors of human history and misery, but the use of their imagery in flippant comparisons of everything from soup choice to stop-light traffic cameras with what was visited upon Germany's Jews, Gypsies, and other such "undesirables" should not be treated lightly. Very few people alive today have had any life experiences that even approach the horror of Hitler's regime as it touched those it persecuted.

Despite that, I chose to title this book with a variation of one of the hallmark events that ushered in the open persecution of German Jewry: *Kristallnacht*, The Night of Broken Glass. I chose that title not to imply that smokers are in imminent danger of being hauled off to camps to be starved, tortured, and incinerated, but to emphasize that what is being done to smokers today is truly not that much different from what was being done to Germany's Jews in the very earliest presages of the Holocaust, the period when virtually no one, either in Germany or elsewhere, would ever have believed in the possibility of what was to come.

When I went to college, I helped build a new field of study that spread to many other schools throughout the 1970s and 1980s. It was an interdisciplinary program called Peace Studies. While there had been one or two programs at small religious schools with similarly named programs, ours was the first to move the concept into a truly secular mode, analyzing the roots of war from the perspectives of every relevant academic discipline available and seeking solutions that would prevent future wars.

We realized that every discipline and its scholars had a tendency to think that it alone had the real answers to the problem. Economists preached that wars had economic bases, psychologists held that they flowed from the Id, biologists pointed to animal behavior and emphasized the analysis of aggression's biological basis, and historians wanted us to believe that learning history was the only way to avoid repeating its mistakes. The Peace Studies program at Manhattan College was unique in insisting that, for a student to be adequately prepared to fight against the evil of war, he or she needed to understand *all* aspects of the sources of human conflict and prejudice – psychological, biological, personal, group, economic, racial, national, and international – and understand them as seen from every field of study that examines human behavior.

It was a grand goal, one that I believe we made very significant progress toward during my years at the start of the program, and Peace Studies programs at Manhattan and elsewhere have continued to build upon that early work.

However, there was one thing, throughout all my study at Manhattan College and on into my graduate work in Peace Science at the University of Pennsylvania, that I could never comprehend: How could the "Good" Germans have allowed the Holocaust to come about? Why didn't they stand up at the beginning, when the first insults to human decency and the incursions upon respect and liberty began to grow? What was going through their minds as they saw their Jewish friends and neighbors being called names, being made the butt of editorial cartoons and nasty jokes, being thrown out of clubs, denied jobs or medical care, or being told they spread disease or couldn't live next to their Gentile friends and neighbors? What restrained those good Germans from action in coming to the aid of those they'd known and cared for before it reached the point where such action meant risking a Gestapo bullet?

That was something I could not understand: the beginnings of it all. But now I do. Even though I do not believe the "War On Smokers" will ever extend much beyond personal animosity and economic, housing, and medical persecution, I have been horrified to see such widespread acceptance of the growth of such persecution. It has been very disturbing to see it met with total complacency amongst a generation that, just thirty years earlier, was screaming high holy hell against any and all behavioral control by authority; the hippies who believed that all people should be loved and practically all non-violent behaviors should be tolerated and accepted as each and every individual was allowed to do their own thing.

I have seen the sea-change in attitude over those thirty years and I have seen how reluctant people are to stand up to authority, no matter how illegitimate, if it dons the robes of acting in the public interest, or for the children, or even just to save tax money.

I have seen people radically alter their views of reality, not because of any sound and rational argument, but purely because they have been hit, over and over and over again, with sound bite philosophies that come to be accepted without thought or question upon the hundredth unchallenged repetition.

I have seen people treat and accept treatment of friends and family members in ways that would have been unimaginable twenty or thirty years ago: tossing visiting grandparents out onto snow-covered porches, evicting elderly patients from long-term care facilities, rudely accosting total strangers who are engaging in "misbehavior" a dozen yards away in the open air, even teaching their children to regard certain sorts of folks as "dirty," while training them to make nasty faces and fake coughing sounds upon the sight of such folks.

I have seen the majority of the population welcome extortionate taxation of a minority simply because they have been given an excuse to vote for taxing someone else without guilt.

I have seen people tossed out of homes they had lived in for years simply because they refused to change their lifestyles to be in accordance with a new fiat regarding their perfectly legal behavior, and

I have even seen people threatened with the denial of needed medical treatments unless they agreed to adopt the current medical ideals and alter their behavior accordingly.

I've seen all that and finally, many years after graduating with a degree in Peace Studies, I have begun to understand just how subtly hate can be built up against a minority with almost no one objecting, and almost no one believing how far such hate can eventually go as it develops step by step.

That is why I decided, despite Godwin's Law, that the title of the opening story and the overall title of this book should ring back to the memories of the early stages of the oppression of minorities in Hitler's Germany. There really should be no need to fantasize about where such things could go in the future; the present-day reality should be more than enough to wake people up.

But it hasn't.

In 2010 the internationally prestigious journal, Tobacco Control, dedicated the editorial focus of its October issue to the concept of "new endgame ideas for tobacco control."[2]

The core actors behind the mainstream of the antismoking movement now believe that they have enough power to actually "correct" the misbehavior of the world's 1.5 billion tobacco lovers. *TobakkoNacht – The Antismoking Endgame* will provide proponents of Free Choice some weaponry to battle against this Final Solution to the Smoker Problem. Unfortunately, it is not a magic bullet or a super-secret weapon, nor does it have the answers to the question of how the world's relatively unorganized mass of smokers, with intermittent, almost nonexistent, and often counterproductive help from the tobacco industry itself, can produce such a bullet or weapon.

What it does provide are some basic tools for attacking the enemy at its weakest point: its lies. Lies are basically what brought the American tobacco industry, once one of the most powerful commercial forces on the planet, to its knees. Just like Richard Nixon facing Watergate, the industry would have done far better if, right from the start, it had made all its research, memos, and commercial efforts public. Can they be badly faulted for not doing so? Not really – try getting the same sort of information out of Big Fast Food or Big Pharma today and see how far you get. While these corporate interests may not have the sheer amount of incriminating evidence in their files as the tobacco companies did, I'm sure lawyers would have a field day if all their secret vaults were suddenly thrown open.

Hopefully, *TobakkoNacht – The Antismoking Endgame* will help pry those doors apart while warning of the dangers of other doors and paths. It's designed to be a toolbox, a mini-armory for those who know the basic layout of the battlefield and the nature of the enemy, and who are looking for weaponry to fight against the Antismokers' endgame plans. Whether those weapons and the will to use them flow from the story at its start, from the jolt to sensibilities from its satires, from its analysis of the statistics and lies that have brought about current injustices, or simply from discussions engendered by its existence as some rail against its inappropriateness and others defend its need, the goal is simple. I am hoping it will remind people of the importance of standing with and supporting any minority that is being subjected to persecution, discrimination, and hatred – even, or especially, at its earliest stages – while also showing how such a thing can be fought.

During the writing of this book, on July 21, 2009, US President Barack Obama went on American television's Today Show and stated, "The only tax change I have made in the six months I've been here has been to cut people's taxes."[3] He said this just a few months after promoting and signing the SCHIP (State Children's Health Insurance Program) bill that hit most of America's 40 million smokers with a 150% federal tax increase; and a minority of smokers, often the poorest of the poor – those forced to roll their own from shreds of tobacco and scraps of paper because they cannot afford store-bought cigarettes – with an incredible 2,150% tax increase! You can see the short video clip and full story online on the Gasdoc website of Dr. Phil Button, a British anesthesiologist who has been quite critical of antismoking extremists.[4]

With that statement, the president effectively denied the very existence of over 40 million Americans as people. According to the president, the smokers who are now paying the SCHIP tax to cover the health care of tens of millions of nonsmokers' children simply do not exist. He effectively erased them as human beings and reduced them to being simply vermin who could be ignored without comment. And no one, not the interviewer, not the mainstream media, and most sadly, not even the general public, seemed to notice or care.

The opening section of this book is the short story, *TobakkoNacht!* The story will speak for itself just a few pages from now and I hope you find it as disturbing to read as I found it to write.

It is followed by a section of lighter fare, *Satirical Smoke,* a selection of satirical pieces I've written over the years to poke pointedly at the foibles of various Antismokers, their activities, and their dubious claims. Satire is an effective way of saying things or exploring ideas that people would not ordinarily be willing to consider seriously; it allows the writer to open the readers' minds to possibly unpleasant concepts in a non-threatening manner so that they can at least begin thinking about the paths to such ideas. Appreciation of some of these satires will be heightened by a deeper background knowledge of the people and claims involved, but the introduction to each one should be sufficient to allow even a totally innocent reader to see the points and get at least a bit of a laugh – although sometimes a slightly sour one – because on the smoking battleground, truth is often uncomfortably close to satire.

The third section, *Stratistics Unbound,* takes a more serious turn and looks closely at the deliberate distortion and misuse of numbers and statistics in pursuit of basic antismoking strategies. It's not highly technical, and it provides some much needed insights into how widely people have been deceived by fear-peddlers pushing bogus numbers and interpretations.

The fourth section, an extensive and very serious one titled *Studies On The Slab,* extends beyond the simple numbers involved and presents critical analyses of the headlined scientific studies that smoking bans, taxes, and the general discrimination against and hatred of smokers have been built upon. While the studies themselves are not reprinted here, they are fully referenced for any who care to check. Their descriptions are accurate and their criticisms are honest and sometimes harsh.

In a standard hearing before a City Council or State Legislature, the very questionable conclusions, or wild approximations thereof, of these studies are repeatedly thrown into the mix as undisputed facts by antismoking advocates. Because they have often been printed in peer-reviewed medical journals, their validity and their all-too-frequently misrepresented conclusions are taken as unassailable.

Opponents of bans rarely have the opportunity to do more than barely give lip-service challenge to the "mountains of studies" that Antismokers claim support the need for bans. Even asking a legislative body to take the time to examine just one or two in enough detail that they will begin to doubt the rest is usually too much to hope for. *Studies On The Slab* takes that time and examines a healthy selection of the major antismoking studies and their copycats that have been repeatedly cited to legislative bodies as solid evidence of the need for ever more stringent bans and ever higher taxes on tobacco. The fact that so many of the most prominent scientific foundations of smoking bans can be so seriously challenged should give any reasonable reader some significant doubts as to the need for such bans.

The fifth section of the book, *Slings And Arrows*, offers a selection of my shorter serious writings – letters to the editor, Op-Ed articles, and such things – along with samples of more formal presentations to three governmental bodies: Philadelphia's City Council, Findlay, Ohio's Department Of Health, and the British House of Lords. It's meant to serve as a ready source of facts and arguments about topics frequently brought up by Antismokers, while serving as an example of how to fight them without funding. The final subsection of *Slings And Arrows* examines some aspects of the battle *In The Trenches* of the Internet bulletin boards and blogs and shares thoughts about effective fighting in that milieu. Overall, *Slings* covers a huge range of topics and the individual entries vary from the deadly serious to the satirical to the highly barbed and political.

The sixth major section, *The Endgame*, takes a look at what the Antismokers see as the upcoming, and in their opinion, near-final moves in their fight against tobacco and explores the best approaches to stopping that endgame in its tracks at this point in time. It doesn't have all the answers, or even anything like a detailed map of the trail ahead, but I hope it will offer pointers, thoughts, and guidance that will help us retain our freedoms to live our own lives by our own choices and values.

The seventh and final main section is quite short but serves well as a bookend to my opening tale. The short story *Breathers* was written at about the same time as *TobakkoNacht!* and offers another chilling, but more succinct, view of an unlikely but thought-provoking future.

Prologue To *TobakkoNacht!*

On one dark night in November of 1938, after years of gradually escalating restrictions and hatred aimed at Jews and Gypsies in Germany and Austria, terror descended upon the Jewish populations of those nations. Jewish commercial and ghetto districts saw entire blocks of store windows smashed, over 250 synagogues destroyed and over a thousand more vandalized, a hundred Jews killed outright and 30,000 more arrested and sent to Himmler's concentration camps by the Nazis. Meanwhile, regular authorities were instructed "not to interfere with the demonstrations" except to protect foreigners and any non-Jewish businesses.[5] Hitler's Aryan gangs intended to teach the "vermin" a lesson they would never forget. That night has come down in history to us as *Kristallnacht* – The Night of Broken Glass.[6]

TobakkoNacht! is not meant in any way to detract from the horror of that night. On the contrary, while writing it and sharing it with others, I discovered how sadly *Kristallnacht* itself has been forgotten by far too many. Perhaps this story will help remind some of its readers of that history and the horrors it presaged.

The story spans roughly 10,000 years in its thirty pages, divided into six separate sections set in different times and eras and told in different styles. The core of it, from the Plaza scene through the end, was written in the late 1990s. When I shared it with friends at that time, the feedback was uniform – the story was just too crazy, too far from anything rooted in reality, to even be acceptable as science fiction. So I just put it in a drawer for a while and went to work on a more serious effort: *Dissecting Antismokers' Brains.*[7]

As you'll see, the passage of just fifteen years or so has made a remarkable and frightening difference, as things once regarded as too unrealistic even for darkest fantasies now regularly grace our news pages and figure in the plots and stories of our TV programs and news. What a sad difference time has made.

TobakkoNacht!

The Tinder

Bear Killer stared into the leaping fire.

It was larger than usual, for the sleep season had come early and the whitewater lay thick on the ground outside the light. Bear Killer had no concept of date or calendar, but the year, in Christian terms, was roughly 8000 BC, and a harsh winter was settling upon eastern North America near what would someday be Washington, DC.

The wolves had been active for so early in the season, as had the bears. Bear Killer ignored their howls outside the circle of light. He was unafraid, for although he was young, with only single tattoos on his toes and most of his fingers, he was the largest and strongest of his tribe. He had the fattest and the most beautiful partner to keep him warm in his winter tent, and he was Chief. But tonight he was a chief who was alone.

He poked the warmth of the life-sustaining fire with a stick. As tribe leader, the fire was his responsibility. It was the same fire that had been given by the rain god to his mother's mother's mother after a giant godspeak storm many seasons in the past. As the god's thunderous voice had lit the sky, he had reached a finger down to share his gift with men. Bear Killer's brave ancestor had taken some of the gift to keep and nurture, and it had made their tribe larger and fatter than any other they had ever encountered. The fire was cared for and carried from camp to camp with as much tenderness as a newborn infant.

A wolf howled nearby and Bear Killer added more branches to build the light that kept them away. His woman screamed and he dropped the branch he was holding before he could place it properly. Sparks spiraled high in the air, an offering to the fire god, but a poor offering because it had not been meant. Several others in his tribe, men and women with all their fingers and some toes tattooed, and some with even two marks on some fingers, laughed and grinned and poked each other while peeking toward the medicine man's tent.

Wolf-Soft-Talker and his woman left the circle and walked into their tent. She would soon scream and be happy and warm too. Bear Killer spat into the fire. It was not right to throw anger from the water god into the fire, but there was also a fire burning inside him and he did not care. His woman screamed happily again in the medicine man's tent.

It was not right! She was his! She was the fattest. She was the best to look at in the whole tribe and she was soon to make another tribe member. Bear Killer did not share the belief of some of his tribe that new members came to women from the gods. His mother's mother with many tattoos had taught him as a child that women brought new life only with the help of the men in their tents. She had shown him how Long Toe was so like Scream Maker and how both looked like Man-Two-Women although they had come from two different mothers.

His woman was screaming and had a new member in her belly, his own new member, and she was in the tent of the medicine man more than in his because of the medicine man's stinky fire. As Chief, he had the fire god in his tent too. But the fire in the medicine man's tent was different. It was foul and stunk more than the places where those who had left forever rotted in the ground. The medicine man put leaves in his fire to make it stink. He called them Toe Leaves and laughed and would not share them.

And Bear Killer's woman loved the stench they made and stayed with the medicine man more nights than with Bear Killer, even when it was warm and the medicine man's tent was hot beyond comfort and burned a fire with no need. And when she came back to Bear Killer's tent, she stunk more of Toe Leaves than of sex.

He had hit her many times once, after she came back in the morning dizzy and happy after many screams, but she still went back to the Toe-cursed tent – by longstanding custom, even a chief had to share his woman with the medicine man. As his last companion by the fire wandered off to sleep, Bear Killer sullenly prepared the fire to be watched over by the moon god.

He heard his woman scream again in the tent with the medicine man and the filthy, dizzying smoke, and something changed inside him. A spark of the fire he tended reached into his brain and grew. When he had met the bear that gave him his name, that same spark had flared. He reached into the fire god and took a log with the gift blazing on one end and ran to the medicine man's tent. He found them, screaming together, and he hit them. He hit them and he hit them and he hit them, again and again amid the clouds of foul stinking smoke polluted with Toe Leaves. The fire-gift from his club created more fire as furs and oils in the tent shared the gift. He did not feel it, but the gift had also come to his long, shaggy hair as the clouds of smoke overcame him and he fell with the two beaten lovers.

As the sun god awakened, it shone down on the sweet clouds of smoke that rose around the first three human beings known to have died because of tobacco.

The Kindling

Nine thousand years later, the Ottoman's Sultan Murad IV prohibited smoking, with first offenders having their noses slit open and repeat offenders, as many as eighteen a day, being executed for their smoky ways. The ultimate headsman's total is unknown, but smoking continued.

Simultaneously, in Merrie Olde England, King James I proclaimed tobacco smoking to be "a custom loathsome to the eye, hateful to the nose, harmful to the brain, dangerous to the lungs, and in the black, stinking fume thereof, nearest resembling the horrible Stygian smoke of the pit that is bottomless."[8] He threatened executions of smokers, to no avail in stopping them.

Three hundred more years would pass before Lucy Page Gaston of the United States founded the Anti-Cigarette League and sought a nationwide ban. She failed, as well.

Just forty years further on saw Adolf Hitler and the National Socialists of Germany creating the concept of *Passivrauchen* (Passive Smoking) as they sought to free all Aryans of the tobacco habit. They lost the war.

But then, suddenly and unexpectedly, the tide turned. The last thirty years of the twentieth century saw a part-time cruise ship dancer named John Banzhaf filing lawsuits to fill American airwaves with antismoking commercials, and a full-time consumer advocate named Ralph Nader pushing legal efforts to have smoking banned from airplanes. They won!

Those victories were quickly followed by the World Health Organization conference (the Godber Conference) in 1975 that revived the Nazi's concept of Passive Smoking and convinced activists around the world to foster the perception that "active cigarette smokers would injure those around them, especially their family and any infants or young children who would be exposed involuntarily to ETS (Environmental Tobacco Smoke)." This time the concept proved golden. Over the next twenty-five years, the new "Great American Antismoking Crusade" grew stronger than any before it, nurtured by billions of dollars and a campaign designed by the best and the brightest of the Madison Avenue advertising firms.

By 1999, flush with their victories in gaining tax funding and banning smoking in California's bars, antismoking activists were claiming that millions had died from the same stinky leaves that a medicine man and his woman had loved thousands of years in the past. The Antismokers, as they became known, wanted a final end to "the horrible Stygian smoke of the pit that is bottomless," and they didn't care who got hurt in the process.

As the new millennium dawned over an increasingly smoke-free country, the leaders of the new crusade smiled. The Master Settlement Agreement (MSA) with Big Tobacco had brought them bounty beyond their wildest dreams. By 2001, US Tobacco Control funding was peaking at well over 800 million dollars per year.

The movement's leaders barely knew where to focus the power of their newfound wealth. There were so many possible attacks on smokers available: banning smoking in movies, airbrushing cigarettes and pipes from historic photos, frightening small business owners and landlords with threats of lawsuits, convincing employers that smokers were less productive and more costly, raising taxes by double and triple digits, and finally, moving to ban smoking at colleges, in cars, on beaches, in condos and apartments, and even in private gardens and row homes sharing boundaries or walls with nonsmoking neighbors.

The push for bans on the home front was the trickiest area to approach due to antiquated American privacy concerns, but the step-by-step approach always worked. Car smoking bans built on images of children unable to protect themselves, and descriptions of "the pink young lungs of infants trapped in mobile gas chambers" were difficult for politicians to vote against. Public housing bans justified by claims of fire dangers soon followed. Condominiums, even condominium balconies, also fell quickly, as images and computer-generated videos of smoke seeking out windows to enter, then winding down hallways or creeping along electrical wires, and eventually pushing open doors to nurseries and attacking babies playing in their cribs were made.[9] Support actually seems to be growing for the idea of giving rewards to neighbors who spy and report on smokers in homes where children might be exposed.

Radical MSA-funded groups pushed for bans all over the country on a scale that neither Big Tobacco nor small groups of resistant smokers, so-called "Free-Choicers," could ever hope to match. A surprising new player gave strength to the radicals as some of the large pharmaceutical companies, seeing almost boundless wealth from their "NicoGummyPatchyProducts," added their own billions to the campaigns.

Small Free-Choice groups battled in city hall hearings and across the Internet, trying to beat the spending of their opponents by exposing how studies had been twisted and manipulated. But without funding, their path was rocky, and their losses outnumbered their victories. Antismoking crusaders now leaped forward with an openly declared push for the total "denormalization" of smokers. By the late teens of the twenty-first century, Las Vegas was virtually the only city in North America that still allowed any indoor public smoking at all, but even there it was confined to sealed slot machine rooms that were derided as "Gas Chambers For Losers." Among the last to fall were Nevada's infamous brothels; in 2019, smoking ban lobbyists successfully argued that the working girls who were provided with condoms to protect them against AIDS should also be guaranteed smoke-free air to protect them against lung cancer – even if they smoked themselves!

The 2020s roared in with a bang as smoking was virtually outlawed almost everywhere outside of private homes, even on the sidewalks of many politically correct cities. Smokers were reviled, with known smokers often denied entrance to universities, care in hospitals, and any but the most menial of jobs. Bags of nicotine-free urine and systems to beat blood tests abounded as aspects of this new Prohibition proved tougher to enforce than first anticipated, but the ostracism of smokers and the hate that grew from it spread through America and Europe like a plague.

However, as things rise, so do they fall. The courts finally weighed in, and by 2025 some of the most flagrant open-air bans and job discriminations began to fall under the weight of court-presented scientific evidence and the sudden re-awakening of the American Civil Liberties Union's awareness of the importance of the Ninth Amendment's unenumerated freedoms.

Smokers heaved a great puff of relief... for a while.

The Spark

Scene: A young mother, carrying an infant, preparing to cross through a crowd of smokers in one of New York's public plazas. Her shoulder-length hair is in some disarray, its black curls falling on her shoulders in tangles as though she's been running. Her eyes seem a little wild, and her breathing is a bit ragged.

She is muttering softly under her breath, but floating security nano-cams augmented with Homeland Security's SUB-VOC* programs record the scene for later broadcast and analysis.

##It was AGAINST THE LAW! Damn smokers' rights groups and commie ACLU got open air bans ruled unconstitutional. Filthy nic addicts! Got to get to the building... They're all in my damn way with their damn POISON!##

The mother is clearly in her fifth or sixth month of pregnancy despite holding a child of less than a year partially wrapped in a rubber raincoat. The morning's rain had just stopped as the security tapes began recording the scene. Smokers were enjoying their reborn freedom during their lunch breaks, sharing benches with smoking and nonsmoking friends, and standing around in convivial groups as the sun broke through the clouds.

---

*SUB-VOC: Statistically Upgraded Brain-Voice Optimized Characterization: a new, flashy, and controversial Homeland Security tool, allowing properly equipped Holocam drones to catch and record subvocalized thoughts of people in public places.

The young woman glares at the scene before her and wraps the raincoat protectively around her infant's head to make a bubble of clean air before starting to rush across the plaza. Her breakneck pace begins propelling her into groups that she pushes through, rather than detouring around, cursing and muttering all the while. Oddly, she's trying to hold her breath, so she doesn't really speak aloud but just keeps jerking her head to the side to tell the smokers they've got to move.

##Why should I have to go around THEM? They should MOVE!!!##

She clasps the raincoat tighter. The wisps of smoke in the air seem to be almost a solid fog when seen through her eyes. She looks down at the raincoat wrapped around her child…

##It's getting in the CRACKS! The Damn Smoke is GETTING IN THE CRACKS! I can feel my baby KICKING! Damn smokers won't MOVE!!!! What's the MATTER WITH THEM?????##

She almost snarls at a group in the way of her progress. The smokers seem shocked and a little frightened by her distraught appearance and wild eyes. She swings an arm at them as though trying to knock them aside with sheer willpower.

##Why are they just STANDING there STARING at me as though I'M the one who's crazy while THEY'RE the ones sucking down poison and blowing it everywhere while my BABY is trying to BREATHE!##

She won't be intimidated. She stops her headlong rush, pulls the raincoat a little tighter to protect her baby, and stands and glowers back at a group right in front of her. Then … horrors! One of them actually starts walking toward her waving a hand holding a lit cigarette. Some critical observers have since claimed that, from the camera view, it looked like he might have been concerned about her and was about to ask her if she needed help of some kind. The SUB-VOC data record that would confirm this has unfortunately been lost.

He is still about three feet away but as he opens his mouth to talk to her...

She sees SMOKE starting to come out of his mouth. He'd had the smoke in his LUNGS and was now going to spray it all over her and her baby and her unborn child! This was TOO MUCH! She couldn't stop herself. She drew in a lungful of the foul smoky air. It almost choked her. She felt it burning down her throat and causing micro cancers in her lungs that would grow into brown seeping masses and kill her but she had to breathe it in because she had to ... she HAD to ... SHE HAD TO

SCREEAMMM!

SCREEEAMMMM!!

SCREEEEAMMMMM!!!

The camera shows her stopped, almost across the plaza. The smokers had stopped as well; the tableau was frozen, but later Holo/SUB-VOC analysis expanded upon what you could see in her eyes and what was going through her mind...

They're not even SMOKING anymore but they're STILL burning their cancer sticks and sending their poison to attack me and my babies! But I'm almost through. I can see the fresh air gushing out of the vents by the doors of the building. There's only one more dirty smoker in the way ... And he's a POLICEMAN!! How can they be ALLOWED to smoke in public where decent people can see them? Where did he COME FROM??? He wasn't there a minute ago!

The camera showed comprehension dawning in her eyes as she realized he'd been in what was left of the old "Smokers' Butt Hutt" built ten years earlier between the office buildings and the main plaza. It was a small, nasty little thing with no light and a depression in the concrete – worn by smokers' pacing feet – where rainwater collected and putrefied in puddles of tar. He'd clearly heard her scream and come running out to see what was happening.

The idiot actually comes out thinking he could HELP ME? With a DEATH STICK IN HIS HAND???

At this point it's clear that the besieged mother has simply lost control. As the officer approached with a hand offered toward her tightly wrapped and protected baby, she kicks out and gets him right in the groin. He drops like a rock and she rushes past into the cool, fresh air from the building vents.

She stops and takes a deep shuddering breath; and then exhales as hard and fast as she can to try to get the poisonous brown foul carcinogenic tarry gobs of smokers' air out of the recesses of her lungs.

She takes another deep breath and hugs her precious pure baby to her bosom. At least she had protected him. His air bubble had kept HIS lungs safe.

She hoists him up in her arm and goes to pull the flap of raincoat free to smile at him and let him know that all is OK...

Damn! The damn raincoat is TANGLED! I didn't pull it this tight! Oh! That's right, I tucked it more at that first knot of stinking smokers! But how did it get so TIGHT???

The camera shows her pulling frantically at the flaps while the knots resist her stubbornly. Her baby has stopped kicking ... isn't even moving. He's lying there loose in her arms as she finally pulls the last corner of the raincoat free.

His head lolls sideways. His eyes are open and sightless and accusing.

SCRRREEEEAMMMMMMM!!!!

THE SMOKERS DID IT!

THE SMOKERS KILLED MY BABY!!!!

She looks up at the smokers in the plaza. The idiots are STILL frozen and staring at her, STILL holding their burning murders between their fingers and blowing their deadly smoke at her.

The cop has gotten up and is limping toward her. He is almost on top of her, reaching out toward her precious lifeless baby. She lunges forward and pulls out his gun. Shoots him. Starts shooting into the crowd of smokers.

BANG!

BANG!

BANG! BANG!

BANGBANGBangBangbangbangbangba....

After a dozen wild shots take down almost a dozen people, the policeman levers himself up from where he'd fallen and shoots her with his backup pistol as he dies. The shot bursts through the head of the child in her womb and severs an artery in her chest, killing them both almost instantly in a spray of blood....

The plaza goes silent.

The Bellows

The evening edition of the *Washington Post* blared in a front page headline:

PREGNANT WOMAN, HER 8 MONTH OLD SON, HER UNBORN CHILD: KILLED BY SMOKERS!!!!

New York's *Daily News* left the details for the inside pages and simply pasted a gigantic headline over a full front-page photo of the mother, taken as the cop's bullet created its fountain of blood:

SMOKER MULTIPLE MURDERS!

The *New York Times*, still the city's Old Grey Lady even in the new millennium, had a more subdued headline, only an inch high:

ADDICTS SHOOT PREGNANT WOMAN

It sat quietly over a picture of her corpse lying in a pool of blood, but was followed by a subhead stating, *"Her Only Crime Was Trying To Breathe."* Nowhere in the following story was it noted that she had actually fired the first dozen or so shots at the officer and bystanders.

The *Times'* Editorial page continued with the omission, suggesting that perhaps it was time for the government to finally get serious about the "threat of killer outdoor smoke pollution" and stating outright that any who voted against such a bill were probably in the pocket of Big Tobacco. Of course, none of the papers mentioned the five smokers in the crowd who had taken fatal hits from the woman's first wild shots, or the four who were still listed as being in critical or guarded condition from the riot slugs that had been loaded into the pistol she'd grabbed.

Predictably, the scattered nuts in the smokers' rights groups whined about how the medical examiner had determined that the infant had died of suffocation, rather than smoke inhalation, but their self-serving mewlings were quickly doused by California Senator Grantz's impassioned plea that *"ANY young life should be treated with respect, and this particular young life should not have been lost in vain; to allow public smoking to continue after this crime would be a crime in and of itself!"*

There were even comparisons drawn between Hitler's gassing of the Jews and what women and children were being subjected to as they were forced to walk in the streets with smokers.

Strangely, things quieted down for a week or two after that. There were letters to the editors in newspapers, and the talking heads on the HoloShows raved on, of course, almost universally in favor of outdoor bans or even the closing of tobacco companies altogether – policies that tax-hungry politicians and tax-dependent antismoking groups both staunchly opposed in private while supporting in public – but the outcry seemed almost subdued compared to the initial head-lines.

The Flame

Three weeks to the day after hundreds watched in person as poor Heather – her name, Heather Mary Katherine O'Malley, had become a common sight on banners and ribbons hung from trees and stretched across porches – was gunned down in cold blood on a public plaza, and after tens of millions had seen those last moments endlessly replayed on the HoloNewsMinutes, the Washington DC corporate offices of Philip Morris were hit with a fiery explosion. Fire examiners later determined it was a bomb, ironically set off with the classic timer of a cigarette burning down to ignite a pack of matches and a fuse.

Unfortunately, the explosion occurred just as a group of 500 third-graders paraded to the front of the building holding signs with such poignant, crayoned messages as "Heather died for our air", "My mother is an addict and I can't breathe", "My little sister died from smoking", and "Are My Lungs Getting Cancer?" One child was just finishing a live interview on CNN about how his 93-year-old grand-mother had finally passed away as a result of her late husband's Vietnam-era habit of smoking an occasional cigar on the front porch. The blast shook the plaza in front of the office building and blasted away the tear that was just starting to trickle from the child's eye at the interview's closing.

In contrast to the evening of the incident in New York, the media was almost silent for a day, attending mainly to the details and minutia of the explosion's aftermath. Two hundred of the children had died in the immediate explosion and another two hundred were in trauma centers, with many not expected to survive the week. The rest were being treated for severe lacerations, burns, and possible tobacco smoke inhalation, as it was suspected that the corporate offices probably had many cartons of cigarettes in drawers and crates.

There were seventy-three other known fatalities, including the schoolteachers leading the children, some unlucky random passersby, and a number of toddlers in a day care center across the street who were sliced by shattering windows. There were, naturally, some tobacco company personnel in the building at the time of the blast, but there was no great concern about them. Their only mention was as part of a story noting that the top executives had gotten away unscathed because they were downtown, enjoying martinis and cigars at an upscale restaurant that flouted clean air laws by enforcing a strict admissions/membership policy. A video making the rounds on the Internet purported to show them laughing at a joke about asthmatics at the moment of the blast, and then rapidly moving into a discussion of insurance coverage as they heard the news.

The story initially was simply too shocking and too gruesome for even the lowest of the media to do more than report on the medical statistics and the tight-lipped police investigation that was getting under way.

The following day was a different tale. The shock had worn off and yellow journalism shrieked at full volume… but the target was unexpected; headlines roared, but not against the bombers. Instead, they pointed at those who had set the stage for the bombing by setting up "A Factory of Death" within a vulnerable city and those who had paid for that factory: smokers. The headlines were fanatical, the stories even more so. There were calls for the complete criminalization of smokers and the execution of all tobacco farmers, manufacturers, and purveyors.

TobakkoNacht!

It's not clear exactly where it started, and it's not clear to what degree it was spontaneous and how much of it was organized by various extremist groups communicating over the Internet, but it's quite clear that the Friday evening following the bombing was the start of what has become known as TobakkoNacht. The carnage of the event can best be seen through the raw numbers and cameos of scenes, primarily in New York City, but also on a smaller scale in many of the more politically correct communities around the country.

– Over 5,000 dead in New York; 50,000+ treated at hospitals for injuries. Over 800 convenience stores looted, primarily of cigarettes;

another 350 vandalized with broken windows and graffiti scrawls of "Death Pushers," "Commie Frogs," and "Stink Smokers" spray painted on what was left. Fifty additional stores, several with the bodies of clerks still inside, were burned to the ground. One clerk had been suffocated by cigarettes stuffed down his throat while "PUSHER!" was carved into the forehead of another with a broken cigar cutter.

– Over 200 "smoke-easies"* had their staff arrested and stocks destroyed by federal agents backed by National Guard troops.

– Over 100 specific in-family (spousal and parent/child) murders took place over smoking disputes on the opening evening of Tobakko-Nacht itself and several dozen more were scattered throughout the weekend.

– One church was burned to its foundations by an angry congregation when a pipe-smoking Irish immigrant priest lit his pipe as an example while preaching a Sunday sermon promoting tolerance, acceptance, and calm. Ironically, it was later determined that he had loaded his pipe, not with tobacco, but with frankincense, historically known as the "peace incense."

– An unknown number of people were left dead in a massive car bomb attack on Nat Sherman's Tobacco Emporium on 42nd Street. The explosion leveled a good bit of the block and heavily damaged the New York Public Library. Nat Sherman's had long been famous for hosting "The Smoke-Free World's Largest Walk In Humidor" with plush leather couch seating and even a few cigar-friendly hot tubs and a spa for high level foreign diplomats and such. The newly elected Russian UN chief was one of the victims... supposedly in one of those hot tubs at the time with a half dozen members of Sweden's stogie-sucking Blond Bikini Swim Team (this final claim has not been fully substantiated).

– The head of the Central Park Zoo was found mauled to death in the lion pit on Tuesday morning along with Sam, the zoo's longest-lived chimp, who had been given an exemption to the public park ban due to the cigar addiction he had picked up in more tolerant times. Efforts to cure him with electric shocks timed to his smoking breaks

---

* Smoke-easies: strict membership bars where illegal smoking was tolerated by S.W.A.P.P. <Smoking Without A Permit Police> officers bribed by cigarette smugglers.

had failed miserably earlier in the year and the zoo official had ended the attempt with a public statement saying, "Let the poor old guy smoke one cigar a day in peace."

– An ex-president from the 1990s, known for his sexual misadventures, was found propped up in Central Park's main fountain, horribly mutilated and with a cigar in place of a missing, but oddly unspecified, organ.

New York's suffering was mirrored on a smaller scale in Boston, Massachusetts; Brattleboro, Vermont; Hartford, Connecticut; Madison, Wisconsin; and dozens of other towns progressive communities had grown in power over the years and where antismoking sentiment and Joe Camel laws had traditionally been both nasty and strong.

Surprisingly, California was a mecca of relative peace. While several dozen independent smoking-related murders occurred statewide, there was little of the burning and destruction that had occurred in New York. Smoking had been effectively outlawed in California in 2017, and although random drug checks indicated that almost 5% of Californians still smoked at least occasionally, it was almost always done in secrecy and with the scent disguised by burning marijuana; the actual sale of tobacco products and paraphernalia was illegal.

There were two sad exceptions to California's general peace. First was the mass execution carried out behind the walls of California's infamous "Smokers' Prison" where illegal smokers, cigarette smugglers, and child molesters were housed away from the general prison population. A career-sniffing warden had decided the time was ripe for a final solution to that segment of the inmate overcrowding problem. Second was the attack on San Fran-cisco's so-called "Smokers' Church" where "priests" burned bowls of tobacco instead of incense during services; not a single known member of the church survived the weekend.

On the darker side, there was the largely unreported story in the last five tobacco-state strongholds: Kentucky, the Virginias, and the Carolinas. The mayhem there was so intense and widespread that regular Army and tank divisions were called in. A complete lid of secrecy was imposed once the B3 bombers arrived from bases in Maryland and Iowa.

The Ashes

Thirteen years P.T. (Post-TobakkoNacht)

President Waxham sat at his desk in the situation room. The fingers on his right hand twitched nervously as he stared at the displays on the wall monitors. The day had been an absolute disaster. From the first meeting in the morning with his cabinet, to a disappointing lunch with his economic chief of staff, through a frustrating evening gathering with the heads of the CIA and NSA drug operations.

It had been over ten years since the enactment of the Uniform Tobacco Control Act. President Waxham had been one of the authors of the Act and one of the main activists in getting it pushed through both houses of Congress and exempted from Constitutional control. For most of the first decade of its existence, he had worked hard, first as a Representative, and later as a Senator, to see it fully implemented and properly supported. Finally, he had run for president of the United States and had trounced his opponent, an old-school Southerner who had misguided notions of returning to the smoky days of yesteryear.

Waxham looked at the full-wall Holoscreen Situation Map and his fingers and mustache twitched nervously once again. The satellite mapping of the Brazilian rain forest showed yet another parcel of jungle cleared and planted with tobacco.

Waxham leaned back in his chair and rubbed his temples. How had it come to this? It had all seemed so clear thirty years ago. Tobacco was destroying our country; it was killing our kids and robbing our treasuries. It was the greatest drug scourge to ever exist in the world.

The work of the dedicated Smoke-Free activists of the last quarter of the twentieth century had been productive. The population of adult smokers had been cut in half through a combination of social pressures, stigmatization, media control, massive tax hikes, and spreading workplace bans. At this thought, Waxham chuckled briefly; the disappearance of relaxed smokers working productively at their desks and the simultaneous appearance of the little groups of "drug addicts" huddling in dirty alleys by the dumpsters behind office buildings had been a key part of the campaign to move public opinion against the smokers. As a pleasant side-effect the campaign had also provided a never-ending source of entertaining snowball and mudball targets in his youth.

Tobacco taxes. Almost the only taxes in the history of the world that were welcomed by the majority of the populace. And the beauty of it was, Waxham grinned, that we didn't even have to call them taxes. They were just "user fees" on the big bad tobacco companies. By the time the spin doctors got through with it, even most smokers believed the price increases were simply a plot by Big Tobacco!

True, some problems had developed. First of all, the hundreds of billions in expected tax revenue simply disappeared as many smokers showed their true criminal orientation and turned to smuggling or growing their own drugs. Larger basement greenhouses were detected by infrared radiation picked up by AntiBac Copters, but small family operations were hard to find unless helpful neighbors, encouraged by generous head bounties, blew the whistle.

Unfortunately, standardized USA NoNic cigarettes had never caught on in a big way with the American public. To the small extent that they were sold, they were usually bought by customers who brought them home and spiked them with concentrated nicotine oils smuggled from South America. The number of children who had died as a result of imbibing these concentrated oils, or through smoking homemade cigarettes pumped up to fifty times the normal nicotine level, made Waxham shiver.

Waxham returned to surveying the Holoscreen in front of him. Brazil was horrible. Colombia was a close second. And Cuba, of course; despite its small size, it was even worse than Brazil – at least in terms of intensity if not in actual quantity of production. The South American governments had laughed at America's concern over wisps of smoke while their own populations were starving and their children were dying by the millions from dysentery and resurgent malaria. They were angered rather than gladdened at the shift in UN spending toward saving their people from the "Grey Scourge of Toxic Tobacco Smoke" while leaving their children to waste away in misery.

Cuba had been the first to blatantly defy the edicts from the United Nations Tobacco Symposium, refusing to destroy its major and most distinctive national crop. Brazil had swiftly followed suit, declaring itself a "Free Tobacco Country" and allowing its farmers to clear vast swaths of rain forest to plant its fields with lucrative tobacco plants for illegal export. Colombia and Venezuela quickly joined the profitable club.

If Latin American countries had been the only ones failing to toe the line, Waxham thought, perhaps we could have controlled it. But when France, Russia, and Japan declared that they would continue to grow tobacco for domestic consumption, it had destroyed the ability of the United Nations to act effectively against the new drug barons of South America.

Waxham studied the papers before him. Not only had tobacco tax revenues failed to materialize as expected, but health costs had spiraled well beyond the increased rate of inflation. As the number of smokers was reduced, healthcare costs actually increased due to higher costs for obesity, geriatric medicine, and depression. The double hit of the massive loss of cigarette tax dollars combined with increased healthcare costs, had dealt the American economic system a blow even more severe than the eternal "Sand Dune" wars in Iraq, Iran, and Afghanistan.

He turned to the next folder. The title was ominous and in huge black caps: YOUTH DRUG EPIDEMIC SURVEY. As laws made it more difficult for teens to get tobacco, youth smoking rates had indeed gone down. Teens then turned to other cheap forms of getting a buzz and having fun – synthetic drugs were rampant and the amount of brain and liver damage caused by huffing spray paint, gasoline, and glue had dramatically increased. Many of these kids ended up on life support, or needed to be cared for, in a vegetative state, for the remainder of their lives. The cost was astronomical. There were now entire hospital complexes devoted to nothing other than maintaining the lives of these empty shells.

The problem had leveled off recently, however, as the black market partly shielded the country from this unintended consequence of the war on smokers. Black-marketeers had made cigarettes easily available once again to America's youth, since they had no concerns about the age of their illegal customers. The actual age for initiation of tobacco use had gone down to below the points reached in the 1990s; it was now quite easy for even third-graders to find a willing and relatively cheap supplier in a back alley on the way home from school – the "White Van Man" or the "guy with the shopping bag" evidently did *not* worry about ID cards. Despite all efforts against it, the combination of bloodthirsty taxes followed by outright Prohibition had served to reopen, and then nail open, this market, and had effectively destroyed all efforts at youth access control. The brain-dead huffers followed by the increase in younger youth smoking should have taught the government a valuable lesson, but governments never have been very smart, unfortunately.

Waxham wiped his forehead and his fingers again twitched nervously. Why had it turned out this way? Where had they gone wrong? It had all seemed so clear. He blamed the tobacco companies for deliberately going bankrupt. If they had stayed in business and used their know-how to produce acceptable, non-nicotine, smokeless cigarettes, perhaps the American public would have made the change more willingly. Perhaps the aging hippies who had so successfully refined marijuana into potent sinsemilla in the 1970s wouldn't have turned their talents to producing pumped-up strains of HiNic tobacco.

A tone sounded. Waxham clicked a button and the air over his desk solidified into the opening symbol for CNN's six o'clock Holo-News. The announcer's face was gray against a background scene of the riot in Chicago. Just last week, an undercover squad had stormed a suburban home when their sensors revealed the inhabitant lighting up a smoke after sex with a sixteen-year-old prostitute. While whoring at sixteen was perfectly legal, exposing a whore to nico-fumes was not.

When a cop demanded a proper inspection of the suspected tobacco cigarette, the smoker started to reach for his bed stand drawer and was shot by the officer. A TV crew on the scene recorded a fellow officer as he opened the drawer and withdrew a packet of perfectly legal, guaranteed tasteless USA Standard NoNic Smokes. The division that had grown between police attempting to enforce tobacco control and an increasingly irate citizenry who protested what they called "the loss of American freedom" had become more deadly in recent months.

Waxham blanked the broadcast and activated the Oval Office's Situation Holoscreen. Brazil had sixteen major tobacco areas scattered throughout the rain forest and countless smaller plots. Colombia had eight huge plantations. Cuba had only two, but they occupied a good 60% of the island. Waxham looked down at the last folder in front of him. This title looked promising: *The Effects of HyperGenetic Tobacco BioDefoliant*. The contents were not nearly as upbeat.

The latest attack against HiNic tobacco had been worse than disappointing. The plan that Waxham had focused the energy and reputation of his administration on for the last three years was a failure. The first full field test of the newly developed TDA* against an illicit operation in the hills of Kentucky had been carried out three weeks earlier. Tens of thousands of tobacco plants had been secretly cultivated by black-marketeers within a million-acre farm devoted to growing FDA medical-grade marijuana. The spraying had killed the tobacco and left the marijuana intact, as planned, but in open field conditions the TDA had quickly mutated. Within three weeks, all food crops within ten miles of the initial spray were dying.

The Army Corps of Engineers and FEMA had just concluded an emergency evaluation: the spread rate was increasing geometrically, and it seemed that there was no practical way to stop it. The Corps had proposed a gargantuan attempt to use two entire Army divisions to create a five-mile-wide burn zone perimeter around a circle 100 miles in diameter centered on the test field by the end of September. But the report candidly admitted there was no guarantee the airborne spread could be halted if a hurricane hit the area before October.

If he did not do something decisive and effective, Waxham's vision of America leading the rest of the world to clean air would be the historic footnote marking the failure of his administration, rather than the headline of its greatest victory. If the mutated biodefoliant managed to reach the Kansas wheat fields...

Waxham shuddered. He reached under the desk and took out the briefcase once playfully nicknamed "The Football" by an earlier administration. As he placed it in front of him, his fingers twitched again. It was almost as if they had a life of their own. He opened the briefcase and entered the necessary codes. He pressed the center button and looked up at the three dimensional image that coalesced in front of him. He hit the button again.

The pre-programmed strike commenced. Waxham watched the monitors as four relatively small and three large nuclear warheads slammed into Brazil. His fingers twitched as two medium and six small warheads hit Colombia. The two large warheads for Cuba brought an oddly wistful smile to his face, along with a single violent twitch of his mustache. Another dozen or so small warheads distributed themselves across Venezuela, Argentina, and Nicaragua.

---

* Tobacco Defoliant Agent.

Finally, just ninety seconds after the last of the South American strikes, its origins hidden in the EMP pulsing haze, a single 50-megaton warhead finished its short drop from an unregistered B3 flight and vaporized both the spreading problem in Kentucky and the brave but unsuspecting bomber crew who'd been told they carried a two-kiloton neutron device pinpointed for tobacco irradiation.

Waxham pushed his chair back from the desk and walked to his private office. As he entered, he turned and carefully locked the stout door behind him. He pushed aside the discreet wall-hanging in back of the oak desk at the far end of the room, and unlocked the unmarked door it concealed. He entered the small chamber beyond, then turned and locked the door.

A week later Waxham sat despondently at his desk in front of the Situation Holoscreen. In some ways, things had gone well. His decisive military move had brought the usual patriotic presidential popularity surge. The Kentucky problem had been solved and the bomb there was blamed on Cuba, as planned; there were no Cubans left alive to say otherwise. However, the bombing in the Brazilian rain forests had gone awry. The intense heat from the three larger bombs had caused the merger of gigantic fires that were now sweeping across 80% of the Brazilian rain forest.

The slash-and-burn farming of the late twentieth century had combined with the further clearing for the tobacco fields to produce an atmosphere as dry as a desert. The top cover of the rain forest was little more than kindling and the fires from the nuclear blasts had spread rapidly, far more rapidly than anyone had imagined they could. The smoke that was spiraling up from Brazil threatened to bring about the nuclear winter that had previously been thought possible only after the release of hundreds of nuclear weapons.

The president rested his head in his hands. It had *not* been meant to turn out this way. The pressures of the Presidential office were beyond what any mortal man was capable of bearing alone. He stood, checked to see that the outer doors were closed, and once again moved the curtain aside before unlocking the back door of the Oval Office and the door to the room that lay beyond.

He took a deep breath and sat at the small but plush desk in the serenely lit room. He opened a delicately carved box, similar to several hundred that lined the walls. Delicious Cuban tobacco aroma wafted toward his nostrils. His nervous fingers relaxed in a familiar and comfortable circle around a cigar as he removed it from its home.

He quickly lit and drew on the illegally pungent smoke. Unlike one of his famed predecessors, Waxham *did* inhale, and deeply. As his body relaxed, he felt a momentary pang of guilt. He reminded himself that he was not like one of those cigarette-sucking addicts outside. He was enjoying a rare Cuban Royal Montecristo Cigar, one of the last few thousand that existed in all the world.

He looked around his humidor sadly. The bombing of Cuba was a necessity, but a sad one. If only they had not so blatantly defied the Unified Americas' Tobacco Control Act and the orders of the United Nations. If only the Kentucky experiment had not gone so horribly wrong. If only...

Waxham had left the Football in the outer office as he entered his inner sanctum to relax. He didn't see the little red light blinking as the briefcase signaled an emergency. France and Russia, convinced that America's madness would soon lead to a strike on their own tobacco fields despite their non-export policy, had launched a simultaneous full pre-emptive nuclear strike against the United States.

The smoke from the burning rain forest would soon be amplified a hundred times over. The first 20-megaton warhead hit the White House just as Waxham was taking a final puff on one of the last Cuban Montecristos in the world. The secret vault and Waxham joined the rest of the Oval Office in a strangely sweet puff of smoke glowing over the hellfire crest that had been Washington.

Ten thousand years after Bear Killer's death, the final peace pipe had been lit and a Smoke-Free World was finally in the offing.

Postscript

Except for a portion of *The Kindling, TobakkoNacht!* is clearly fiction. But that portion is important precisely because it is not fiction. Smoking has been seriously attacked by autocratic powers in the past, and is being seriously attacked by an autocratic health establishment today, one which proclaims "Nanny Knows Best" for us. The ostracism and denormalization campaign aimed at smokers is real, as is the undercurrent of hate that is being built up against them. Readers with Internet access need only take a quick visit to smokersclub.com or forces.org to find multiple examples of divorces, beatings, murders, and even torture resulting from the paranoia that has been created around the fear of curls of smoke rising from a few leaves burning quietly in a small tube of paper. While I've decided not to detail them here, they're not hard to find on the Internet, and later, in *Slings And Arrows*, you'll be seeing some of the language and thinking that creates such things.

Is smoking harmful to the smoker? Yes, I believe it can be and would not argue otherwise, although I might take issue with the degree of harm often claimed, and might argue that an individual can rationally balance the pleasurable benefits of smoking against such potential harm.

Is smoking generally harmful to those around the smoker? No, the evidence shows no meaningfully significant harm to any normally healthy nonsmoker exposed to smoke in decently ventilated situations, even in lifelong daily cases of such exposure; to say nothing of the occasional passing exposures that currently have people holding their breath and crossing streets out of fear. Such reactions, objectively, are far closer to mental illness than to any rational assessment of reality.

But such concerns have become the norm rather than the exception in recent years because of massive media expenditures that have drilled those fears into our brains through a constant barrage of press releases and Madison Avenue TV ads. These fears run so deeply that any actual public questioning of such mantras as "Secondhand Smoke Kills!", "There Is No Safe Level Of Exposure!", or even "Secondhand Smoke Is More Deadly Than Smoking!" has become as unlikely as characters in Orwell's *1984* publicly questioning "Freedom Is Slavery!"

or "War Is Peace!" The meanings of words have mutated so much that rational discussion sometimes seems almost impossible. Concepts that would have been considered outright ridiculous just ten years ago – danger from wisps of smoke outdoors or a threat from invisible toxins left on floors after smokers leave a room ("thirdhand smoke") – are now treated with seriousness by respected media and even by some reputable scientists and medical authorities.

The Antismoking Crusaders are powerful, heavily-funded, and have tools far beyond the imagination of a Sultan Murad, a Lucy Page Gaston, or even a Joseph Goebbels. They started out as loose collections of idealistic or neurotic individuals, but as huge sums of money poured into their hands, they grew into a movement that now hosts multi-million-dollar conferences where thousands of devoted activists gather to plan strategies extending years into the future.

The $883 million US expenditure on Tobacco Control was a minimum figure cited in the annual report by the American Medical Association in 2001.[10] It did not include the money raised and spent by advocacy or extremist groups like ASH (Action on Smoking and Health) or GASP (Group Against Smokers' Pollution). It did not include lobbying and grant money from the Big Pharma purveyors of NicoGummyPatchyProducts (usually referred to as "Nicotine Replacement Therapy," or NRT, by its advocates), or the vast amounts of public service airtime and advertising that fan the flames of irrational fear on our TV sets. And it most certainly did not include the lobbying power commanded by those who take our children into their care and turn them into political pawns, parading them in front of City Councils in bright uniform shirts before sending them home to berate their mommies for killing them with deadly smoke.

The use and abuse of our love for our children has been the cornerstone of some of the bloodiest wars and dictatorships in history. Attila the Hun promised to spare women and children only if the men would surrender – and to rip them to bloody shreds otherwise. Hitler frightened the "Good" Germans with tales of Jewish ceremonies where the blood of Christian children was sacrificed. George Bush Sr. heightened American hatred for Saddam with tales of Iraqi soldiers dumping Kuwaiti babies from incubators onto cold hospital floors, and Saddam responded by "lovingly" ruffling the hair of a young American hostage in front of TV cameras.

Today's antismoking advocates consistently raise the flag of protecting the children whenever an extra emotional push is needed for an unpopular policy. The SCHIP bill for children's health insurance sailed through Congress and was immediately signed with fanfare and in front of TV cameras by incoming president Barack Obama despite his firm pledge of "no tax increase for anyone earning less than $250,000." Not only was the tax increase the largest such absolute federal increase on commercial cigarettes (a 150% hike equaling sixty-one cents per pack) in American history, but it carried an additional nasty sting by raising the tax on roll-your-own tobacco by over 2,150% – a hike from about $1.10 per pound up to $24.78 per pound. That hike hit hardest at impoverished, and often elderly, smokers who had long depended on the price break of making their own smokes. But it was signed with pride and fanfare since it was "for the children" – although afterwards, as we saw in the *Author's Preface*, Obama went on TV and denied having raised taxes on any "people."

Even worse than the simple political playing of the children card is the wider-reaching effect of the massive advertising and educational effort aimed at young children. "Raising a generation of nonsmokers" may be a laudable goal, but the long-term results of such intensive childhood indoctrination are yet to be seen.

If the only effect was to reduce the number of young smokers, few would argue against it being a good thing. But children do not easily make fine distinctions. A fear of and aversion to smoking all too readily becomes a fear of and aversion to smokers, and that fear and aversion will all too readily translate into an avoidance of, and eventually a hatred of, the people, rather than the custom. And that fear and hatred may well carry over into adult lives and attitudes.

Are we likely to see *The Ashes* coming true in our world? No. But are we likely to see *The Kindling* that is now being gathered and stacked grow into *The Flame* of commonly sanctioned violence against smokers? That's something we're already seeing hints of in the printed news and on the Internet; isolated stories of attacks, beatings and even outright tortures and killings – supported by open expressions of hate expressed without shame on public news boards and applauded by observers; and it's something that becomes more likely on a larger scale with every passing day that the extremists of the antismoking movement remain in power.

Epilogue

Over ten years after the bulk of *TobakkoNacht!* was written, I put it up on Amazon's Kindle in January 2009, and just a month later the story saw its climactic plaza scene take a baby step toward reality as *The Scotsman* ran an article by Lyndsay Moss with some of her observations briefly excerpted here:

Time to Reclaim Our Streets From Persistent Smokers

[Our smoking ban has brought us] A FUN new game ... pavement hopscotch. Imagine ... walking down Glasgow's Sauchiehall Street ... strewn with annoying obstacles – or smokers as they are more commonly known.

To negotiate your way past this smouldering flotsam and jetsam, you must do a little jig, ... repeatedly jump[ing] on and off the pavement, hopefully avoiding cars, buses and students being sick in the gutter. It is not a pretty sight. ... pavements clogged with smokers ... Don't get me wrong, the smoking ban has been a fantastic success [and] removed the stinking and cancer-causing fumes from our public places. ...

Knowing what to do with the remaining smokers is [tricky] Personally, I would set up smoking outposts on desolate wasteland, ... Sadly, I know this is not a plausible suggestion and, anyway, what has the desolate wasteland done to hurt anyone?[11]

As noted, it was a baby step; simply a somewhat irrational and intolerant characterization thought up by a journalist for a minor article in a Scottish newspaper. But baby steps are always how things start – we have only to look at the history of smoking bans over the last thirty years to see how that works and only to look at current news stories to see *The Kindling* being arranged.

The hopscotch article does not stand alone. In February of 2010, the tale of the zookeeper and his poor old cigar-loving chimp moved a skip closer as Reuters reported on a Russian chimp thrown out of his home at a local zoo, and sent 500 miles away to be put through rehab because of his nasty smoking and drinking habits.[12] And in February 2011, almost fifteen years after the bulk of *TobakkoNacht!* was written, more than the *Scotsman*'s baby step toward the plaza scene was taken as New York's Mayor, Michael Bloomberg, signed into law a ban on outdoor tobacco smoking at all of the city's "parks, beaches, and public plazas."[13] Friends who had read drafts of *TobakkoNacht!* back in the late 1990s and criticized the concept of legally mandated outdoor smoking bans as being too silly even for science fiction were clearly being a bit too short-sighted.

Meanwhile, the problem of international tobacco smuggling has taken on a grim reality as prohibitive taxes in countries such as Ireland and the United States have created a black market in which tens of millions of dollars' worth of cigarettes are actually *caught* by customs enforcement agents each year – with hundreds of millions still leaking through and fueling the coffers of organized crime and terror groups. There's even been an instance of a group of thirty tobacco smugglers being the victims of a rocket attack by Turkish military aircraft – although the military later claimed it had merely mistaken them for a wandering group of Kurdish separatists.[14]

And while my 1990s' predictions of inhalant abuse may not have come to pass to the degree pictured, my concerns about concentrated nicotine oils poisoning children has come a lot closer to reality with both Big Pharma's Nicotine Replacement Therapy and the advent of the e-cigarette with its little liquid nicotine capsules and refill bottles. While I don't know of any misadventures with e-cigarette liquids yet, there's been at least one instance of a fourteen-year-old student taking near-fatal advantage of a school's NicoGum giveaway: he tried chewing the equivalent intake of 400 Marlboro Light cigarettes during a lunch-hour break. Fortunately for the student, he passed out in public and was rushed to a hospital rather than simply dying. The school and the antismoking group sponsoring the giveaway defended themselves by pointing out that the gum can be freely and legally purchased by twelve-year-olds at local pharmacies, and by claiming they'd taught the students about its potential dangers and correct dosage.[15] Meanwhile, students at middle schools can find themselves strip-searched if they are merely suspected of "cigarette possession."[16]

Remember those government marijuana fields with the illicit tobacco in them? The little municipality of Colorado Springs is now on track to be collecting well over a half-million dollars a year in taxes on government-approved medical marijuana.[17] As for the "aging hippies" Waxham fretted over in *TobakkoNacht!* – those who had turned from growing sinsemilla to engineering more potent strains of tobacco – I just recently traded a copy of *Dissecting Antismokers' Brains* to Bill Drake in return for a copy of his newly published *Cultivators Handbook of Natural Tobacco*.[18] Mr. Drake is generally better known for his 1970 authorship of the *Cultivator's Handbook of Marijuana*.[19]

Marijuana, inhalants, and NicoGummy products aren't the only concern our schoolchildren may have in the brave new world though. On March 1, 2013 I was surprised to read that the government of Norway had announced plans to move toward legalization of smoking heroin![20] After checking to make sure that I hadn't pulled a Rip Van Winkle and awakened on April 1st rather than March 1st, I discovered that the powers-that-be in Norway seem to think this is a good way to cut down on the HIV problem. It remains to be seen if schools develop a problem of students mistakenly smoking deadly tobacco cigarettes in the boys' room when they're merely taking a healthy heroin smoke break between classes.

As for Waxham's plans to eradicate tobacco, I have discovered that, under Governor Jeb Bush, Florida had at one point seriously considered using a strain of the fungus *fusarium oxysporum* to eradicate the state's marijuana fields with mycoherbicides. Florida's Department of Environmental Protection, however, had a concern: "It is difficult, if not impossible to control the spread of *Fusarium* species. The mutated fungi can cause disease in large numbers of crops."[21] They succeeded in stopping the governor, but I wonder how well they would have done against President Waxham?

Finally, for any sheltered souls who might feel that the extreme attitudes of President Waxham and Heather O'Malley toward smokers would never be found in real life, I would urge you to just take a few minutes to read some of the comments sections after the news stories featured on the Internet. The fires of hate there may not have spread to the general population in any prominent way, but they've been growing and getting nastier year, by year, by year. Don't go there now, but when you eventually reach the last sections of *TobakkoNacht – The Antismoking Endgame* you'll see just how far those expressions of hate have gone.

And if those fires of hate *do* spread further? Never fear, the groundwork for *The Flame* will be quite well prepared. On October 4, 2010, the *Stanford University News* reported on a new research project titled "Cigarette Citadels."[22] A Stanford anthropologist named

Matthew Kohrman has mapped out the locations of more than 300 of the world's largest cigarette factories. "The goal of the project is neither to agitate nor defend … The point of the project is to share information that he hopes will motivate people to think in new ways." The article notes that this research "shares information that could combat the single largest cause of preventable death." One has to wonder just how knowing the addresses of cigarette factories will aid in "combatting" smoking in any way *other* than *The Flame*, particularly when the author makes a special point of noting such fine details as the disturbing fact that some of the factories are "just a short walk from tidy residential neighborhoods," while emphasizing the number of deaths supposedly caused by those factories (e.g., one Dutch factory was labeled as "responsible for some 80,000 deaths"). Somehow the assurance that his research is not meant to agitate seems to ring a bit hollow.[23]

And finally, when the bulk of *TobakkoNacht!* was first written in the 1990s, there was nothing at all like edicts from a "United Nations Tobacco Symposium," but today we are seeing pressures brought to bear from the United Nations and the World Health Organization as they attempt to enforce the WHO's *Framework Convention on Tobacco Control*. The FCTC has been signed by over 150 countries and requires member nations to do such things as increase tobacco taxes and mandate smoking bans if they don't want to lose out on UN handouts and face the possibility of sanctions and health funding.[24] The May 6, 2011 issue of the *British Medical Journal* ran a story castigating Germany for its failure to crack down on the freedoms of its member states, many of which were refusing to properly enforce the federal ban.[25] And Hans Stam, CEO of the Netherlands Heart Foundation, has labeled Edith Schippers, the Dutch Minister of Health, the "Minister of Death" because she has supported the idea of small pubs allowing their customers to smoke if they wish to do so.[26]

Most recently, in April of 2012, the World Trade Organization informed the United States that its ban on importing clove cigarettes was illegal. Oddly, while American antismoking groups seem to feel that the FCTC is akin to Biblical script, the Campaign for Tobacco Free Kids urged government leaders to keep enforcing the ban despite the WTO decision.[27] So far, however, no country has gone to war over the tobacco policies of another. So far.

The Kindling is pretty much already here; the stage for *The Spark* has been set; and *The Flame* may be only a hopscotch away. We may not see a true *TobakkoNacht!* on the scale of *The Ashes,* but if we continue to move in the direction we're heading we'll certainly be seeing more of a taste of it than any sane person would ever want.

Factory of Death

{For Research Purposes Only! Via Matthew Korhman's interactive map.}

Nestled comfortably between Fox Trail's family-friendly housing and the Trinity Evangelical Church. Nicely accessible via Benjamin Franklin Pkwy. & Limerick Road. A nice afternoon's bicycle ride from culturally rich Philadelphia along the scenic Schuylkill River Trail. Welcome to Philip Morris, USA! Bring the kiddies! Do not agitate!

Satirical Smoke

Words ought to be a little wild,
for they are the assault of thoughts on the unthinking.

- John Maynard Keynes

The dozen or so pieces included in this section may not all fall under the strict definition of satire, but they all use an element of humor or irony to point to various excesses in claims or actions of the antismoking movement. Some of them are short – little more than notes that might be printed as a humorous letter to the editor – and some are fairly long and actually try to impart some real information about the details of arguments used for and against smoking bans and taxes. Some are just playful, but some will be almost as disturbing as Jonathan Swift's solution for the excess of Irish babies. Each satirical piece has a brief introduction placing it in the proper context in which it was written, and all, hopefully, will be enjoyable to those who appreciate somewhat pointed humor.

I'll start with a piece inspired by a 2008 article that actually tried to draw a serious comparison between allowing someone to smoke in an apartment and allowing someone to shoot random shotgun slugs through the walls into neighbors' abodes. This appeared at roughly the same time as another article which argued that diet sodas represented the next great health crisis and needed to be as heavily taxed and perhaps legally restricted for youth as tobacco products.

Put the two together and you get...

Love, and a Coke 45

It was a dark and stormy night. The glaring lights of the 7-11 Qwickee Mart on the corner sputtered, spangled, and sparked, casting an eerie glow over the surrounding dreary neighborhood.

A hooded teen pushed in through the doors. The place was empty aside from two store clerks who were jabbering at each other in Lithuanian or Zimbabwum or French or Vietno-Arab-Mongolianese and the teen headed back to the Soda Safes.

He pulled out a cold can of Coke 45 and ran up to the clerk at the cash register. The clerk started to ring it up while calling out the price when the kid suddenly started shaking the deadly can, priming it to blow before the befuddled bejabberer could reach under the counter for his sawed-off shotgun.

"HANDS IN THE AIR! NOW!!!" the kid shouted! The merchants of death reached for the sky and backed away as the cold-blooded psycho-killer, a veteran of the Fizz Wars of the '90s, reached into the open register and grabbed for the cash.

At that point the braver of the two clerks reached for a Camel, intending to light it and cast a toxic cloud toward the armed teen but the kid was fast, too fast, and the top of the loaded can went off with a BANG!

Frothing, Fizzling, Foozelling Fantails of Coke 45 suddenly spewed sizzlingly across the intervening space and sent both clerks reeling toward an early grave as the marauding murdering miscreant mustered back out into the night.

Another sordid tale from the Naked City.

Antismokers have been getting, quite simply, nuttier and nuttier with every month that goes by as they proclaim the deadly dangers of even the shortest and most fleeting exposures to secondhand smoke in hopes of new, attention-getting headlines.* The public presentations of these studies never mention that they involve sticking smoke-avoiders of the most extreme variety into smoke-choked chambers that would make a sealed airplane smoking section seem like the fresh air on a mountaintop by comparison; chambers literally 2,000% as smoky as those airplanes of years past.

The funny thing is that the researchers – those who jump up on a podium at a moment's notice to spread the fear that even healthy people might meet the Grim Reaper at the mere sight of a smoker – feel no qualms about paying innocent subjects $100 to "risk death" in these chambers. Indeed, it seems to be routine for antismoking scientists to claim, "No risk is expected to volunteers in collecting the data," when subjects are sent to spend hours measuring smoke with little backpack sniffer-boxes in smoky bars and restaurants.[28]

---

* Antismokers prefer the term "Secondhand smoke" rather than the more accurate "secondary smoke" because it evokes the image of smoke that's been "used" and cast off as waste by smokers, although most of it comes from the burning end of a cigarette. As Campaign For Tobacco Free Kids puts it, *"Secondhand smoke seems like the most unappetizing name for smoke inhaled by nonsmokers, and using the most unappetizing name possible should probably be the goal for those of us who are working to prevent and reduce smoking."* Unfortunately, the term has been so widely promoted in the public mind that it's hard to avoid without introducing confusion. Discussion available for viewing at CTFK site at: http://tobaccocontrol.bmj.com/content/8/2/156.abstract/reply#tobaccocontrol_el_20.

The Island of Dr. Michaelious

In his groundbreaking experiments of early 2012, *"The Health Conse-quences of Involuntary Exposure to DHMO,"* Dr. Michaelious McFadden found that even rather brief exposure to relatively small amounts of DiHydrogen MOnoxide can be a killer.*

In a study hailed as being ahead of its time, Dr. Michaelious paid 15 students $50 apiece and then held their heads in five-gallon buckets of water for just five minutes. Upon laying them out on mats after-wards, he was appalled to note the drastic effect that even such brief exposure to relatively small quantities of DHMO had upon their circu-latory and respiratory systems. Uniformly, by every modern scientific measurement available, the subjects generally expired.

One student, a member of the Swim Club and founder of Students For a Smoke-Free Campus, showed a profoundly atypical reaction and began spontaneously reviving after several minutes on the mat. Dr. Michaelious duly noted the odd reaction and immediately repeated the experiment to see if it would recur. According to his meticulous notes, it did not. Autopsies revealed that DHMO exposure was definitely the culprit in every case, despite Big Water's predictable claims to the contrary.

The Michaelious Experiments met with great academic success and approval, and his conclusions laid the groundwork for many new laws and workplace regulations. Despite his good works, however, Dr. Michaelious is currently being held by authorities pending investiga-tion for intent to defraud. Evidently he'd paid the students by check without actually having funds in his checking account to cover those payments. The subjects themselves never complained, but some disgruntled family members (probably DHMO abusers) were not so charitably inclined.

---

* The good Doctor can't claim full credit in his concern about DHMO. There's an entire website devoted to its dangers at: http://www.dhmo.org/

Over the past several years, I've tried to help some of the bar owners and workers in Hawaii whose lives and livelihoods have been devastated by the smoking ban there. Even though Hawaii's climate is ideal for outdoor patios, many Japanese tourists find it unbearably rude to be told they cannot sit at the bar while enjoying a smoke with their beer and have evidently stopped coming back. Those who have found themselves subjected to grossly limited hotel smoking room availability and the specter of open-air beach bans are even more deeply insulted. Those tourists then return home to spread the tale that smoking is allowed virtually nowhere in the Hawaiian paradise. The result, naturally enough, is that in following years many Japanese tourists have sought other vacation spots where they'd be made to feel more welcome.

The loss of Japanese tourism dollars has dealt those business owners and the larger island economy a significant blow, but the Hawaiian antismoking army is a strong one and they insist that the bar, restaurant, and hotel owners, and even Hawaii's Board of Tourism are all simply making up tales of losses for some unknown, but surely nefarious, reason! Those claiming that the smoking ban has caused economic problems are simply written off as "Big Tobacco and its Allies."*

Thus was born the tale of a fanciful next step in Hawaii's efforts to attract only the "best" sort of tourists.

---

* Actually, from what I have heard on an anecdotal level from my Hawaiian contacts, many of the bars that have survived have done so by ignoring the ban in various below-the-radar fashions.

COME TO HAWAII!
The Land of Sex-Free Hotels!
{Leaked from the Hawaiian Future Tourism Bureau}

Under a new Hawaiian Health Ministry program, Hawaii's family-friendly tourist hotels will be going SEX-FREE!

No longer will your children be kept up late at night listening to "banging" noises, breathless grunting, and animalistic howls from the room next door. No longer will you find suspicious stains on poorly cleaned bed sheets or gag on the lingering musky odors of fornication. Nevermore worry about your toddler choking to death after finding a used condom under the bed and exploring it with innocent lips!

Under the proposed new rules, evidence pointing to clear violations of the sex-free policies will result in violators being fined $500 to $5,000 and barred from future visits to our pristine shores. Evidence may take the form of testimony from sharp-eared neighbors in nearby rooms, physical evidence such as used condoms or lubricant bottles, laboratory evidence of stains on sheets or UV fluorescence in tubs, or in hard-to-prove cases, blood tests for sexually induced endorphins.

➔ SPECIAL BONUS for diligent travelers ⬅
$50 finder's fee for confirmed reports of Unbridled Bacchanalia!

If you HEAR sex, you are AFFECTED by sex! Do not feel bashful about reporting lascivious behavior or observed hints that a couple may be planning such behavior back in their room. The new "heat-flash" sensor technology currently used to monitor school bathrooms – see http://CatchASmoker.com – is now being adapted to detect the rapid increases in body temperature that accompany sexual activity. Remember: there is no safe level of sex or exposure to it. You *cannot* be "just a little bit pregnant."

End the "Sextel" industry in Hawaii
{The preceding notice was co-sponsored by the Voluntary Human Extinction Movement (http://VHEMT.org) and ASH – Action on Sexing and Health.}

The following satire has a few in-joke references that people not involved in the day-to-day battle with the Nicotine Nannies may miss, but it was fun to write and it showcases several of their wackier claims in one grand piece of Silly-Pseudo-Science.

It is based on real studies and their misuse by ban advocates who blithely overlook the dating discrepancies involved so that they can claim that smoking bans produce amazing decreases in lung cancer and heart attacks before the bans even come into effect! Such claims were actually used successfully to promote smoking bans in the United States and abroad in the 2000s, with the media never picking up on the temporal discrepancies and almost no one being any the wiser.

Except now, dear reader, you.

Magickal Tobacco Smoke!

BREAKTHRU! BREAKTHRU! BREAKTHRU!!!!

The newest **PARADIGM SHATTERING STUDY** from the *Time&SpacePortal* (Reg. TM by Roger Wood's Junky Foundation) of the WorldWide Antismoking Lobby has hit the airwaves and the world of Epidemiological Temporal Physics will never be the same again!

New research has smashed through the old *Post Hoc Ergo Propter Hoc* fallacy (*After this, therefore because of this.*) and is leading us along the Mobius Highway at warp speed into the new *Pre Hoc Ergo Propter Hoc* world.

As Dr. Michael Siegel has noted while examining the latest Italian smoking ban, "The data clearly show that the decline in heart attack rates among adults in these two age groups began prior to the implementation of the smoking ban."[29]

In a just-published study to be published in 2000-and-something, researchers will discover tomorrow that a smoking ban implemented next month will cause a fall in heart attacks THREE MONTHS AGO!

The implications of this breakthrough are *ENORMOUS!* It has long been recognized that tobacco smoke exhibits strange and wonderful Magickal properties heretofore unknown to the minds of mortal men. This was first noticed in 1998 when a rogue "Health Physicist" prominent in antismoking circles, the renowned Jamie Disgrace, argued that the merest whorl of smoke could maintain frightening integrity and fight its way through hurricane gales to find and destroy children.

His work was confirmed in the midst of the ravages of Hurricane Katrina when Louisiana's Superdome was filled with unfortunate children who had been ruthlessly assaulted by the toxic fumes of scumbag smokers while they were looting Luckies from the local Lollapaloozas.

Disgrace's landmark work is now being extended beforehand by others who will discover that even solid steel walls are no match for insidious tendrils of smoke as they wend their evil way along electrical cords only to sneak out of light sockets to attack innocent lungs cowering in their owners' apartments.

The Temporal Breakthrough!

Marlboro's Magical Molecules (Reg. trademark: ***MMM*** by Evil Big Tobacco) shredded osmotic boundaries in the early years of the new century as they learned how to sneakily degassify out of the air onto walls and cabinetry, and then regassify and launch themselves at any infants who were inadvertently brought into a death chamber that would soon be occupied by a smoker just a few months in the past. The carnage has been unspeakable and CNN's advance-delayed broadcast footage has had to be heavily censored to avoid creating panic. Reports of **Toxic Teddy Bears** (**TTBs**) stalking nurseries may be thought by some to be exaggerated flights of macabre fantasy, but the graphic footage from Canadian antismoking ads and the ever-growing piles of bloody corpses are gradually convincing even the most skeptical historians.

It is only tomorrow, however, that Greco-Roman researchers discovered the **Chronological Invertedness of Magickal Tobacco Smoke**: the ability of the unburned leaves to cause disease before they are even seedlings in a stamen's wet dream! This had previously been hinted at when New York's 2002 smoking ban was promoted through the use of a 1996 study claiming a 14% decrease in California's lung cancer rate caused by that state's 1998 bar and tavern smoking ban, however tardy researchers had largely ignored this obvious truth at the time.

But Science has now **TRIUMPHED**, and the silly superstition of temporal cause and effect shall bother us no more. Magickal Tobacco Smoke may soon bring back amputated limbs while heralding our glorious past aspirations into our indeterminate future memories! Even now (or then) it is rumored that the Italian government is conducting secret experiments deep underground in the Roman Catacombs where they are makingg use of hundreds of smoker s who have beenn enslaved and forc d topuff thous&nds s o f cigarettes at t th t0mbx of Juli`us Caes@rr in %^ n attefmpxt o x$reincarN8&% thh late xemper0rr @v Rom3#.#`~af 235f.a*(@^$XD-

3~7&$^@)(*#@|
#$(*^&$@(&*#)=<!
+(~ x
3#)%$(@#...

(We are sorry. This broadcast has been interrupted by the Pentagon. Military applications of Magickal Tobacco Smoke have been granted National Security Status. With the ending of the start of the War in Iraq tomorrow by yesterday, the new research HQ of Philip Morris, RJ Reynolds, and the National Security Agency will take up past residence in the newly Puffed-Up™ buildings of the World Trade Center without previous delay.)*

---

* The various claims about lung cancer and heart attacks being reduced by smoking bans *before* the bans were even implemented, as well as the claims of the inability of hurricane-force winds to waft away curls of smoke are pretty much all straight from legislative hearings and the news. I was actually present at a New York City Council hearing in 2000 where the claim about California's 1998 ban affecting their 1996 lung cancer rates was made, although at the time neither I nor, apparently, anyone else at the hearing, was aware of the temporal discrepancy involved.

When I first came up with the beginnings of the following piece in 2006, "secondhand fat" was nothing more than an outright fantasy. Yes, people poking fun at antismoking extremists sometimes talked about the problems of being squeezed between two heavyweights on an airplane, but airlines hadn't even begun to broach the idea of charging people by weight and the concept that fat could be contagious was as laughable as... as... well, as secondhand smoke had been not that long ago.

Times change, eh?

By 2007, not only were airlines and insurance companies eyeing the pocketbooks of the generously endowed, but researchers had actually produced a study that claimed teenagers could "catch" being fat from their parents and that even adults might suffer from a form of contagion as they associate with obese friends who persuade them that "Chubby is Chic!" and "Fat is Fun!" and urge them to join groups like the National Association to Advance Fat Acceptance.[30]

Despite the development of such contrary advocacy groups, the world is beginning to look at the McWhopperies and Chipperdoodles with the same jaundiced eye that was once reserved for Evil Big Tobacco. On the website of SmokersClub.com there is even a short video of a Pennsylvania Burger King with more smoke pouring out of its chimney than a thousand smokers could produce.[31] On YouTube there is a video seriously urging that pictures of obese people with bedsores be applied to sugary and fatty supermarket foods, a minimum purchase age set for soft drinks, and courts start giving serious consideration to removing obese children from their parents and putting them under proper care by the State.[32] There is another YouTube presentation, featuring an expert from the University of California, where a medical doctor proclaims that "We are in the midst of the biggest public health crisis in the history of the world!" – and then goes on to explain the "pandemic metabolic disease" caused by our fondness for sodas and cookies.[33]

People ask me at times where I think the Great American Antismoking Crusade is going next. I used to feel confident in pointing at alcohol, but in light of some recent developments I have to honestly say that I simply have no idea. No one can really predict what direction a parade of lunatics is going to take at the next intersection.

Fork You!
{Ancient Chinese Proverb}

Addiction experts worldwide have begun to realize that forks and spoons are basically Western-style "enabling tools" for the morbidly obese and the greedily gluttonous who drain our taxes and degrade the quality of our lives with their fetid folds of floppy flesh.

The Eastern use of chopsticks, however, makes each morsel a contemplative experience while pleasantly slowing overall consumption. There are few obese native Asians. Those few who do have weight problems often own Western culinary implements; Sumo wrestlers are particularly proud of their deadly cutlery collections. (Oversized tablespoons are current eBay favorites.)

Forks and spoons are clearly neither necessary nor intrinsic to healthy eating. Billions of people throughout history have survived and thrived using only fingers, twigs, chopsticks, and such. In the US, the FDA could easily be empowered to stem the Utensil Problem's contribution to the Obesity Epidemic despite industry lobbying. Big Food is almost always pulling the strings behind the Naysayers and the giggling Chubby Cheerleader types.

Morbid obesity costs all of us real money, from increased taxes to carry the failing life support systems of waddling lumps of fat, to increased wear and tear on our mass transit systems, to increased air fares borne by the thinner and more socially responsible passengers who are squashed between the seats of the over-corpulent.

It is now known that the use of Western utensils is the real root of the problem and it's one of the easiest contributing factors to control. Real Americans may not carry a lot of weight in this fight, but our cause is just, and we *will* be heard! Removing sporks from stores and restaurants simply makes good health sense. Out of sight, out of mouth! Eliminating them from the media via creative air-brushing and CGI retouching of old-fashioned dining scenes will protect our children from even thinking of forking bloated masses of congealed fat and charred carcasses down their gullets. Forkless dining is the first step toward a healthier, happier, Chopstick-Centered future for all of us as Useless Eaters are banished from the new, joyous, Fork-Free Society!

You'll thank us later.

The following piece was written shortly after the publication of a study by a Dr. Neil Klepeis that seemed clearly biased and specifically designed to support the concept of banning smoking in areas such as outdoor patios.[34] In order to achieve anything even approaching the experimental results that would be needed to justify such bans on health grounds, the researchers had to jump through some very convoluted hoops.

The situations described in the following piece are really not far from the reality a study that generated hundreds of mass media headlines about the New And Deadly Threat Of OTS: Outdoor Tobacco Smoke! (Actually, the concept of worrying about OTS was not totally new; as will be described later in the section *Studies On The Slab*, a Mr. James Repace had already grabbed the prize for that particular honor while enjoying a sunny cruise vacation as his research venue.)

Fun With Stan And Jamie has a rather unique claim to fame in its very first sentence. In 2007 a Canadian reporter whose stories often seem to have a pronounced antismoking scent to them wrote a story about smoking bans on outdoor patios. In that story Andre Picard repeated a statement that I had made to another reporter, Liz Szabo from the Washington Post, as if I had made it directly to him, and then followed it up with the following statement: *"In a posting on the website of Smoker's Club Inc., [McFadden] wrote that anti-smoking groups had "come up with some kind of scientific sounding research to justify these outdoor smoking bans."* He made no note that the quote was not actually from me, but instead was simply lifted from an antismoking character in a clearly presented piece of satire... as you will now see.[35]

Fun With Stan And Jamie!
Main Personas: Stanley Rantz, Jamie Disgrace, and Nell Klaptrap*

::curtain rises: Personas seated around table::

Disgrace: We need to come up with some kind of scientifical-sounding research to justify these outdoor smoking bans. Too many people are laughing at us and my students keep giving me raspberries behind my back when I'm drawing funny pictures of smokers on the blackboard!

Rantz: Hmm. OK, how about this? We've got a few million dollars still sitting in this year's budget that we haven't been able to give away to ANYONE because we're all getting pushed into that nasty top tax bracket. But if we DON'T use it all we won't be able to demand a bigger budget next year! Soooo... let's do an experiment with some really expensive new measuring equipment and show how bad it is to be around outdoor smokers!

Klaptrap: Er, one problem, Boss. The machines aren't expensive enough to use up all the money.

Rantz: OK... look, there are five different KINDS of machines. Let's get one of EACH!

Disgrace: Good idea! I *LOVE* scientificky tekkie stuffs!!

::scene cut::

Rantz: OK! We're all set now! Nell, you get to wear the little machine feed sniffer-thingie right here on the end of your nose while

---

* Any resemblance to antismoking nuts, living, dead, or undead, is purely coincidental.

you sit at this table. Don't mind if it pinches a bit: it makes you look sexy! Now we'll have a cigarette sitting at this table over here five feet away from you and see how it goes....

Disgrace: Damn! Zero reading.

Rantz: Hmm... maybe it's the wind direction.... OK, Nell, try to move to a table so you're sitting directly where the smoke will blow in your face. OK... that's good.

Disgrace: Damn! Still can't get a reading big enough to measure.

Rantz: ::sigh:: It's all this AIR moving around out here causing the problem. Nell, pretend you just got to an outdoor diner and there are five smokers all sitting in a circle around one empty table in the middle. Now go sit there and let's see what we measure.... Ahhh! Finally! We're getting a reading showing that if you sit in the middle of a filled up smoking section that you might accidentally breathe a trace of smoke!

Disgrace: Hmm... it's still not enough to say anything bad about though. Look, let's move all the smokers' tables together so they're all touching each other in a teensy-tiny circle. Now Nell, climb over the tables and sit on that little chair in the middle. Just pretend it's the last available chair in the whole dining area and you don't really want a table of your own, OK?

Disgrace: Darn it! The readings are STILL too small! These tables are just too BIG! The smelly icky smokers are still too far away.

Klaptrap: Well... er... the tables are just standard thirty-inch tables, boss.

Disgrace: I DON'T CARE! Get me some SMALLER tables! Twenty-inch ones! Fifteen-inch ones! Ten-inch ones!!! We'll lick this problem yet!

::shuffleshufflescrapebangdrag::

Rantz: Nell, OK, just stand in that little circle. Now, you smokers: you see where Nell's nose is? I want each of you to light a cigarette and hold it just 10 inches from her nose. Yes, you, right in front! Ten inches! You two on the sides, about eight inches from her ears. And you two guys in the back: just about three inches away ... Goooooood! Now everybody Spark Up!!!

::Mix Rantz and Disgrace with Klaptrap screamings::

AARRRGGHHHH! HELP!! HELLLLPPP!!! Get the FIRE EXTIN-GUISHER!!!! Nell's HAIR is on fire!!!

::runrunrun::
::spraysprayspray::
::soaksoaksoak::
::bandagebandagebandage::

Rantz: (Will someone PLEASE give Nell a tranquilizer?) OK Nell, we need you to do it one more time. Now this time you don't HAVE any hair so you don't have to worry about it catching on fire. You'll be fine! Disgrace? Disgrace??? Where ARE you Disgrace????

Disgrace: Whoops! Sorry... I thought I saw a tiny tornado over in that flower pot and I was chasing it with some smoke. OK, let's see what the dials show now ... Ahh! At last! We're getting some decent readings on what could happen to an innocent nonsmoker outdoors in a typical situation standing in the middle of a tiny circle of people chain smoking. We'll make Nell stand there for ten whole minutes, measure what she breathes, and call this a "Cigarette Event."

Rantz: Now ... hmm ... let's do some figuring. OK! If Nell came to this diner nine times a day and always stood in the middle of tight little circles of chain smokers, she'd be breathing something that would nearly be sort of like an atmospheric mix kind of similar in a way to the air of a place like New York or Philadelphia or Cleveland or something.

Disgrace: All Right! Time to write up the press release! Hmm.... hmm.... Darn it Stanley! These numbers STILL aren't all that impressive. Look, let's move this all into the living room and close up the windows to get rid of all this goddam AIR that's fouling us up.

Klaptrap: Er, Boss? Isn't this whole study about OUTdoor smoking?

Rantz: Look, twit, if we wanted your opinion we'd give it to you. We'll go indoors and just SIMULATE outdoor smoke with some fans, OK? Sheeeesh, it's the same damn thing and simulations are ALWAYS more like scientifical modeling anyway! I used to simulate being a cardiologist, ya know!

Disgrace: OK! Nell, stand in that circle! Gentlemen, FIRE UP THOSE CIGGIES! Hold them close! Closer!! *CLOSER DAMN YOU!!!* Ahhh! OK! Good! We're getting some fine readings now. Wonderful! Magnificent! So True To Life!!

Klaptrap: Look, Boss, I don't mean to be disrespectful, but it would really help our press release if you could show readings like this from being around just ONE smoker instead of five.

Rantz: ::sigh:: Nell, you are SUCH an annoying stickler about things! You spend too much time over on Siegel's denialist blog! OK. Look, Nell, just sit here. Now I'm going to burn a cigarette fifteen inches in front of your face here for ten minutes with this little fan blowing the smoke right at your nose. Yes, I know it's uncomfortable but we'll give you an oxygen bottle and you'll be fine ... just squint your eyes a bit and DON'T MOVE!!!

Disgrace: GREAT! WONDERFUL reading! Look at those numbers! Wow!!! Just from being exposed to ONE SMOKER in a simulated verisimilitudinalacious outdoorsy-type environment! Nell, you ready for one more reading? Move in just a little bit closer…

:: *BOOOOOOMMMMM!!!!!* ::

::Screams::

::Mix of all::

OH MY GODDD!!! Nell's oxygen bottle just EXPLODED! I ***TOLD*** you to be careful with those cigarettes! They're deadly weapons you know! Help!! *HELLLPPP!!!* Someone call the Surgeon General!!!!!

::scene cut to St. Joan of Arc Cemetery setting::

Rantz: Well, let's look on the bright side. We've now proven that secondhand smoke kills people and we FINALLY have a death certificate for evidence!

Disgrace (nodding): Yep! OK boss, can we go to Kansas now?
I hear there are lots of nice clean air tornados there at this time of year!

::fade to smoke::

Another Great Peer-Reviewed Scientificacal Study by Rantz, Disgrace, and Klaptrap!

In mid-2007, the Mayor of Belmont, California, Coralin Feierbach, came out in favor of a plan to ban indoor *and* outdoor smoking within twenty feet of all workplaces.[36] While it may be hard to believe in today's world, such a proposal was almost unheard of at the time. An image sprang to my mind and resulted in the following unpublished submission to a newspaper out there.

Alice Down The Worm Hole

{a.k.a. "Alice the Health Inspector"}

Reading about Belmont's smoking ban, one that would ban smoking in all indoor and outdoor workplaces as well as within 20 feet of any place where smoking is not allowed, reminded me of a favorite childhood tale, *Alice Down The Worm Hole.*

Alice walks up to Mr. Caterpillar and asks a question. "Mr. Caterpillar, since your job is giving answers and you do it from atop that mushroom, isn't your mushroom a smoking prohibited workplace?"

Mr. Caterpillar considers, and then admits that it is. He slowly climbs down and walks 21 feet away with his hookah.

Alice again walks up to him and stands 19 feet from the mushroom.

"Mr. Caterpillar, you will note that I am again standing in a place where smoking is not allowed. You must move 20 feet farther away!"

Mr. Caterpillar scratches his head, but her logic is unassailable.

Alice now walks 39 feet from the mushroom and confronts him again. "Mr. Caterpillar, I am still standing in a smoking prohibited zone since I am 19 feet from the area where smoking is prohibited around your cruddy old mushroom. And you are only 1 foot away from this prohibited spot. I am afraid you will have to walk another 20 feet away."

No one has seen Alice and Mr. Caterpillar for quite a while now, although it is rumored that the Hubble Space Telescope glimpsed them heading in piecemeal fashion toward a gigantic black hole near the edge of the universe where it is rumored that Mayor Coralin Feierbach has a condo.

The following was originally written in 1999 as an AOL Hometown Page complete with pictures of cute, innocent, adorable little children. It's a bit long and is quite definitely a rant; but it was meant to be such. I wanted to appear as a bit of a raving lunatic while also warning of the road ahead.

I knew the path because I used to help write the road signs. I often say that one reason I understand Antismoker psychology so well is that, as mentioned earlier, I used to be an Anti of a sort. In the 1970s and 1980s, I worked with a radical transit collective, promoting bicycles and mass transit as automotive alternatives. We developed and used primitive versions of antismoking style propaganda and wielded similar weapons: make the driver feel guilty and outcast, emphasize the harms to children, and take any quote or reference from anyone who supported our case and repeat it as Biblical verse.

Some of us even advocated direct action techniques similar to those pushed by some Antismokers – putting up fake no parking signs or blocking off "play streets" at random in order to make driving less convenient and make drivers walk farther, pushing for higher gasoline taxes, orchestrating letter-writing campaigns under multiple names, playing on the needs of the disabled and the vulnerability of the very young and the very old, advocating seatbelt laws (ostensibly just to save lives but really because they made driving a little less comfortable while also reminding folks of just how dangerous the "Detroit Death Machines" actually were), and generally criticizing drivers for being an unhealthy drain on our medical system while polluting our lungs.

We even played up the idea of people being "auto-addicted" and trumpeted news (quite real, actually!) of the conspiracies of Big Auto and its allies to rip up trolley tracks in the 1930s![37]

I know the motivations, methods and tricks of the Antismokers all too well.

This web page was a bit of an experiment. I wanted to see how far down the tubes the politically correct folks had already taken us. The response wasn't too bad – only about 20% responded at all positively – although 10% actually wanted to join up. Do you think the numbers might be higher today?

S.A.F.E. 4 Our Kids!

As we enter a new millennium, America leads the world in recognizing the impact of addictions on our lives. Opiate and hallucinogen users, sadly tolerated in the 1960s, have been pushed back behind closed doors and their numbers diminished. Tobacco users can still smoke in some public places but acceptable venues for nico-junkies are disappearing and they'll soon be limited to stand-alone homes and smokers' adult trailer parks, far from kids and non-smokers. Lawsuits, taxes, and a responsible media correctly portraying smokers as addicts, lowlifes, and criminals have worked together to create this new reality.

The time has come to move to the next level in creating a society truly free of substance abuse. S.A.F.E. (*Substance Abuse Free Environments*) is an organization dedicated to raising awareness of the addictive and destructive nature of brain altering drugs. Tobacco and the other hard drugs are already being successfully dealt with. It is now time to address the touchier subjects of alcohol and caffeine.

Alcohol is touchy because of the historical mistake of Prohibition. Eighty years ago, those seeking to free mankind from the yoke of demon rum thought it could be done by simple fiat. The disaster that followed laid the foundation for decades of backlash in which public drunkenness became accepted as long as it wasn't "excessive" and didn't involve the use of a motor vehicle.

Families that drank "responsibly" had no fear of social workers despite the ready access of their children to a poisonous and addictive drug. Parents openly drank in front of their children, often using the excuse of "company" or "a glass of wine with a meal" to satisfy their cravings while setting an example for receptive and imitative minds. This must and will change in the new century. Just as tobacco users have begun to have their kids removed by courts, in the future alcohol users will also find their children moved to safer quarters.

Skillful use of media can reduce the number of active drinkers to the same manageable levels as smokers. We are tired of seeing alcohol on TV and in sports arenas. Fans will happily pay higher ticket prices or cut some playoff events rather to save their children from exposure to alcohol ads. Drug pushing does NOT belong on TV or at sports events where much of the audience is under twenty-one. We do not let Big Tobacco target our children in these media; nor should we allow

Big Alcohol into their minds. Network TV generally forbids portraying smoking as "normal or acceptable behavior" and should apply the same rule to characters using alcohol. Programs such as *Cheers* and *M*A*S*H* should never again be allowed to pollute the airwaves!

In addition to using the media, we will follow the legal path blazed by our clean-air allies. Ninety percent of all alcoholics began their addiction as under-21 children. Alcohol consumption is widely tolerated on college campuses. Those who produce and promote this deadly drug have gotten away scot-free with the mass murder of our children. It is time the Alcohol Cartel and their media lackeys feel the full legal fury of the millions whose lives they have destroyed.

Such legal actions take time though; until cultural re-education has reached a level similar to that enjoyed by tobacco, it will be difficult to get juries to unite in destroying Big Alcohol. But persistent class-action lawsuits and thousands of cases will make it real. Eventually a verdict will be won and the Liver-Killers will face destruction.

Do your part in the fight! Make an impression when you eat out – ask to be seated in the alcohol-free area. If there is none, express deep dismay or leave outright. If someone at a table near you is drinking, complain openly about the smell or the loud voice of the drunk. If a restaurant lush becomes obnoxious (as many do), simply get up and leave without paying; there is no reason for you to jeopardize your life because of a restaurant's greed for drug money! Restaurants are supposed to be places where people eat, not get high. And remember: Alcohol is a carcinogen that evaporates into the air faster than water! Alcoholics should not be able to force carcinogenic vapors on those around them! There is no safe level of exposure to alcohol.

Use the power of your children! Complain that they are being forced to watch drug use and may wander from you to grab a glass of poison from a careless drunk's table. Innocent young lips don't know the difference between soda and a rum and coke! We have fortunately left the era in which children were allowed to sip foam from dad's beer or light mom's cigarettes. We've even largely moved past the point where a child visiting from college or military service would be offered a Bud by dad during the Super Bowl. But children are still being corrupted by public examples and state regulation is sorely needed.

We are not trying to ban drinking in bars while they exist. Bars, after all, are set up for the express use of drunks who want to hang out in each other's sodden company while shooting up their drug (Why do you think they call them "shot glasses"?). But restaurants are for serving food to sustain health and should forgo alcohol's blood money. Those that can't survive on a healthy menu are better off closed. Those that remain will be safer for our kids and be happier places to eat.

While not focusing as strongly on the problem of caffeine abuse at this point, S.A.F.E. recognizes that caffeine is also a mind-altering drug very similar to nicotine in terms of brain chemistry and addictive behavior.[*38] Research has shown how "soft-drink" companies addict our children by injecting carefully measured doses of caffeine into their beverages. Coke and Pepsi freely sponsor not only sports, but all sorts of events specifically aimed at very young children; they know that when the caffeine habit is introduced early it will ensure a lifelong source of profits. Remember: Coke costs more than gasoline, and Coke, just like Big Tobacco, refuses to publish its "secret" formulae!

While health problems from caffeine are less than those from alcohol and nicotine, it's still a mind-altering addictive drug. Children need only pure water and fruit juices. Treats of non-addictive soft drinks are sometimes OK, but the gateway drug of caffeine should *never* be given to those under twenty-one. Caffeine-drugged drinks may be acceptable in adult-only settings, but adding a $1/liter tax will help keep them from the young and gladly be borne by adult users.

A final note on caffeine: that morning coffee "aroma" wafting through the house contains thousands of poisonous and carcinogenic substances. There are over 700 volatile oils in a cup of steaming coffee... and if you know the meaning of the word "volatile" you'll readily see the danger of the drug to others when used in this particular form.

Support politicians who'll make the world S.A.F.E.r for our kids!
Write us at SAFE4OurKids@xyz.com

---

[*] Note: on February 28, 2013, almost fifteen years after the above was written, the UK's *Daily Mail* ran headlines warning "Caffeine is so dangerous that it should be regulated like alcohol and cigarettes... Sales to children in particular should be restricted." The stories gave special attention to "Fizzy Drinks." Two weeks later, on March 18, a new story hit the papers about a 31-year-old woman who died after years of drinking ten liters of caffeinated soda per day.

The Senior Problem was inspired by a 2003 news story about all the negatives associated with smokers. I did not distribute it very widely because – even though I felt its satirical point was valid and strong, showing how even our parents (or ourselves when older) could be targeted in many of the same nasty ways as smokers – I felt it was almost *too* nasty.

I shared it at one point with a woman who'd written me from a nursing home where she and others were facing possible eviction, in addition to being made to hike fifty yards through a deserted parking lot to stand on a curb if they wanted to smoke. Sadly, though I continued to do my best to help her and her friends, I always felt as though a bit of a wedge had been driven between us after she read this piece. Even knowing it was meant as a satire in the wider effort to help her and her friends, I think it hit just a bit too close to home.

Seniors who smoke are perhaps the hardest hit of all when their children deny them access to their grandchildren unless they literally strip and shower to cleanse their "contamination" and as those grandchildren are warned not to get too close because they might breathe the leftover "poison" coming through Gran's dentures.

Eventually, one of the greatest joys of grandparenting – the unrestricted love and adoration of one's children's children – is destroyed as the prejudices of the parents are injected into the unknowing grandchildren. As one mother said of her mother-in-law on a parenting blog, *"I know my children do not like to be held by her since she smells like an ashtray."*

Another, whipped up by fears over recent thirdhand smoke scare stories, spoke of her restaurant-working husband, *"I know it is difficult for him especially since our son is more aware now and usually goes crazy with joy and wants to be held right away as soon as he sees Daddy coming through the door after work. He can't understand why Daddy has to wash hands, brush teeth and change clothes before he can give his hugs and kisses."* [39]

The Senior Problem is nasty... but no nastier than what smokers are hit with every day.

The Senior Problem

{Inspired by Charles Arlinghaus' *Cigarettes and Candy Bars*}[40]

After a decade of smoker-funded spending sprees our states are now running into a new crisis: the well of smoker taxation is running dry. After a certain level of taxation, the costs of imprisoning smugglers and convenience store robbers begins to become prohibitive. If we add the costs of future terrorist attacks financed by bootleggers' profits, further taxation truly becomes problematical.

We need a new minority group. Preferably one that's defenseless, has lots of money, and is about the same percentage of the population as smokers. How about Senior Citizens? After all, they are a minority group that supposedly shares many characteristics with smokers. Seniors tend to be a much greater drain on our healthcare budgets than younger folks, and while smokers more than offset their own increased health costs with tens of billions of dollars of targeted taxes, seniors actually get various forms of tax breaks! This is despite the fact that with their compromised immune systems and heavily medicated healthcare they often serve as breeding grounds for virulent strains of mutated microbes that take aim at the young and healthy who are sometimes forced to share a breathing space with the old and coughing.

It's hard for companies to know which employees to fire to avoid the pension-draining seniors that live into their 90s, but if the health authorities are to be believed, firing nonsmokers could save companies untold billions in pension payouts. As an additional measure, separation of older employees into negatively pressurized and separately ventilated work units could help stem the spread of office illness and disease. Break areas and cafeterias present a bit of a problem, but there's surely no harm in simply asking the seniors to step outside for their snacks and rest breaks. Most of them will enjoy the fresh air and seasonal changes.

In terms of living arrangements, seniors, like smokers, tend to present a somewhat greater fire-hazard in apartments and condos as they forget the pots on the stove or the candles on the mantle in much the same way some smokers forget their smokes in an ashtray. They also tend to dangerously overload their extension cords with remote-controlled gadgets they barely understand and leave lots of papers lying about. Antismoking groups spend money educating landlords about the disadvantages of renting to smokers, but landlords also need some solid education about the desirability of renting to young couples!

Antismokers like to say that smokers smell bad. Well, ask any little kid with a sharp nose who's just been visiting the grandfolks: give them the right sort of prompts and they'll probably tell you seniors smell bad too! And if grandmom happens to smoke, well, that's just a bonus as you can smile and wrinkle your face and say "Icky Grandma smells like dirty cigarettes, doesn't she?" as you leave their house. But aside from the smell, there's no reason children should be forced to breathe scads of aging, flaking skin cells from grandparents; healthy and happy non-touching visits can always be arranged outdoors.

The big problem with discrimination against seniors (aside from temporarily inconvenient constitutional and legal proprieties) is that they're a powerful lobby. They're actually a far more self-aware and powerful lobbying force than unorganized tobacco smokers ever were, and vote overwhelmingly in their own selfish self-interest. The only way to pull the false teeth out of AARP's grinning skull will be through a double punch right to the geezers' guts.

First a skillful use of "Divide and Conquer" (as used so well by the Antismoking Lobby with their "level playing field" gambit pitting restaurants against bars against casinos). In the case of the Seniors, we simply need to temporarily redefine "Senior" as anyone over seventy-five rather than over sixty. Suddenly, their political numbers are sliced by 80% and their political power carved up as well!* Once their numbers are further reduced through skillful allocation of limited medical resources, the bar can be moved gradually back downward if younger folks are feeling generous.

---

* I don't know exactly when it happened, but at some point while I was napping in the 1980s, the official definition of "middle age" was extended to 69 years old!

Second, we need a concentrated media campaign portraying the hoarding tendencies and selfishness of Seniors while slamming the shameful manipulation of our political system by their front groups and allies. Cruise lines, care homes, and resorts love to steal the inheritances of needy middle-class offspring by supposedly offering "creature comforts" to those too old to tell the difference between a bath-spa and a bedpan. When it comes to campaign time, politicians are all too willing to go to bat for their tottering "friends."

Tobacco Control groups spend hundreds of millions of dollars annually demonizing Smokers and encouraging their segregation and ostracization. Imagine how our thinking about Seniors could be changed with just a few years of similar spending! MTV audiences would be a prime vehicle for this campaign – instead of ten minutes of antismoking ads every day, they could be forced to run ten minutes of antisenior ads!

Ultimately, if Antismokers are to be believed, the final solution to the "Senior Problem" may simply be to encourage more of our youth to become smokers. Voila! The Social Security Crisis will dwindle to a puff in a cloud of Alzheimer's!

Several years ago I ran across an article in favor of a universal restaurant smoking ban. The article included a statement from a local Antismoker named Ms. Andresen. *"The smoke is more toxic than mainstream smoke. Second-hand smoke travels 50 feet to land in a plate of food that I just paid $25 for."*[41]

Her concern brought about the following image:

Evil ETS...

Eeeeviiiiilll ETS skulks down the aisles, past the waiters' table and the salad cart, past floral bouquets and innocent diners and their children, looking for Andresen... searching for Andresen... the ubiquitous Andresen... Andresen with the big quivering querulous nostrils... Andresen with the oozing oogling outrageously orange olfactory orifices....

AHAHH!!! FOUND HER!!!!!!

And **THERE'S** her FOOOOOOOOOOOOODDDD!!!!!

::jumping into Andresen's $25 plate of GreaseFries(TM) and rubbing orgasmically all over her oils::

Ahhhh....

Nothing like a good meal after a smoke!

Signed,

Eeeeviiiiilll

Ships Travelling in British Waters
Face Complete Smoking Ban!
{August 9, 2008 headline in the London Times}

"William Gibbons, director of the Passenger Shipping Association, said that announcements would be made to let passengers know when smoking was permitted. *The rules will apply to all ships, whatever the flag.'* "[42]

I couldn't ignore the challenge and sent the following note (which somewhat crosses the line from satire to serious political letters) to the *Times'* editors. Unfortunately it sank without a trace....

Rule Britannia!

Dear Editor,

Your August 9th story, "Smoking Ban in British Waters," notes that "Ships travelling in British waters face a complete smoking ban next year," and quotes the director of the Passenger Shipping Association as saying, "The rules will apply to all ships, whatever the flag."

As an American who might travel your waters, I am curious. If a lookout on a passing destroyer spots me enjoying a cigar on the bridge one morning, will the Royal Navy first fire a shot over the bow as a warning or will they simply sink the ship on sight?

It seems the psychotic fringe that has taken over the British antismoking movement has finally dragged the government as a whole to join them in their smoke-free dance through the daisy fields of insanity. Her Majesty's Government seems to have forgotten one of the primary rules of psychotherapy: you do *not* join in the dance... it only makes the psychopaths more dangerous. The near destruction of the traditional British pub has been only an appetizer. Your homes (and, as you can see, even your floating homes) are the next target.

The following satire, although at the very end of this section, was actually one of the very first I wrote, with the original version being scribed in 1998 when California was first instituting its bar smoking ban. At the time, it was still uncertain whether the nuttiness would take hold even in the LaLa land of hippies and movie stars, so I wrote a little play about what it might be like during a bar raid by the Smoke Police!

This satire also holds a special place in my memories because in the first year or two of the 2000s, I showed it to several bar owners in Philadelphia, along with a warning that if they didn't start getting active we'd see a similar ban hit here. They laughed and told me in a fatherly way that I really needed to concentrate on some other topic before people concluded I was simply a nutcase; obviously, such a ban would *never* happen in Philadelphia.*

Here's the story with only one significant change made since its writing: After the New York City ban, I changed the name of the top cop from Captain Grantz to Captain Bloomy. It was meant to be performed as a play, a "guerrilla theater" for the benefit of television cameras and reporters, but in pre-ban New York and Philly the bar owners were in deep denial and uninterested in such silliness (what I've come to call the Ostrich Syndrome), and in post-ban New York and Philly they simply gave up (the Borg Syndrome).

---

* Philadelphia's ban arrived in 2007, seemingly taking many bar owners, including several I'd spoken to in quite some depth beforehand, completely by surprise. With just a little bit of effort, several hundred bars could have gotten exemptions, but as it worked out, only about a hundred applied. Too many simply didn't want to bother with the paperwork and figured the law would never be enforced. Bad move.

S.W.A.P.P.

It's a peaceful and fairly quiet night at the neighborhood bar. Several regulars mix with a few wander-ins and the bartender keeps a wary eye for empty glasses. Suddenly the front door slams open:

(Team of four S.W.A.P.P. [Smoking Without A Permit Police] officers march into the bar, all attired in exaggerated cop-like uniforms with big hats and HUGE tin "Cigarette Police" badges.)

(Captain BLOOMY blows his whistle. Officer ODDBALL takes out a citation book. Officer WATERS presents his SuperSoaker at the ready. Officer CUTTER waves a huge pair of scissors/garden shears in her hands. Sergeant KLINK has pairs of handcuffs on his belt and holds up a pair in each hand.)

CAPTAIN BLOOMY: This is a RAID! Everybody FREEZE and put your HANDS IN THE AIR!!

CAPTAIN BLOOMY: *(Walking up to first smoker at the bar.)* Sir, I regret to inform you that you are in violation of the Crisply Clean New York Air Ordinance. Please put out your cigarette immediately. Officer ODDBALL! Give this man a ticket: $1,000 just like in California!

SIR: *(Gulps, looks down, nervous and guilty while putting out his cigarette.)*

OFFICER ODDBALL: *(Walks over and starts writing ticket.)*

SIR: *(Looks up, his eyes hardening, and says loudly…)* But Officer, this is America!

CAPTAIN BLOOMY: *(Glares.)* Sergeant KLINK! Arrest this man! He's inciting a riot!

(KLINK slams man's head on the bar while expertly applying cuffs.)

CAPTAIN BLOOMY: *(Walks up to female smoker at the bar.)* Young Lady, how can you sit there and spew pollution into the clean fresh air of our city like that?

YOUNG LADY: *(Squeaks, mumbles.)*

CAPTAIN BLOOMY: Aren't you concerned about the example you're setting for the CHILDREN? *(Gesturing around…)*

YOUNG LADY: But… but Officer, this is a bar. There ARE no children!

CAPTAIN BLOOMY: Nonsense! Everyone knows that tobacco companies put cigarette machines in bars so that every day thousands of school children will flock into the taverns to buy cigarettes!   I certainly hope you're not a MOTHER!

YOUNG LADY: Er…. Actually, I have two sons, Officer.

CAPTAIN BLOOMY: WHAT???   Sergeant KLINK! Place this woman under arrest for child abuse, notify Child Protective Services, and see that her children are taken and given to a proper God-fearing nonsmoking couple! *(KLINK slaps cuffs on woman.)*

CAPTAIN BLOOMY: *(Accosts woman sitting at a table with an ashtray on it. Picks up ashtray and ostentatiously smells butt in it.)* Madam, this cigarette is OBVIOUSLY yours, or you wouldn't be SITTING near the repulsive thing! Officer ODDBALL!! Write her a ticket and take her into the back room for a full body search for cigarettes!

(At this point another woman sitting at the bar screams, clapping her hands to her face in obvious terror. As her scream dies she tremblingly points across the bar at a man who'd been concentrating on a TV ball game and who has JUST lit the cigarette in his mouth. He is still holding the lit match.)

WOMAN: That … That … *MAN!* He just LIT A CIGARETTE!!!
(She then collapses on the bar.)

MAN: *(After a shocked second the man hastily waves out the match. The cigarette, lit, is still in his mouth.)*

CAPTAIN BLOOMY: SIR! You are under arrest for assault with a DEADLY WEAPON! Officer CUTTER! Disarm the suspect!!

OFFICER CUTTER: *(Strides over with her garden shears and cuts the cigarette in half.)*

OFFICER WATERS: *(Runs over and squirts the cigarette <and the man> thoroughly with SuperSoaker.)*

SERGEANT KLINK: *(handcuffs smoker suspect to barstool.)*

CAPTAIN BLOOMY: *(Looks around the bar in satisfaction and accosts the bartender – who himself sneakily puts his own cigarette and ashtray under the bar as he's approached by the unseeing officer.)*

CAPTAIN BLOOMY: There now, we, the Smoking Without A Permit Police have saved you from these smokers! Aren't you happy? They'll never come back to trouble you again.

BARTENDER: *(Stammers, obviously fearful.)* Oh! Er … yes! Thank you, Officer! But what will my boss say?

CAPTAIN BLOOMY: Don't worry, he'll be grateful to us too for improving his business. By the way, here's the total fine for your establishment: 27 smokers at $1,000 apiece, plus a $5,000 bonus fine for the bar smelling like smoke. $32,000! Just like in California. Always remember, as a conscript of the New American AntiSmoking State, YOU are responsible for enforcing the law in here!

CAPTAIN BLOOMY: *(Looks around again, raises hand, then barks…)* Officer CUTTER!!

OFFICER CUTTER: *(Raises Shears overhead and snaps them open and closed while shouting.)* S.W.A.P.P. TEAM MOVE OUT!!! ON TO THE NEXT BAR!!!

(All line up to head out, with ODDBALL hastily returning from the back room while pulling his pants up. Officer WATERS squirting his Super-Soaker all around while marching, KLINK rattling his handcuffs, CUTTER snapping her shears, and ODDBALL fanning at invisible traces of smoke with his ticket pad. The parade moves outward into the night while BLOOMY pauses at the door for a last look around. He nods in satisfaction at another job well done, and lights a Tiparillo cigar before rejoining his troops.)

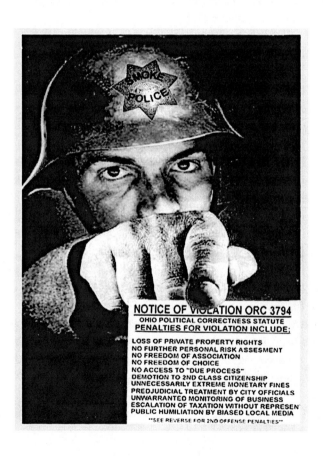

Stratistics Unbound...

Statistics are like a bikini. What they reveal is suggestive, but what they conceal is vital.

- Aaron Levenstein

In later sections we will be seeing many references to the statistical fantasies spun by Antismokers as they seek to lure more innocent passersby into their web of paranoia. *Stratistics Unbound* serves as an introduction to some of the more common tricks they use and, hopefully, will keep the numerically faint-of-heart from fleeing in panic at the sight of a correlation claim. Statistics have a real and valid use in science and public health, but when it comes to using social engineering techniques toward the end goal of creating a smoke-free world, they have been destructively abused to create fear and resentment far more than they have been constructively used to share information and enlightenment.

Stratistics

What are "stratistics"? Basically, they are statistics bent to a particular strategic and ideological purpose by researchers or advocates seeking to advance a belief or to secure future grant funding. They are what advocates of an idea develop and use when the simple facts and numbers in a situation prove to be less supportive than desired.

The best examples of stratistics and their use in today's world come, not surprisingly, from the antismoking movement, and often involve the basic foundational pillar that has driven smoking bans since Godber's 1975 World Conference on Smoking and Health. The overall guideline from the panels and speakers at that conference was one advising Antismokers that to successfully eliminate smoking, it would first be essential to foster a perception that would "emphasize that active cigarette smokers injure those around them, including their families and, especially, any infants that might be exposed involuntarily to ETS."[43]

ETS, environmental tobacco smoke, is the combination of the smoke coming from the end of the burning cigarette and that exhaled by the smoker. The claim that exposure to ETS is deadly was to become one of the antismoking movement's most powerful weapons, but at the time it was first seriously proposed as a major tool at Godber's conference, the claim had very little real scientific support.

A Bit Of Background

Before getting into specific "stratistical" examples it might be good to look at a bit of general history. Fifty years ago, there was a segment – a very small segment – of the population that was seriously bothered by their normal encounters with tobacco smoke. For the most part, even within that small segment, their concerns focused almost entirely upon specific situations in which such smoke was more concentrated than usual and in which they were less able to avoid it.

Airplanes were a prime example of such an unavoidable and concentrated situation, though particular experiences, such as being stuck between two smokers at a diner counter while trying to enjoy a breakfast of bacon and eggs, stood out as well. The relative rarity of the concern can be seen in the total lack of any enterprising entrepreneurs trying to capitalize on the niche market by offering occasional "No Smoking" flights between busy hubs. Today, if there were not already a total ban on airplane smoking, such flights would be snapped up faster than the eye could follow!

Sensitivities were particularly acute in the mornings, a time when one's olfactory senses are sharp and highly functioning after a night's respite from the social assaults of perfumes, body odors, and car exhaust; Hungarian, Australian, or Upper Mongolian cooking odors; or even a co-worker's unfortunate fondness for too much Old Spice aftershave or pine-scented air-fresheners. Bars were simply written off as being smoky dens that would never change; the idea of banning smoking in those dens of iniquity was pretty much simply unimaginable in the world of the 1970s. Indeed, even as late as 1998, Clive Bates, the head of Action on Smoking and Health's UK branch, declared that rumors of plans to ban smoking in pubs and restaurants was just "scaremongering" by the tobacco industry. Less than ten years before every pub in the UK would be forced to throw its smokers into the streets, ASH UK reassured the public that, "No one is seriously talking about a complete ban on smoking in pubs and restaurants."[44]

Outside of rarely encountered situations, it was relatively unusual for anyone to be bothered enough to think of complaining about another's smoke, and even such a non-confrontational adjustment as changing one's seat was uncommon unless a particularly offensive five-cent stogie happened to be set afire. And it was pretty much unknown for anyone to claim they could not tolerate a work situation because of the amount of smoke in the air, despite the fact that probably about 90% of work situations involved varying but, relative to today, quite intense exposure.[*]

How can I say with surety that such reactions to smoke were highly unusual? Simple: I was alive at the time and I noticed such things because as a young boy, I was one of those oddballs who hated

---

[*] Two well-known exceptions to the above rule involved, not surprisingly, two of the people who have gone on to become high-profile antismoking advocates today: Martin Pion, the idealistic and hard-working founder of Missouri GASP, and James Repace, a self-styled "Secondhand Smoke Consultant," of whom we'll be hearing more about later.

cigarette smoke. I remember how totally and completely alone I was in my desire to avoid people smoking. I was seemingly unique in my willingness to complain about or fan smoke away from my face at a restaurant counter if a passing dining acquaintance of my grandmother happened to light up. I remember several such incidents where people looked at me in puzzlement, concerned about what might be wrong with me, while my grandmother reassured them that I was just a bit strange in that way. They'd usually respond with a smile and a comment about how most kids had little likes and dislikes that they grow out of. And then, of course, they'd light up another smoke!

I was the only kid among all of my friends and classmates that I ever knew to feel strongly about smoke in those permissive times. Indeed, smoking was so totally accepted that I remember my future kindergarten teacher smoking at the registration desk in the basement of the local church while talking to my mother the day before school started. My teacher didn't usually smoke during the bulk of our day at kindergarten, thank goodness, but she did always take a two-cigarette break while sitting at her desk and watching us all as we lay on our mats for our daily half hour of nap-time. So I most definitely noticed other peoples' reactions, or more usually, their lack of reactions, to smoke and smoking.

Is there other evidence I can offer that such strong feelings were rare, aside from my personal memories? Again, it's fairly simple; if you go back to the records of our culture before the 1970s, you will find almost no mention at all of anyone with such concerns unless they were heavily involved with fringe religious or temperance groups. True, Big Tobacco may have had some influence over cinematic portrayals of smoking, but that strength did not really build until the era of World War Two. Yet there are virtually no instances of anyone waving smoke away or making strangled coughing sounds or complaining about smoke during the earlier cinematic period when such pressures had little, if any, significance. Other than the very occasional scene of a harassing wife bitching about the boys' cigars smelling up her house on poker night, such a concept was essentially absent from the silver screen.

Now, think about it: if the reality was the opposite, wouldn't we have seen it portrayed as such? What would have been the point of hiding or twisting such a basic facet of life in a universally smoky world? Wouldn't five, or ten, or even twenty percent of movie directors, writers, and producers have objected to the constant assault of smoke strongly enough to put at least some passing statements about it into their films? Evidently not, and I strongly doubt that the evil minions of the tobacco companies swarmed through the movie vaults of Hollywood in the 1950s and systematically burned all the "offensive" antismoking instances ever recorded on film.

Even if they *had* succeeded in such a cultural cleansing, even if such firebrushing of silver nitrate stock *had* taken place, wouldn't it have been noted somewhere, anywhere, in the fiercely pro-freedom, anti-censorship political movements that soon followed? Yes, portrayals of sex and drugs were censored, and that censorship has been duly noted and roundly examined in history... but censorship of negative tobacco attitudes? If any such censorship did exist in the film world, it was too rare to have ever been recorded in any source I've seen.

Going beyond visual depictions, what about literature? Of the tens of thousands of authors penning novels and stories, wouldn't there be hundreds of expressions of such feelings? I've always been a fairly prolific reader of fiction of all kinds in my life – though with some bias toward science fiction and bad vampire novels – and I can recall only one instance of an Antismoker appearing prominently in a story before the "politically correct" invasion of the 1980s and 1990s. It was a 1950s story by Isaac Asimov, one of the most respected and prolific science fiction authors of the twentieth century, and one who was known to have at least some degree of personal aversion to smoking in his private life. Despite those feelings, it was an unusual enough "bug" for someone to have that he rarely mentioned it in any form in his writings outside of one particular short story, *The Dead Past*.[45]

In that story, a scientist was striving mightily to perfect a machine that would allow archeological historians to view the distant past. As the story progresses, there are several instances where he reacts with strong and notable negativity when another character would take out or prepare to light a cigarette. It was an unusual enough reaction that other characters in the story were fairly surprised by it. One character took enough note of its oddity that much later, right near the end of the tale, he realized that the scientist's affliction must have deep psychological roots.

Sure enough, it turns out at the end that the scientist tries to scuttle his entire life's research into his time viewer when he discovers it would primarily only be useful for viewing the recent past of the last few decades. Why the sudden switch? Because, twenty years earlier, his child had died in a house fire that he always guiltily suspected he'd caused with one of his cigarettes. His reaction and guilt after that fire had caused him to become something almost unique in the world of the 1950s: an Antismoker. The unusual strength of his antismoking feelings brought about the ultimate realization by the other character in the story that something along those lines must have happened in the past and the man's guilty secret was exposed!

Aside from this one tale, though, there's almost nothing in America's post-Prohibition general literature before the mid-1970s that shows a strong antismoking leaning. While censorship of *Fanny Hill* and Mao's *Little Red Book* were well known, I have never heard of a novel or political tract of that time meeting the knife of a smoky Winston Smith* because of an antismoking scene or piece of dialogue. Practically the only places outside of Asimov's short story where such writing seemed to exist were venues that were religiously, rather than societally, driven. In particular, publications like the *Christian Herald* and the remnants of groups like the Women's Christian Temperance Union were by no means abashed about letting their feelings about the tobacco habit be known. But even those groups spent most of their energy railing against the physical and theorized moral harms of the smoking habit on smokers themselves, with only occasional adjectival slams at the "stink" of smoking on the side.

One of the first such instances that I'm aware of in post-Hitler times was in the *Christian Herald* itself, as presented by foot-massaging reflexologist Don Matchan in his unusual 1977 antismoking book, *We Mind If You Smoke*.[46] Matchan presented a reproduction of an article written in the October 1964 issue of the *Christian Herald* where the author lamented his travails of traveling to a hotel in New York City. Along every step of the way, from train cars to train stations to taxis to

---

* Iconic character in George Orwell's *1984*. Smith's job was to rewrite old news articles to conform with current, constantly-changing, approved political beliefs, erasing and changing facts to match up with current political dogma.

the hotel lobby and even in the elevator, he was assailed by clouds of smoke; until finally he hurries "to antisocially seek the privacy – and clean air – of my room. That room is the one place I can find with pure air."[47]

The story is notable not just because of the rarity of having someone express such extreme concern about random encounters with smoke at that point in history, but also because of the author's vast relief at finally getting to the pure, clean air of his hotel room – a room almost certainly surrounded by other rooms filled with people smoking, and a room in which thousands had most definitely smoked in the past! But in the 1960s, even an extremist, a fanatic, a likely candidate for psychological counseling in that era, even such a person would never have thought of worrying about people smoking in other rooms or fretted over fears of previous smokers in the room they were currently occupying. With very few exceptions, the concept of being concerned about such things, or even noticing such things, was simply beyond rational thought even by crazy people in that period.

That is simply a reality, a reality with no practical denial, and a reality that almost anyone over fifty with a clear and fair mind will quickly confirm. It's a reality attested to by the fact that in all of Google's billions of pages, there are less than a dozen pre-1980 hits on the phrase "smoking in the jury room," after references to *12 Angry Men* are deleted.* This is despite the fact that hundreds of thousands of juries throughout the United States have deliberated amidst clouds of smoke over the years without any difficulty. It was simply, for most normal people, a complete nonissue.

And yet today, the antismoking movement would have us believe, and has largely convinced the majority of the population to believe, that this reality never existed. They would have us believe that the ordinary nonsmoker has been choking, gasping, pleading, and literally dying from fogbanks of poisonous gas for decades. Some of the favorite antismoking sound bites to gain popularity in the 2000s have revolved around the concept of the "Right To Breathe" trumping the "Right To Smoke," as though the two could never, and have never,

---

* Googling ("smoking in the jury room" -"angry men") in May of 2013 produced just ten hits, with four of them about a single Antismoker in a 1988 jury.

coexisted. How have they managed to create that belief, that false world which existed almost nowhere outside of a very small number of fairly warped minds fifty years ago?

The answer lies largely in the world of politics and idealism, money and prestige, propaganda and statistics, and yes, stratistics. Stratistics created the fertile soil of fear in which people could grow conditioned to taking instant notice of the barest hints of smoke in the air, either by sight or by smell; a soil fertilized further by silly warnings along the lines of, "just like asbestos and radon, there's no safe level of exposure,"* and statements like, "Your nose isn't lying. The stuff is so toxic that your brain is telling you: 'Get away.' "[48]

Without the constant drumbeat of the stratistics served up by morning talk shows, news articles, and TV stories on new research that seem to appear practically every day, such words by themselves would be laughed at. But people respect and fear numbers, particularly when those numbers are surrounded by scientific-sounding terminology and are presented with an aura of unquestionable exactitude by supposedly responsible and respected authorities.

The power of using such stratistics to build fear around smoking actually predates Godber's 1975 conference. It was originally applied, not to fears around secondhand smoke, but to fears aimed directly at smokers themselves in hopes they would be driven to quit.

Smoking And Death

One of the earliest examples of popular scare stratistics created to reduce smoking was the claim that 85% of lung cancer deaths occurred in smokers. There were two clear problems with this claim. The first problem was simply that during that period of history, the proportion of the population meeting the formal definition of a "smoker" – basically anyone who admitted to having smoked more than five packs of cigarettes in their entire life wasn't all that far below 85%.   Given

---

* Such warnings nicely overlook the fact that the same could be said of an errant sunbeam sneaking through a curtained window, or even the carcinogenic alcohol fumes wafting out of a wine glass at a neighboring restaurant table.

that fact, it's not so surprising that 85% of lung cancers might have occurred in smokers. Pure statistical chance, even if smoking had *no* effect on lung cancer, could easily have resulted in producing at least some studies finding such percentages of lung cancers in smokers, particularly once the question of possible confounders was considered.

The second problem was that the 85% figure was really only half a statistic; the second, and more meaningful half of that statistic – the percentage of smokers who might expect to get lung cancer – was almost totally invisible to anyone outside of the dedicated medical research community from the 1950s through the 1980s. That second half of the statistic would have stated that a lifelong smoker's chances of eventually getting lung cancer generally ranged from under 5% up toward a maximum of 15% or so – quite significant as a relative epidemiological risk when compared to nonsmokers, but not nearly the virtual death sentence that was commonly being pictured. Of course, if there were additional clear and unusual aggravating factors, such as long-term, heavy exposure to asbestos, that 15% figure could go even higher, but for most smokers the fears were clearly being exaggerated for a behavioral effect.

The constant repetition of just the first half of the statistic is what made it into a stratistic. It was a statistic that had been deliberately misrepresented in a form calculated to produce a specific socially desired effect: scaring smokers into quitting by causing them to make the mistake of assuming that the vast majority of smokers would die young and horrible deaths from lung cancer during a time period when cancer was even more feared than it is today – almost superstitiously so. Back in the 1950s and 1960s, people seemed almost to fear saying the word and those who died from cancer were almost always noted as simply having passed away "after a long illness" or some other related euphemism. Sometimes a brave soul would say about a passed one, "Yeah, he had the Big C," but even that was uttered with mannerisms that might put one in mind of crossing oneself when speaking of vampires.

Another stratistic favored by Antismokers since the 1960s and still in wide use today is the statement "Smoking kills 400,000 Americans a year!" A close variation is "Smoking kills 1 in 3 smokers!" and this is sometimes increased to 1 in 2 or "50% of all smokers." If you go back in time, you'll see how the claims gradually increased from the initial twenty to thirty percent as the computer formulae were juggled to produce more impact. I would never argue that smoking is harmless, or even attempt to downplay it as having the potential for a significant and concerning health impact over the course of one's life, but some of the numbers produced by antismoking stratisticians deserve serious reconsideration – particularly when they are extended into the craziness of microscopic or sub-microscopic secondhand and thirdhand smoke exposures.

But to stick to the effects of smoking itself for the moment, most people assume that these 400,000 deaths are "real" numbers – in the sense that doctors have actually determined that 400,000 corpses each year would still be walking around enjoying happy and healthy lives if they hadn't smoked. In reality, this stratistic, and hundreds of smaller derivatives for local areas trotted out at smoking ban hearings, are just creations of a computer program family known best as SAMMEC – Smoking Attributable Mortality, Morbidity, and Economic Costs.

SAMMEC is based on a set of made-up formulae fed into a computer that wildly estimate how many smokers will eventually die from various diseases if no further medical progress is ever made. The total number churned out at the end depends on how many smokers are put in at the beginning and on what numbers and percentages and variables are plugged into the various sub-formulae within the overall program. The resulting number can be made to go up, down, sideways, or completely around the bend just by adjusting some of those formulas or ignoring/including various confounding variables. Even when researchers supposedly correct for such things, those corrections may sometimes be included simply because they help shift the outcomes in the desired direction. Finally, about half the "smoking-related deaths" spewed out by SAMMEC actually occur in people over seventy-five years of age. Almost 20% occur in individuals over *eighty-five* years of age![49] Think for a moment: If you'd had a lifetime of enjoying your after-dinner cigars, would you *really* regret dying at age 87 because you'd smoked them?

A primitive stratistical contortion of one of the concepts above appeared in a newspaper several years ago: "In schools across our area, students have been telling their peers to stay off the stuff because it kills one in five Americans each year." For the mathematically chall-enged, that currently equates to over 60 million Americans a year meeting the grinning reaper. If these numbers were true, America would be populated by almost nothing but healthy, high-jumping, smoke-free rabbits within five years!*[50]

MaxiMiniMicrotizing

MaxiMiniMicrotizing is a stratistical way of taking advantage of people's ignorance of scientific terminology. Yes, people are generally aware that prefixes like kilo- and mega- mean something big while micro- and nano- mean something small, but most people don't know much more than that. If you ask them how many micrograms or nanograms are in a quart of tap water, few would come near the correct answer of about a billion micrograms or a trillion nanograms, and if asked about picograms, even fewer would guess a quadrillion. Probably only about one person in twenty would recall ever having heard the word picogram, and I doubt one in a hundred would be familiar with the term femtogram – much less have any real idea what one is.

But a good antismoking stratistician can terrify a roomful of soccer moms by warning them that their babies may absorb hundreds of femtograms of a deadly neurotoxin through their skin if smoky grandmom is allowed to hold the child without first having been put through a bio-warfare style sterilization chamber. A clever propagandist can warn of a single cigarette producing 30,000 picograms of something like arsenic without anyone in the room knowing that a single, almost invisible, grain of table salt masses about a hundred million picograms.[51]

By pretending that such small amounts of almost anything have any real physical meaning outside of a laboratory, stratisticians lie to and mislead the scientifically innocent, and they do so in a very clear and conscious attempt to frighten them and engineer their behavior in a desired direction. To drive this point home, let's look at an example, the story of "The Baby And The Deadly Arsenic Smoker."

---

* OK. Strictly speaking, successive statistical reductions would still leave us a hundred million or so pleasingly plump rabbit-eaters. Nice stratistic though!

A standard cigarette produces about 30 nanograms of arsenic.[52] A reasonably sized, standard living room might contain about 50 cubic meters of air that is changed about three times an hour, for a total of 150 cubic meters for the cigarette's smoke to disperse through during the course of that hour. A child breathes somewhat faster than an adult, but his or her lungs are far smaller, so whereas an adult engaging in moderate activity might breathe a full cubic meter of air in an hour, a small child would likely breathe only about half that. Now let's say that you're the type of mom who would let her little girl play happily with some blocks on the rug as you enjoy a cigarette while watching a rerun of Oprah in the afternoon.

The 30 nanograms (i.e., 30,000 picograms) of arsenic from your cigarette are going to disperse through those 150 cubic meters of air. Your poor little architect-in-training will inhale at least a half cubic meter of that air, potentially absorbing 1/300th of those 30 nanograms (i.e., a total of 100 picograms) while hefting her building blocks. An Antismoker might confront you with the following screamed accusation: *"You are poisoning your child with RAT POISON! You're forcing the poor little innocent to breathe 100 picograms of arsenic from your cigarette!"*

Sounds bad, eh? Even though you kinda figger that picograms aren't that big, still, your poor kid is going to gulp down a hundred of them, all filled with arsenic, which you *know* is poisonous and is indeed used as rat poison! And you are deliberately blowing them out into the room with your innocent little child who is only trying to gasp some air in order to stay alive! What the hell kind of parent are you anyway???

You've just been stratisticized. There are several ways to look at this that will make you less vulnerable next time.

First of all, a hundred nothings is still nothing. And a hundred times a hundred nothings is also still nothing. Now to be fair, a picogram is not exactly nothing. It's a very real quantity and your smoky cigarette is indeed resulting in your child arguably ingesting a hundred picograms of indisputably real and deadly arsenic.[53]

But how real is it? How real *are* those 100 picograms of arsenic in terms of anything that should have any meaningful impact upon your approach to life and your love for your child? Well, let's say you decided to spend your entire life collecting arsenic, and you collected a picogram every single minute of every single day, both day and night, for a hundred years and piled it all up in the middle of New York's Time Square.

What do you think it would look like? If you do the multiplications, you'll get 60 minutes x 24 hours x 365 days x 100 years as being the total number of picograms collected in an enterprising lifetime: a grand total of about 50 million picograms! Would it be a big pile? A small pile? A truckload? Maybe just a teaspoon or two?

Nope. It would be a speck. A speck about half the size of a single grain of table salt (which weighs in at roughly 100 million picograms).[54] And if you ate or inhaled *all* of it, would you die? Heh... nope again. You could happily eat a thousand times that amount, collected after a thousand lifetimes devoted to nothing but arsenic collection from smoke, and still go out for your morning jog and die by getting hit by a truck. Arsenic is actually an essential ingredient for healthy human metabolism. The average, healthy human body has about twenty milligrams of arsenic in it – about two hundred million times the amount you were supposedly "poisoning" Little Miss Architect with.

The overall point here is that these quantities and measurements are absurdly small. No ordinary people would (or should) ever have to think about such things as femtograms or picograms. They ordinarily have no meaning at all in normal human experience.[55]*

So let's return to you watching Oprah every day of the week, rain or shine, and the warning that you are exposing your little block builder to 100 picograms of arsenic each time. You would have to sit there and smoke with your child for roughly 3,000 years before reaching that grain of salt level. But, if you were truly a psychopathic parent and actually *did* want to poison your little one, you'd have to reach a level of about ten *milligrams*. It would take roughly 300 billion years to accomplish your dastardly deed – about 30 times as long as the

---

* Another quick example: Would you give your child a nice little alcohol cocktail to get them going on a cold morning? No? Well, if you give them an eight-ounce glass of healthy, vitamin-filled orange juice, you'll also be giving them roughly 100,000,000,000 (one hundred billion) picograms of pure, rotgut, white-lightning grain alcohol! It's just a totally natural part of orange juice!

entire universe has existed. As for Oprah, if she lasted that long, just think of all the Botox involved!

Of course, a normal human child would usually excrete most of that arsenic along the way,* so a ground rule would have to be set: no changing of diapers allowed for those 300,000,000,000 years. Hopefully at the end of it all you'd be able to find an antismoking researcher to take over that rather unpleasant task.[56]

Arsenic is not the only element in smoke to be played with like this. Polonium, formaldehyde, and cadmium are also antismoking favorites. Polonium 210 will be examined in some detail in the later section on *Thirdhand Smoke,* and formaldehyde gets a once-over on the *Of Vapors and Vapors* Slab of *Studies On The Slab,* but let's take a quick look at the heavy metal cadmium here. Cadmium poisoning can cause symptoms like headache, anemia, fatigue, muscle and joint pain, and a generally rundown feeling called the "cadmium blues."[57, 58]

Well, the town of Alexendria, Louisiana was celebrating its first anniversary of a smoking ban with a January 31st, 2013 dinner at a local restaurant with a local actress, Faith Ford, as a special guest. Ms. Ford told attendees her scary story of having been diagnosed by her doctor with cadmium poisoning which the doc attributed to her living in a house where people smoked. She was very concerned, she said, because she put a lot of effort and attention into "healthy living," which, in all likelihood, included a nice healthy diet with an emphasis on wholesome snacks like sunflower seeds. Ms. Ford drew approval from the crowd as she recounted how thereafter she threw her roommates outside to enjoy their nasty habit.[59]

It should be noted, however, that the diagnosis seemed somewhat soft – based largely upon vaguely imagined symptoms and arguably supported by an anomalous blood chemistry reading. A kindly and fatherly doctor, wanting to "help" people stop smoking, would naturally take the opportunity to not only reassure his patient that there was a solution to her problem, but also rejoice in the fact that

---

* Our bodies are actually pretty good at getting rid of all sorts of nasty stuff that they manufacture or that enters them through various routes in everyday life. We have defecation, urination, perspiration, and respiration – all throwing out life's natural toxic waste products that would otherwise quickly kill us.

the solution might result in encouraging some smokers to quit! Who could find fault with that?

Well, the smokers might not be too happy. And they'd be even less happy if they did some research. Yes, the doctor was correct in noting that Ms. Ford was exposed to cadmium from her roommates' smoke. Where he failed was in not telling her that her likely exposure level of .003 to .03 micrograms per day was about ten thousand times less than her likely exposure if she was a hearty muncher of all-natural-organically-grown-and-watered-with-pure-raindrops sunflower seeds. A good sized serving of sunflower seeds can contain up to 100 micrograms of cadmium.[60] So was her doctor a victim of stratistics? Or just a victim of good intentions? Hard to say, but in either event it seems likely that Ms. Ford and her likely relationships with her room-mates were victims of the war on smokers.

The "Commander Almost Zero Fallacy"

The best example of this stratistical weapon can be seen when the presence of an expected element in a smoking environment (such as "smoke" or "nicotine") is compared to the presence of that element in a nonsmoking environment. It is then "revealed" that the smoking environment has five times, or ten times, or even 53 times the amount of that element as the nonsmoking one.

If a smoker's home contains 53 times as much of a deadly toxin as a nonsmoker's home it seems like a good reason not to bring your precious young one into such an environment. But once you realize that the amount in the nonsmoking environment is, quite literally, almost zero, then you might also realize that 53 times almost zero is still going to be ... almost zero. It's like trying to frighten people into never taking showers by telling them that homes with shower-takers have 53 times as much deadly chlorine gas in them (evaporating from healthily chlorinated, bacteria-free tap water) as homes where nobody but a bunch of grubby, long haired, non-showering hippies live.*

---

* I've been called a grubby longhaired hippie at times in my life, but I *do* take occasional showers – whether I need them or not.

We saw this trick used in 2009 when the Smoke-Free Campus movement was being generously funded by such folks as the NicoGummyPatchyPushers at the Robert Wood Johnson Foundation (RWJF),[61] as well as being supported by programs and organizations like TFU, TTAC, CTPR, BACCHUS[62], GAMMA*[63] and other alphabet soup groups, shells, and fronts springing from such mega-sources as the Master Settlement Agreement's "invisible tax"† on smokers. These groups were pushing colleges around the country to follow up on classroom and dormitory smoking bans with bans covering the entirety of their outdoor campuses – even to the far corners of parking lots enshrouded in clouds of engine exhaust! The background, usually unstated, justification for such pressure was that the bans would "foster campus and community environments that promote healthy lifestyles…"[64] or "The hope is that those who desire to quit smoking or desire to quit chewing tobacco will take the opportunity to do that."[65]

As part of that effort, the University of Georgia produced a study supposedly showing that simply being around smokers would boost your blood nicotine (actually cotinine, the nicotine metabolite found in blood) level to over 150% higher than the levels of those who avoided such exposure. It isn't until you read the study itself that you'd find out that it compared a group of test subjects who sat around outside in a location nowhere near any smokers to a group that sat right in the middle of crowds of smokers in smoke pits outside of smoke-banned bars for six hours straight on busy Friday nights.

I took the figures from the Georgia study and computed what the level of exposure would actually be if a waiter worked every Friday night on the crowded smoking patio of a college bar where indoor smoking had been banned. It turns out that the waiter would have to work in such conditions for about a hundred years to get the equivalent exposure of smoking a single pack of cigarettes.[66]

---

* Alphasoups Translated: Tobacco Free Universities, Tobacco Technical Assistance Consortium, Center for Tobacco Policy Research, Boosting Alcohol Consciousness Concerning the Health of University Students, and Greeks Advocating the Mature Management of Alcohol. {The Endnote nicely documents BACCHUS as a creature of the CDC's Office on Smoking and Health co-funded by state MSA tobacco control monies -- despite its name only mentioning alcohol. GAMMA meanwhile specifically notes it "is not against underage drinking!"}

† The MSA is often called an invisible tax because, although it's money collected to be given directly to the government, it was never passed as a legally legislated tax.

Levels of exposure that were a bit more normal on a campus, say, walking through clumps of smokers at doorways or maybe sitting on a bench several times a week while a couple of smokers smoked on a bench nearby, would be far lower. A hapless student might have to wander around such a "smoke-filled" campus for almost a thousand times as long – a hundred thousand years of Friday nights – to enjoy the equivalent of smoking a single pack of cigarettes or a couple of marijuana joints. *

Now if the above concern about momentary passage by smokers near a doorway seems a bit far-fetched to you, you're clearly not a card-carrying Antismoker. In early 2011, a sadly brainwashed student wrote an article in support of a campus-wide smoking ban and expressed concern that "toxic chemicals from cigarette smoke leave harmful residue" on campus benches that might poison nonsmokers unwise enough to sit there later.[67]

Later in this book (*Slab IV*) I will examine the theoretical danger posed to an infant who obsessively licks "thirdhand smoke" off ten square feet of smokers' flooring every day of the week. I show how it would take literally trillions of years to absorb the amount of "poison" touted as a deadly threat by Antismokers. On an outdoor campus, assuming the typical college bench surface is roughly ten square feet with a smoke deposition rate (in an outdoor environment with normal breezes etc.) of roughly 1/1,000th of that which settles on an inside floor, a student would have to extend their educational opportunities for roughly 3,000,000,000,000,000 (3 quadrillion) years while licking an entire bench nice and clean every single day before dying. As in the case of the arsenic-eating baby mentioned earlier, they'd have to refrain from "going potty" all that time. If they delayed their doctoral thesis until the end, they'd most certainly win the Commander Almost Zero Prize for a thesis more full of a certain aromatic substance than any produced in all of previous academic history.

---

* I've always thought it rather funny that while Smoke Free Campuses so stridently pushes their full campus smoking bans they never seem to consider the students who smoke marijuana rather than tobacco. After all, if a college can expect to successfully halt tobacco smoking then it would seem that they should also have full legal responsibility for any pot smoking that they allow to occur under their watch.

A final and very simple example of a mismatched comparison is found in the common statement seen on the Internet and heard at legislative hearings that "Breathing secondhand smoke is more deadly than smoking." Sometimes the claim is even explained "scientifically" by pointing out that the filter protects the smoker. Obviously it's a silly statement, or researchers would have concluded decades ago that smoking protected people from lung cancer in smoky environments rather than caused it – as their own cigarette puffs filtered out the deadly secondhand smoke in the air surrounding them. But how did the comparison come about?

The answer lies in the fact that the true scientific statement would be along the lines of "Smoke taken directly from the burning ember at the end of a cigarette contains more X (where X is just about any substance that exists in smoke) than the smoke that has been filtered through the length of the body of the cigarette and inhaled by the smoker." The trick that provides for the misunderstanding is that nonsmokers do not suck on the lit ends of cigarettes and inhale lung-fuls of that concentrated smoke. The average concentration of smoke that the nonsmoker inhales in most reasonably ventilated situations today has been diluted by a factor of at least a thousand times. A very simple distinction, but one that has been completely lost in the constant retelling as reality has morphed into a stratistic.

Stratistical Mutations

Unfortunately for antismoking researchers, their stratistical games can be exposed by Free Choice advocates or fellow researchers who are a bit more honest in their approach to science.* Of course, they can always claim that part of their statements were true, or that they just made a simple error, or that some sort of innocent misinterpretation of terms led to later corrections – but hey, everyone's got an excuse, right?

---

* Sadly, only a very small segment of the general public ever sees these rebuttals. They may get play on the Internet, but technical arguments and computations don't exactly make for pulse-pounding excitement on TV. The scary headlines promoted by the supposed experts in the original story simply become the story itself.

There's no need to delve into ancient history. One of the largest international case-control studies ever done on secondhand smoke exposure and cancer, the 1998 World Health Organization study by Boffetta *et al.* provides a perfect starting example.[68] This massive, and intended-to-be-definitive, international study examined ETS exposure among spouses, in work situations, in social situations, and among children to see if such exposure eventually resulted in an increase in lung cancers.

The study was finished, the results tabulated and calculated, but the findings, for some odd reason, were simply not released. And then suddenly, on March 8, 1998, the British *Sunday Telegraph* broke a headlined story titled, "Passive Smoking Doesn't Cause Cancer – Official," where they revealed that this be-all-and-end-all of scientific studies on passive smoking had found basically ... nothing.[69] Not a single measure of exposure, whether at work, from a spouse, in social situations, or in childhood, was able to meet even the bare minimum statistical standard of significance showing any increased risk of cancer at all from exposure to others' smoke at work, at home, or in bars.

Indeed, although within the formal abstract of the study itself the finding was nicely downplayed with the phrase "no association," the only true statistically significant finding of the study was that children growing up and developing their immune systems in homes with smoking were actually *22% less likely* to develop lung cancer later in life than matched children growing up in nonsmoking homes!

Now, how does the concept of stratistics rear its ugly head in this study? In two ways: first, as pointed out, the abominable characterization of the one significant finding – the one on children – as showing "no association" in the hope that the media would overlook the actual numbers right next to it; and secondly, by the equally abominable press release put out by the WHO itself, with its headline blaring, in bold capital letters, "Passive Smoke DOES Cause Lung Cancer; Do Not Let Them Fool You,"[70] despite the fact that the statistics of their own master study concluded nothing of the kind.[71] The stratisticians in question can seek some refuge in the defense that their findings for adult exposures pointed to a small and statistically insignificant correlation similar to some found before, but such questionable findings are not normally meant to drive actual political policy and decisions. Unfortunately, in the realm of antismoking science, such findings are used that way far too often.

A second and rather closely related example lies in the claim that exposure to others' tobacco smoke will increase your risk of lung cancer by 19%. Statements like that have probably resulted in real and indisputable deaths as innocent, stratisticized nonsmokers walk across streets and end up getting hit by cars simply because they'd been fooled into believing there was a real and significant risk involved in passing by a single SOAS (Smoker On A Sidewalk). They don't realize:

(1) That the 19% increase claim is loosely taken from the 1992 US Environmental Protection Administration (EPA) Report and is based on eight hours of exposure every day for thirty to forty years or more;

(2) That the claimed increase relates to a disease with a natural occurrence in nonsmokers of only about four in a thousand, so it works out to only about one extra chance in a thousand, even after such constant exposure;

(3) And, that even the 19% increase has been highly disputed and criticized as being exaggerated.

For the moment, let's accept the EPA figure as accurate for this discussion since, even if it were true, there'd still be no real risk involved in the normal meaning of the word. What kind of risk is someone actually taking when they walk by that "cloud of smokers" every day as they pass the corner bar? Any real exposure is likely to last, at most, about ten seconds,* and is highly unlikely to average more than a hundredth of the intensity of the average exposure of the indoor

---

* Based upon a defined "elderly" walking speed of two miles per hour – actually slightly below the three to four miles per hour "average" walking speed determined at the 3rd Urban Street Symposium (2007) held in Seattle, Washington – and assuming significant exposure within ten feet of the center of the smoking party.

1950s-era worker. Overall then, it's about equal to a single tenth of a second of the levels of exposure spoken of in the EPA Report. Now forty years of constant exposure at the heavy levels examined in the EPA Report supposedly produced about one extra lung cancer in a thousand (see (2) above), so it's fair to characterize that as, on the average, one extra lung cancer per 40,000 worker-years of exposure.

So how many times, on average, would you have to stroll by that rowdy bar crowd to get lung cancer? Well, you'd be getting only about one-tenth of a second of equivalent exposure per day (Remember: ten seconds at 1/100th the intensity.), rather than the 28,800 seconds in a working day (60 seconds x 60 minutes x 8 hours). This means that on the average you'd have stroll by that cluster of nasty smokers every day of the year for somewhat over eleven billion years (11,150,000,000 years, to be exact) to suffer, on the average, a lung cancer from your exposure. Crossing the street to avoid that "risk" would most certainly result in a far greater and far more immediate chance of death under the wheels of a speeding car.

On a somewhat different level, we can see simple stratistical mutations in such things as the widely claimed instant reductions in heart attack rates after the implementation of bar and restaurant smoking bans. The first such study to make headlines came from an isolated northwestern American town: Helena, Montana. In April of 2003 the Helena researchers claimed a miraculous 60% reduction in heart attacks during a six-month smoking ban in 2002, and a frighteningly counter-miraculous "bounce back" to regular heart attack rates once that ban was lifted.[72, 73]

However, when the study was eventually published in the British Medical Journal, that 60% drop had somehow dropped by a full third, down to 40%.[74] And months after that, when a certain intrepid but nameless researcher (well, ok, not nameless – it was me) tracked down the original PowerPoint presentation which had been conveniently erased from the original location, it turned out that most of the loudly proclaimed bounce back actually occurred during the smoking ban, not after it. Helena gets the prize for double-barreled stratistical research – a rarity even in the antismoking community!*

---

* The Helena study is given a detailed examination in *Studies On The Slab*.

Another time jump, this time up to 2009, brings us to a similarly mutated claim from Scotland where researcher Jill Pell published research showing an astounding nationwide drop in heart attacks of 17% following Scotland's smoking ban.[75] Pell's study was characterized as "virtually flawless" by Tom Glynn, speaking for the American Cancer Society in USA Today.[76] By the time analysts such as British historian Christopher Snowdon and longstanding antismoking advocate Dr. Michael Siegel (an MD from Yale now teaching at Boston University) got through with their analyses, the "flawless" nature of the stratistics disintegrated.

Pell's flawless 17% drop fell to 7% – which happened to be just about the same level of reduction that had been going on for a number of years before the smoking ban. Dr. Siegel went on to point out that national Scottish hospital data showing only a 7.2% decrease wasn't much different from the 6% decrease reported the year before the ban. Siegel also caught the fact that Pell's famous 17% drop actually seemed to begin three months before the Scottish smoking ban was actually implemented – another example of the time-mutating properties of "Magickal" tobacco smoke as satirized earlier. And then, most amazingly, in the second year of Scotland's ban, government figures showed the first increase (by 7.8%) in heart attacks that Scotland had seen in close to ten years.[77, 78]

Figure 1 shows how Snowdon exposed the falsity of Pell's claims in a simple bar graph format of the raw data. Even a casual glance makes it obvious that the heart attack rate in the year after the Scottish smoking ban simply continued in a slow drop related to dietary and medical changes as though the ban had never happened. Note that unique increase in heart attack numbers in the second year of the ban!

Similar graphs and the stories they tell will be examined in detail in Studies On The Slab. As for this particular case of Scotland with its annoyingly unexpected increase in heart attacks in the second year of their ban... well, that's a statistic that probably gave the antismoking stratisticians heart attacks of their own.

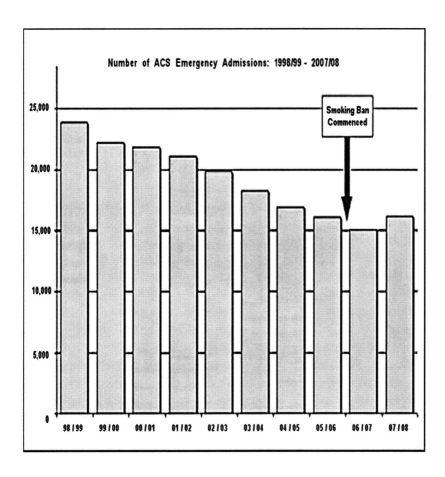

Figure 1

Hospitalizations for acute coronary syndrome. Source: ISD Scotland

Conclusion

These analyses may seem to be over-simplified criticisms of statistical methodology, but simplicity can be a strength rather than a weakness when legitimate statistics are involved. My own statistical training was far more intense than most. I studied statistics not just in high school and college, but also under a full fellowship at the University of Pennsylvania's Wharton School as I worked on a doctorate in Peace Science. Peace Science was a blend of propaganda analysis, statistical theory, and general social-economic mathematical modeling. I'll be quite honest: statistics was never my strongest academic skill, and I left Penn without completing the degree; but I certainly learned enough to understand such basic concepts as correlation, multivariate analysis, confidence intervals, and statistical significance.

One of the strongest lessons I learned at Penn was to never simply take statistics and their claimed meanings on faith. Anything beyond the simplest of statistics can quickly and easily be warped to produce almost anything a study's funder desires. And if you are an academic researcher dependent upon continued grants to rise in your department, pay off your mortgage, and feed your family, you can always find an excuse to justify the warping and salve your conscience, particularly when you know in your inner soul, with all the fervor of a religious martyr, that there's a good side (reducing smoking) and a bad side (the ultimately and quintessentially evil tobacco companies that you think you are fighting against).

Such a combination of monetary, prestige-seeking, and career-advancement motives mixed with the drive of a strong idealistic belief makes a deadly stew as far as truth is concerned. I believe that when history looks back and examines this period, it will find that stratistical chicanery and even cases of outright fraud were far more common in antismoking research than in any other scientific field. The global warming skeptics seem to think that mass perversion of science toward ideological ends is something new in post-World War Two history, but it's not. The path was well-trod long before anyone worried about melting icebergs. There actually aren't all that many scientists willing to risk their good name and scientific principles for mere money, but when the belief that such risk is also for the betterment of humanity, that number seems to multiply like ribald rabbits in rutting season.

Studies On The Slab

They're alive! They're ALIVE!!!

- Baron Von Frankenstein (paraphrased)

The "Mountains Of Studies" that Antismokers love to rhetorically wave in the air are more like "Mountains Of Cotton Candy" once they are examined in detail. Yes, there are some fairly good studies out there linking smoking and various smoking-related diseases;* but no, there are most certainly *not* mountains of studies justifying extremist smoking bans.

This section of *TobakkoNacht – The Antismoking Endgame* will examine, dissect, if you will, many of the primary ban-justifying studies and their clones that you are likely to have run across on the front pages of your newspaper or heard about on the evening news. I think that by the end of *Studies On The Slab*, you'll agree with my evaluation.

---

* These are not all as solid as they have generally been made out to be. Even the strongest foundation block in the antismoking castle, the causal link between smoking and lung cancer, has started to be questioned as monetary motivations of some of the early researchers in the area and defective study designs have been revealed. It's not an area I have examined closely myself (As you'll see in the next 150 pages my hands are quite full with just second- and third-hand smoke concerns!) but the BoltonSmokersClub.com is a good place to start some alternative research and reading in that area if you're interested.

Setting: Charon's Foyer

The Medical Examiner stands in the foyer of the morgue, surveying the ranked slabs basking in cold fluorescence. Only six cadavers tonight, all victims of the war on smokers – sad and mute, but each with a tale to tell – lie before him.

He sighs. He's gotten so much older in the last decade or so as the victims have gotten ever more gruesome. Every night it is the same and more of the same ... only worse. Always worse. Worse with each passing night and each passing year. Truth is such a simple thing, but there's so little of it that comes before him. Sometimes he can see it hiding just under the cold flesh, sometimes he has to dig deep, but it's always a struggle to pull it out from where it has been butchered and buried.

He sighs again, dons his gown and gloves, and crosses the threshold...

The lunacy that resulted in America's "Noble Experiment" with alcohol prohibition in the early part of the twentieth century had some small amount of scientific and medical basis at its foundation, but the main structure was almost entirely the result of moral fervor and a rabid belief in a central doctrine: Alcohol was evil, alcohol was bad, alcohol was the devil's brew, and it made men sad.

However, while our first steps toward a new and very different Prohibition today may have had some roots in morality, they were carefully planted in the supposedly objective soil of medical science. Throughout the twentieth century, evidence had been growing that smoking was connected to, and probably caused, lung cancer. Under Adolf Hitler, the mainstream German medical establishment produced research tying the two strongly together and even invented a new term and concept – *Passivrauchen* – passive smoking. Hitler used that information and concept as the basis for a national smoking Prohibition stronger than anything that had been seen in Europe for centuries.[79]

But with the fall of the Reich to the victorious smokers — led by Roosevelt with his ever-present cigarette holder, a chain-smoking Josef Stalin, a stogie-chomping Churchill, and the hard-smoking scientists who created the atomic bomb — the antismoking movement fell into disfavor. Despite steadily accumulating evidence linking smoking to lung cancer and other health problems, little more was said publicly about the health effects of smoking for more than a decade, and our television sets were flooded with ads sporting doctors endorsing one cigarette brand after another. Unfortunately for Big Tobacco, though, by the late 1950s the pile of studies pointing to smoking as a major cause of lung cancer and other ailments simply grew too huge to ignore, despite criticisms of them by such prominent and respectable figures as Sir Ronald Fisher, the father of modern statistics,[80] and Dr. Philip Burch of the Department of Medical Physics at Leeds and author of *The Biology of Cancer: A New Approach*[81, 82]

Then, in 1964, the US Surgeon General unveiled his famous *Report on Smoking and Health*,[83] and by 1970, faced with ruinous competition from free TV antismoking advertising made possible by Banzhaf's Fairness Doctrine lawsuit, the tobacco companies agreed to halt all broadcast advertising of their products under the Public Health Cigarette Smoking Act of 1969.[84] But after that victory, the Great American Antismoking Crusade fell into a rut. The smoking rates that had dropped steeply during the 1960s began leveling off, with smoking even increasing among some segments of the population. A new approach was needed, and it was found when 1975's World Conference on Smoking and Health was conducted under the guidance of a well-known British Antismoker, Sir George Godber. Before 1975, Godber was best known in antismoking circles for comparing the situation of living with smokers to living in a home infested with head lice.[85] The 1975 conference resulted in a new guideline for a successful War On Smoking, or, as it was now becoming, a War On Smokers.

Godber's conference promoted the recognition and acceptance of the idea that to eliminate smoking it would be important to "emphasize that active cigarette smokers injure those around them, including their families and, especially, any infants that might be exposed involuntarily to ETS."[86] It sounded like a great idea. Peer pressure was thought to be why people often began smoking, so maybe peer pressure would make people stop. The only problem was that there was no real scientific evidence to support the claims of injury to others.

In the late 1970s, the antismoking movement had to content itself with word games and misrepresentations that would never have stood up to even cursory examination. Let me illustrate with two examples.*

---

* Unfortunately, I do not have a proper reference for the first one, but I remember it clearly from writing about it informally in the 1980s.

The first simply involved wordplay. Experimenters would seal several people into a heavily smoke-filled environment for an hour – a favorite was a tightly sealed Volkswagen Beetle, well known for its ads claiming it to be so airtight that it was "The Car That Could Float." The researchers would have two nonsmokers and two chain smokers sharing the rather limited air, and then question and examine the nonsmokers when the doors were opened and they crawled out. The subjects would have red eyes, stuffy noses, itchy throats, and coughs that would probably have done justice to a terminal tuberculosis ward. After all, they'd just been subjected to conditions far smokier than Humphrey Bogart ever endured in Casablanca.

Unfortunately, just saying "lots and lots of really concentrated smoke might give you a stuffy nose" clearly wouldn't generate the sort of fear and concern needed for people to break social mores and start poking around at other people's behaviors. So… what did groups like ASH and GASP do with these study results? Simple: they produced reports and factsheets that made no mention of intensity when speaking of "Nonsmokers exposed to smoke," and then used strange and scary-sounding scientific and medical terms to describe the results. Red eyes suddenly became "scleral vasodilations." Eye-watering could be described as "conjunctival lachrymal hypereffusions." A stuffy nose grew into "acute rhinoidal congestive discharges." Complaints of an itchy throat and cough transmogrified into "severe bronchiolar pruritic irritation and upper thoracic spasming."

The summation then looked something like this:

> Nonsmokers were exposed in commonly encountered situations to second-hand smoke for just 30 minutes. After even such a limited exposure, a medical examination found that the nonsmokers were suffering from pronounced scleral vasodilations, extreme conjunctival lachrymal hypereffusions, acute rhinoidal congestive discharges, severe bronchiolar pruritic irritation, and upper thoracic spasming.

All that from just being exposed to a few wisps of smoke in a supposedly common scenario.

When they wanted to look even more scientific, they did studies like the one that appeared in *The Lancet* in 1973.[87] That particular study stuck twenty people in a sealed room just 43 cubic meters in size (about the size of a small child's bedroom) for 78 minutes *while burning eighty cigarettes and two cigars!* The scientific part, aside from being surprised that the subjects were still alive at the end, was the finding that their blood carbon monoxide levels had increased. Amazing, eh?

The propaganda worked. Not very widely, but for a limited number of individuals and within certain sensitive subcultures, these frightening phraseologies, images, and selective antismoking presentations created an atmosphere in which nonsmokers who simply didn't like smoke or smoking could suddenly feel justified in demanding that smokers not smoke around them. The situations were fairly limited at first – a nonsmoker's own home, a car set aside on a train, an occasional nonsmoking section in a restaurant or a nonsmoking room or ward in a hospital – but they were something that had been almost nonexistent in the smoky 1950s and 1960s.

Just as important as the development of scientific studies was the use of effective sound bite psychology. Such messages might have very little substance and very little hope of standing up under real examination or challenge, but they can be accepted, remembered, and repeated easily by most people without much thought. Basic anti-smoking sound bites were developed and spread by activist groups:

"Secondhand smoke is worse than first hand smoke." Only true if you hold your nose a half inch from the burning end of a cigarette and inhale deeply for ten minutes.

"A non-smoking section in a restaurant is like a non-peeing section in a swimming pool." Sort of true if you ignore the fact that the pool only changes its water once a year while a decent restaurant changes its air about 50,000 times a year.

"There is no safe level of exposure to smoke." Almost completely true, but the "almost" qualification is very important here. As Class A Carcinogens, cigarette smoke, ethyl alcohol, radon gas, sunshine, sawdust, and asbestos all share the same characteristic of having no known absolute "safe level" of exposure, if – and this is a very important scientific "if" – you accept what researchers call the "no threshold" model of carcinogenicity. But even if you accept that model, it still makes no sense to whisk your child out of a room with a mild scent of smoke or away from a well-curtained window where an errant sunbeam might sneak through.

"Cigarette smoke contains arsenic." Fully and completely true, but you'd have to sit in a smoky bar for roughly 165,000 hours to get the same amount of arsenic you'd get from drinking a single pint of government-approved-as-perfectly-safe tap water.[88*]

These sound bites and others like them worked to some extent, but mainly in vulnerable populations that were already susceptible and open to their messages. Anyone who took even a few moments to ask about the basics of the studies or who looked even slightly beyond the surfaces of the sound bites would quickly see through their misleading nature.

More evidence was clearly needed if the newly minted Great American Antismoking Crusade was going to extend beyond the province of a few small niche populations and a few noisy extremists. Simply put, not enough of the general population was buying the secondhand smoke theory in the 1970s, despite the efforts that followed the Godber conference. People were willing to admit that tobacco smoke might be more annoying for some folks than had generally been recognized, but few believed their health was actually threatened in any real sense simply by being in a room with a few smokers. Those who preached such a concept were still small enough in number and lacking enough in resources to be dismissed as neurotics and cranks.

In the mid-1970s, even the American Cancer Society itself summarily dismissed the suggestion that it ban smoking in its own offices, stating that their organization was "not authoritarian" and that they did not believe in such a "dictatorial approach."[89] But the 1980s threw a life preserver to the antismoking movement. Researchers in Japan and Greece studied wives of smokers and nonsmokers and claimed that women married to smokers seemed to develop lung cancer at a significantly higher rate than those married to non-smokers.[90, 91] Takeshi Hirayama's 1981 study was particularly large and appeared to be fairly well done, although serious questions were raised

---

* A nonsmoker sharing a decently ventilated room with smokers would inhale roughly .00032 micrograms of arsenic over the course of an hour. Government "safe" tap water standards allow about 5 micrograms per pint.

about such things as the misclassification of supposedly non-smoking wives hiding their smoking from smoking husbands in the strict Japanese culture. Still, both studies were trumpeted uncritically in headlines around the world.

To the frustration of antismoking advocates, the rest of the 1980s saw follow-up studies that generally failed to replicate those findings.[92] They often showed some level of nonsignificant correlation, but at that point in history, such nonsignificant findings were still generally dismissed as practically worthless from a scientific point of view, other than in offering some support to the null hypothesis that there was no real connection to be found. Epidemiological researchers were very aware of the dangers of possible unknown or inadequately corrected confounders and felt uncomfortable speculating about small increases in population risks. Nevertheless, such concerns did little to slow down antismoking extremists, and in 1986 the United States Surgeon General's Office came out with its milestone *Report On The Health Consequences of Involuntary Smoking.*[93]

Given the strong belief held by tobacco control advocates that fear of other people's smoke would be a key card in getting support for the levels of punitive taxation and regulation they believed necessary to drastically reduce smoking, no one was surprised when the Surgeon General concluded that environmental tobacco smoke (ETS) was likely a significant cause of lung cancer. Then, as a follow-up, in 1993 the Environmental Protection Administration threw caution to the wind and published its own report concluding that a lifetime of workplace exposure to secondhand smoke could increase the risk of lung cancer by 19%.[94] The news stories fed by those reports made faint curls of smoke sound more deadly than wartime mustard gas attacks.

In reality, if those findings had been presented fairly, they would not have been all that terrifying. The actual findings of the EPA Report indicated only that if nonsmokers had thirty or forty years of intense daily exposure to smoke in the poorly ventilated and smoky conditions common in workplaces from the 1940s through 1970s, their chance of ultimately getting lung cancer would increase from about four in a thousand to about five in a thousand[95] – basically, one extra lung cancer for every 40,000 worker-years of exposure. A recent large study in Europe indicates that even the four in a thousand risk may be too high an estimate: Norwegian researchers pegged general nonsmokers' rates at about two in a thousand as a baseline, which would then work out to one extra lung cancer for every 80,000 worker-years of exposure.[96] Even the claim of that small level of risk increase would have disappeared if the EPA had stuck to the normally accepted scientific guideline of using the standard confidence interval set at 95% rather than lowering it to 90%.* It was only through lowering that confidence interval that they were able to claim that 19% increase over the small base cumulative lifetime rate.

Statistical chicanery of that sort should have gotten the results laughed out of the scientific community right at the start, but the EPA was a big enough player to make up its own scientific rules and feed them to a willing press. By ignoring the possibility, small but real, that exposure to low levels of smoke could activate bodily defensive mechanisms that would result in a reduction of lung cancers, they were able to justify using what is called a one-tailed statistical test under something called the bioplausibility theory† – effectively doubling the acceptable margin of error. All they needed was a scientifically naïve press that was willing to put aside inconvenient questions and run with the alarming headlines.

The press was indeed willing. They ignored the scientific impropriety of changing the 95% goalpost and never explained how the scary-sounding 19% increase meant an average of just a single death per 400 centuries of intense daily exposure. By the time reporters and anti-smoking advocates got through writing the stories and supplying hyperbolic quotes, many Americans simply assumed that being around a smoker for a few minutes would practically doom one to an early and cancerous grave.

---

* This change from 95% to 90% was basically what allowed the researchers to declare their findings were statistically significant and meaningful. If the traditionally accepted scientific standard of 95% had been used, their findings would have failed that all-important test. Without that fundamental and highly questionable jiggling of standards, the EPA Report would have had to report *no* real findings for lung cancer and secondhand smoke exposure.

† Bioplausibility asserts there is a reasonable and well-acknowledged biological basis for expecting a certain result – thus a lesser need for proof. Unfortunately it denies the possibility that low exposures to something like a vaccine could ever offer any sort of immunity boost that could bring about a protective effect.

The EPA Report was heavily criticized by several respectable sources — including the research arm of Congress, the Congressional Research Service[97] — and in 1998 it was also judged as being worth little more than garbage in a landmark ninety-three-page decision by Federal District Court Judge William L. Osteen.[98] While Judge Osteen left the EPA's findings on childhood respiratory disease intact, he thoroughly ripped apart the most heavily publicized findings on the relationship between secondhand smoke and the development of lung cancer.

> EPA's theory was premised on the similarities between MS, SS, and ETS.* In other chapters, the Agency used MS and ETS dissimilarities to justify methodology. Recognizing problems, EPA attempted to confirm the theory with epidemiologic studies. After choosing a portion of the studies, EPA did not find a statistically significant association. EPA then claimed the bioplausibility theory, renominated the *a priori* hypothesis, justified a more lenient methodology. With a new methodology, EPA demonstrated from the 88 selected studies a very low relative risk for lung cancer based on ETS exposure....
>
> In this case, EPA publicly committed to a conclusion before research had begun; excluded industry by violating the Act's procedural requirements; adjusted established procedure and scientific norms to validate the Agency's public conclusion, and aggressively utilized the Act's authority to disseminate findings to establish a *de facto* regulatory scheme intended to restrict Plaintiffs' products and to influence public opinion.
>
> In conducting the ETS Risk Assessment, [the EPA] disregarded information and made findings on selective information; did not disseminate significant epidemiologic information; deviated from its Risk Assessment Guidelines; failed to disclose important findings and reasoning; and left significant questions without answers. EPA's conduct left substantial holes in the administrative record. While so doing, produced limited evidence, then claimed the weight of the Agency's research evidence demonstrated ETS causes cancer.[99]

Stop here for a moment and read those excerpts from the ruling again. Osteen's work was truly a scathing and scholarly indictment of

---

* MS stands for Mainstream Smoke – the smoke the smoker inhales. SS stands for Sidestream Smoke – smoke from the burning end of the cigarette. ETS is the combination of sidestream and exhaled mainstream smoke created when someone smokes.

the EPA Report, and it has never been seriously challenged on its substance. None of the points he made above have ever been invalidated. Antismokers instead criticize it on the basis of the fact that, twenty-five years earlier, while in private law practice, Osteen had been hired as a lawyer by a group of North Carolina tobacco farmers to go to Washington and present their case to the Secretary of Agriculture.[100]

This extraordinarily weak association was used by antismoking advocates to dismiss the judgment as worthless because it came from a "Big Tobacco Judge," while they conveniently ignored the fact that just a year earlier, in 1997, Osteen had handed down a devastating ruling *against* Big Tobacco when he upheld the authority of the US Food and Drug Administration (FDA) to regulate tobacco. Osteen's earlier ruling against the tobacco companies had even been described as "the biggest court victory ever against tobacco," by William D. Novelli, president of the National Campaign for Tobacco-Free Kids.[101]

Osteen's 1998 ruling against the EPA should have spelled the end of the line for antismoking efforts based on fears of secondhand smoke except for two unfortunate circumstances. First of all, the EPA had very carefully presented its report as being "advisory" rather than "regulatory." That seemingly small difference was later ruled to be crucial as Osteen's decision was legally vacated. It was not vacated because of any defects in its findings, but purely on the administrative technicality that his court did not have proper jurisdiction over "advisory reports," but only over "final regulatory actions." As we saw above, Osteen had argued that the societal impact of the EPA Report was such that it was indeed a *de facto* regulatory paper, but the higher court disagreed.[102]

The second factor that weakened the impact of Osteen's finding was that it came out six years after the EPA's conclusions hit the headlines; the Antismoking Crusaders therefore had six solid years under their belts during which they could proclaim, without reservation, that an official EPA Report had determined secondhand smoke to be a killer. In the 1990s, the EPA was still riding the wave of 1980s' environmentalism – it was regarded as the golden-haired boy who could do no wrong and whose every word was Gospel. By 1998, a view that very few Americans would have shared six years earlier had become accepted folk knowledge for a substantial portion of the population simply by dint of official endorsement and constant media repetition. Smoking was now banned in the vast majority of standard workplaces around the country, and in most family restaurants as well. In California, all restaurant smoking had been banned since 1995 and, previously unheard of, even most bar smoking (except for places with fewer than six non-family employees) was banned in 1998.

Unfortunately for the state of public knowledge, the news of Judge Osteen's landmark decision was never trumpeted in the media. While the *New York Times* had headlined the original EPA Report and followed it with a major editorial in their Sunday edition, they buried the dismissal of that report in a small column on page 14 of a Monday edition.[103]

In the interim between the production of the EPA Report and Osteen's 1998 ruling, antismoking activists used the excuse of "protecting the workers," buttressed by millions of dollars from earmarked tobacco taxes, to bring about the near-universal California workplace smoking ban. California sat alone in the bar-banning niche until powerful Governor Ruth Ann Minner forced one upon little Delaware in November of 2002.[104]

The real turning point in the war on public smoking came within a year of Delaware's move when popular New York City Mayor Michael Bloomberg and New York State Governor George Pataki joined them in March and July of 2003.[105] In the meantime, a smoking-ban referendum was won in Florida, where the smoke-banners outspent their opponents by a huge margin while portraying the vote as pitting the lungs of the children against the profits of Big Tobacco executives. Florida, however, at least had the sense to exclude free-standing bars from their ban.[106]

But after this little cluster of successes, further victories proved to be a grindingly slow and frustrating process for an antismoking industry that had grown with the Master Settlement Agreement into a behemoth commanding well over 800 million dollars a year.[107]

Despite their resources, ban advocates found that, by and large, the hospitality industry and the general public remained stubbornly resistant to widespread government-mandated smoking bans. Smoking was still generally accepted as an enjoyable and fairly inoffensive activity that people often engaged in while socializing with friends in public gatherings.

People felt that going to a bar or restaurant was a voluntary activity and most saw those locations as recreational facilities rather than workplaces. Antismokers, however, emphasized the concept that bartenders and wait-staff worked in those places and called up the imagery of the company towns of the early 1900s. Playing up that image of forced labor, they claimed that the workers in such places had no choice but to work there and endure the deadly threat of second-hand smoke in order to feed their families. The classic sound bite image portrayed the teenage mom forced to choose between her life and a paycheck. This overblown rhetoric was largely met with well-deserved scorn and generally did little to help the antismoking cause in the public eye, though it may have swayed a few votes in legislative chambers.

Something more was needed to terrorize the general public into demanding the wide-reaching smoking bans required to support the emerging (although rarely admitted) goal of a total denormalization of smoking and smokers.[108]* That "something more" came about in the form of dozens of studies from the late 1990s into the late 2000s that twisted science into pretzels not seen since the sealed Volkswagen gas chambers.

Scientific studies published in peer-reviewed journals and then reported on in such papers of record as the *New York Times* and the *Washington Post* are the standard by which truth is usually judged in today's world. Regrettably, truth has been one of the greatest casualties in the War On Smokers. Far too much of what people today believe to be indisputable fact is little more than a mishmash of exaggerations, distortions, and outright lies.

That may seem like strong language, but it's justified when one examines how ban-supporting studies have been done by researchers, how they've been presented by the supposedly responsible medical

---

* Denormalization was rarely mentioned in public by Antismokers until recently. Googling (denormalization + smoking) in five-year periods from 1990 yields occurrences of 2, 93, 264, 1,100, and, from just 2011 - 2013, 2,220 hits. I examined its potential for social engineering while writing *Brains* in 2003 despite some criticism that I was being "too tinfoil hat." Its more recent open admission may be due to a greater public acceptance of social engineering as being a good, rather than a bad, thing, or may simply be due to its increasingly important role in the Antismokers' endgame planning.

authorities, and how they've been uncritically reported on by a media in search of headlines that will bleed and lead newspaper sales and Nielson ratings. Truth is a very relative term in the hands of propagandists, and it's easy for them to construct claims that distort reality while still hanging on to a tiny shred of fact – just in case they're ever publicly challenged.

The shred of fact defense is an old one in political gamesmanship and one that I wrote about in *Dissecting Antismokers' Brains*. Let us suppose that I earn $500,000 a year as a fast-talking lawyer. I decide to run for public office and, to boost my image, I proclaim my belief in the importance of giving to those less fortunate. I adopt an earnest and heartfelt look for the TV cameras and claim that I give regularly to both organized charity and to the poor homeless folks I see on the streets. I go on to emphasize that I maintain my giving year in and year out, no matter what my financial situation might be, and pledge that I will carry similar dedication and selflessness into my future career as an elected official. I am roundly applauded by my audience and am eventually elected by a landslide on my platform of "Honesty And Charity."

In truth, my "regular contributions" consist of my dropping one shiny new penny into a Salvation Army bucket each Christmas, and then tossing a grubbier one at a homeless guy sleeping on a vent (while being careful not to get too close).

Did I tell a lie during my campaign speeches? Technically, I did not. I *do* contribute regularly to organized charity and the homeless, and I said nothing about the amount of the contributions. But anyone in their right mind would certainly argue that I had not been very honest.

This sort of playing with truth is what we have seen repeatedly with regard to many of the studies that are used to support increased smoking bans and taxes. David Kessler, past head of the FDA, in commenting on questionable claims about drug effectiveness, drew the distinction between a statement being "accurate" and a statement being "true." This distinction applies perfectly to our pseudo-philanthropic barrister, as well as to what we are now seeing in the world of medical science and reporting concerning tobacco control topics. Much of what we read about the health and economic aspects of smoking bans may indeed be accurate, but would certainly not be true in Kessler's wider sense.[109] As we'll see in *Slab I*, even something as simple as the misuse of the three-letter word "and" can make all the difference in the world.

How much of what we're told goes beyond that thin edge of technical accuracy? There's no real way of knowing. There's been no brave "Insider" willing to share the hidden papers of the antismoking organizations and no court orders blasting open secret vaults protected by attorney-client privilege. However, we can get at least a sniff of the dirty laundry simply by looking at what happened when Congress held the famous hearings where tobacco executives were dragged across the coals on national TV. While tobacco executives' testimonies under oath received wide coverage, an outright refusal to testify under similar oath by then-FDA Commissioner David Kessler and Surgeon General C. Everett Koop got almost none... after all, they were the "good guys," so why should *they* have to take an oath? When asked to testify before Congress, they simply said, "We see no reason for the committee to suggest that our testimony about tobacco now requires that we be put under oath or treated akin to tobacco executives."[110]

Such an outrageous refusal by two of the nation's leading Antismokers to be held to even a minimum standard of honesty before the United States Congress should have made banner headlines all over the country... but, of course, it did not.

Studies On The Slab searches for the truth behind the studies and shows how it is often misrepresented or even consciously and deliberately covered up by those claiming to represent science. The structure of those studies will be dissected so that the methods of their creation and strengths and weaknesses of the actual substance of their results can be better appreciated. The presentation of those results to the media via statements, fact sheets, summary posters, and press releases – and the media's almost universally uncritical acceptance, exaggeration, and reporting of those results – will be closely examined as well.

I'd like to add one final introductory note to the *Slabs*: the studies examined here have not been specially selected (cherry-picked) for their weaknesses. On the contrary, the ones I've chosen for the most detailed analysis have generally been the flagship studies, the ones most heavily presented to the public and legislators as groundbreaking and noteworthy in their support for the urgency of smoking bans. These are also the studies which have often inspired copycat studies;

those that repeat the same formulae in a slightly different style or in a somewhat different market for different paymasters. These copycats are then presented to the public, over and over and over again, as being yet more "new" studies finding yet more "new" threats from second-hand smoke. The copycats serve to corroborate and support the original results while generating continued headlines about the sup-posedly new and deadly dangers of even the most casual moments of encounter with threads of smoke.

I will admit however, that in some of the subsections of *Studies*, I have selected a few of the stranger headlines that have made it out there. Even though they may not be the studies that political decisions are ultimately based upon and justified by, they still form an important part of the substratum of attitude and feeling that ends up being translated into the bans, taxes, discriminations, and other negative stimuli designed to shock the "smoker rats" into the desired and proper behavior patterns.

The next time you read a news story about some new study examining smoking, think not only about how it might be twisted, but also why it would be twisted. Think about just what effect hearing of or reading about this new piece of multi-million dollar research will have on the overall psychological base that has allowed us to so blandly accept the treatment of people who smoke as if they were second-class citizens – or less.

**An Antismoker's Brain about to be
Dissected...**

... by a Smoker.

-   Sam Ryskind

Slab I

The Twisted Economics

We do not talk to say something, but to have a certain effect.

- Goebbels

The classic studies used by antismoking advocates in their push for smoking bans in the 1980s and early 1990s focused on long-term epidemiologic research examining lifetime daily exposures to home and workplace ETS. Even though their results were highly mixed and could not show any causality, Antismokers argued that they often found at least small statistical increases in lung cancer and heart disease mortality among lifelong spouses and co-workers of smokers. Appendix A of *Dissecting Antismokers' Brains* lists over a hundred results from lung cancer studies stretching from 1981 through 1998, with the vast majority of them failing even the most basic epidemiologic criterion – statistical significance.*

Whenever one examines the results of these sorts of studies, it is important to remember that even full statistical significance says nothing about how well or poorly designed a study is, or whether the resulting relationships between the variables measured truly indicate any sort of causality. It simply means that if you repeated exactly the same study thousands of times, you'd expect the results to lie outside of your claimed confidence interval less than one time in twenty. A

---

* Statistical significance exists only when the 95% confidence interval does not include the null value of 1.0; if the value of 1 is included in the confidence interval, it means the study found that no effect—a zero correlation between the variable being studied and the observed result—was included as a possibility in the study's range of findings.

study of sunrises occurring after roosters crow in the morning would show very great statistical significance despite the fact that there's obviously no causality. Or, to look at an example somewhat less obvious, there is probably a strong correlation between how many times people eat ice cream and how much skin cancer they get – not because eating ice cream causes skin cancer, but because both are likely to be more frequent occurrences in hot and sunny climates. Very few Eskimos die from malignant melanoma. In that last case, if we were totally ignorant of the relationship between sunshine and skin cancer, we might actually conclude that the ice cream, specifically its high fat content, was indeed the culprit.[111]   We might even defend our conclusion by citing the bioplausibility of fat's relationship to other cancers or go off on a different tack altogether by claiming a relationship between temperature stress and skin cancer.[112] Without previous knowledge, or in the face of a massive public health campaign aimed at eliminating ice cream, the role of solar radiation could be underplayed or ignored outright for decades.

Since 1998, there have been several more studies examining the relationship between ETS exposure and lung cancer, most notably a very large 2003 study by James Enstrom and Geoffrey Kabat, using American Cancer Society data, that showed virtually no effect at all from lifelong exposures to secondhand smoke.[113]   The evidence that even decades of fairly concentrated daily exposure to secondhand smoke has any significant effect on health is far from overwhelming. The Surgeon General's Report of 2006, *The Health Consequences of Involuntary Exposure to Tobacco Smoke,*[114] seemed to reach its major conclusion with regard to lung cancer by claiming that Enstrom and Kabat's 2003 study simply came out "too late" to be included in the Report. This was despite the fact that several studies published in 2004 and even 2005 supporting the Surgeon General's desired conclusions were somehow included in that 2006 Report. Even with such transparent juggling, the evidence that there is any threat at all from the much lower exposures generally found in today's well-ventilated and air-filtered Free Choice environments is simply and literally nonexistent.

How did the antismoking movement get from being the preserve of a few extremists who were widely perceived as cranks to being the powerhouse it is today? The influx of money from taxes and the pharmaceutical companies eager to market nicotine replacement therapy (NRT) products played an important role, but money without a plan could not succeed on its own. Antismokers knew they could take what they wanted if they built their position step by step, and if they could accomplish the dual goal of reassuring people that the transition to a smoke-free society would be both economically painless and produce instant and strong health benefits – not just for smokers, but for nonsmokers.

Unfortunately, the relative weakness of any real scientific basis behind the antismoking arguments has been overshadowed by the availability of billions of dollars in tobacco control money to create a thriving market for studies that are extremist in their claims, biased in their designs, and focused not just on indicting ETS in terms of health, but also on defending smoking bans as not being harmful in economic terms.

Back in the 1960s or 1970s, most employers would have agreed that banning smoking would lead to economic difficulties as smokers would leave to seek employment elsewhere or would demand or sneak excessive amounts of free time for smoking breaks. When bans were initially phased in and smokers objected, they were generally reassured and pacified by relaxed smoke-break time provisions. They were misled into thinking they were actually getting a pretty good deal, never realizing that once the bans were firmly in place, the extra break time would simply be eliminated. Al Gore's warning allegory about the frogs who happily let themselves be boiled alive as long as the water temperature is increased slowly proved all too true when it came to implementing smoking bans.

A second allegory, coming from a time when Al Gore was still in diapers, speaks of a similar concept. During World War Two, Italy was ruled by the velvet-gloved iron fist of Benito Mussolini. His method for removing freedoms was modeled on the behavior of a dishonest butcher shop employee. If that employee tried walking out with an entire salami all at once, he would be fired. But if he simply sliced a little bit off and popped it in his mouth each afternoon, he'd have the whole salami sooner or later, and the boss would never notice! Mussolini's method of slicing bits off the "Freedom Salami" has been given new life in today's antismoking world.

Medical studies tend to be intimidating to most people because they involve obscure medical terms or are based on data that are difficult to obtain and interpret. Economic studies are less frightening because almost everyone has had the experience of having to balance their budgets and do their taxes. In addition, unfortunately for the Antismokers, the basic economic data for such studies is also usually more available for critical examination since it isn't hidden behind the veils of doctor-patient confidentiality or buried in protected hospital records.

The Klein Study

To start off our little dissecting class, we're going to begin with a simple economics study that seemed to make an indisputable claim that smoking bans have no effect on bar and restaurant employment. The clarity of the bias and seemingly clear intent to mislead the public in pursuit of grant money will make the later dissections of medical studies a bit less daunting.

In May of 2009, Jean Forster of the University of Minnesota and Elizabeth Klein Ohio State University joined with several other researchers to publish a study titled "Does the Type of CIA [Clean Indoor Air] Policy Significantly Affect Bar and Restaurant Employment in Minnesota cities?"[115]

The study's conclusion, as stated by the researchers in media headlines throughout the country and overseas, was that, quite simply, "bar and restaurant employment" is not significantly affected by smoking bans. Within weeks of its publication, a Google search for news stories and references to it brought up thousands of hits. A sampling of quotes from the researchers clearly shows the message they wanted to deliver. The top two quotes below are from different primary sources, but both quote the main corresponding researcher, Elizabeth Klein. The third quote is from Jean Forster, the study's lead researcher.

> We didn't find that there was any significant effect on employment in bars and restaurants. [Enacting comprehensive smoking bans] provides the greatest public health benefit, and didn't produce any sort of significant economic effect.[116]

> Opponents to clean indoor air policies tend to say that having a partial policy, with bars exempted, will be less painful economically

for the community. They say people who work in these businesses that are dependent on alcohol sales would experience a catastrophic effect. ... We certainly did not detect anything close to the dramatic claims that opponents make based on the concerns that they have for bars ... bars do not need to be exempted from clean indoor air policies to protect against severe economic effects.[117]

Hospitality industry representatives have argued that smoking bans cause declines in bar and restaurant businesses because of the link between smoking and drinking behaviors. ... They have used that argument to push for exemptions for bars and bar areas of restaurants. We wanted to see if their argument was valid. ... Neither full nor partial bans have a negative economic effect on business, as measured by employment.[118]

In the interviews and articles that flowed from their study, Klein and Forster spoke glowingly of how their findings could reassure communities considering smoking bans that their bar and restaurant employees would not suffer financially because of those bans. But hidden within those assertions, by the clever use of the word "and," there was a linguistic trick. Most reasonable people, seeing the way these claims were presented, would assume that the research had shown that neither restaurant *nor* bar employees had been hurt by the imposition of smoking bans. In reality, this was not how the word "and" was being used, nor did it fairly represent what the raw data showed.

When the research was published, I emailed Dr. Klein and asked her, if the main purpose of the research had actually been, as stated, to examine whether establishments with heavy economic dependence upon their alcohol service were harmed by smoking bans, then why did the researchers decide to deliberately lump two sets of very different data together for their analysis. The data given to the researchers had come to them clearly separated into two distinct categories under the North American Industry Classification System (NAICS). NAICS lists one specific category for full service restaurant employees and another specific and separate category for bar employees. While the raw data for the study period examined did not show a statistically significant decline for employment in restaurants, or even in the combination of bars and restaurants, it clearly showed a strongly significant decline in real terms for bar workers. The researchers had obviously made a conscious decision to combine the two categories for their research presentations and I wanted to know why that decision had been reached and how it could be justified. Dr. Klein's response to me was simply that they had felt adding the two together was the "most appropriate" thing to do.[119]

Klein then pointed me toward studies that seemed specifically designed to imply that any who claimed economic harm from smoking bans were likely connected to Big Tobacco.[120, 121] The reading she directed me to seemed to write off most findings that bans hurt business by implying that bar and restaurant owners have secret and scurrilous motives for wanting to lie about bans harming their businesses and prefer to have studies that will mislead them rather than offer them real guidance in their businesses – something I would think rather unlikely.*

Klein's research was done at the University of Minnesota. The editors of the university's own newspaper had just recently interviewed Andrea Mowery, the vice-president of ClearWay Minnesota, the state's official MSA-funded organization that had provided $516,568 for Forster and Klein's research.[122] In that article, titled "Conclusions For Sale," they asked Mowery "if any of the studies funded by ClearWay have concluded against the organization's politics." She responded by saying, "I can't think of one off the top of my head."[123]

Think for a moment: which would you trust more? Research funded by the bar owners whose own pocketbooks depend on its validity? Or research funded by ClearWay Minnesota whose vice-president could not recall a single study ever funded by them that contradicted their predetermined and desired antismoking outcomes?

Rather than be deterred by Klein's veiled inference about my integrity and motives, I continued my examination of the basic data. Although the researchers refused to publicly identify most of the cities studied, the two largest – comprising roughly two-thirds of the total

---

* When a bar/restaurant activist, Keep St. Louis Free's Bill Hannegan, contacted Dr. Klein and pointed out contrary findings in studies by some economists, her response was, "Based on your argument, I assume you possess doctoral credentials in economics, otherwise you would not be qualified to render judgment in this material...." Klein seems to have overlooked her own lack of "doctoral credentials in economics" when she wrote that.

population data – couldn't be hidden because of their uniquely large sizes: Minneapolis and St. Paul. I tracked down the raw NAICS data for those two cities and examined it just as it was given to Klein and Forster: separated into two distinct groups, restaurant employees and bar employees.

What I found was astonishing. The researchers had been given, in a very clear and simple format, data that clearly showed that bar employment in the two largest cities they'd studied had been literally *decimated* after the imposition of a smoking ban. The bar employment in those cities had dropped by roughly 11% when bans had come into play! It should also be noted that this drop reflected only the jobs that were absolutely lost. The raw reality of the workers' situation was likely even worse due to remaining jobs being downgraded from full time to part time. Additionally, even many of the remaining full-time workers were likely surviving on greatly reduced tips, the real core of income for most bar workers. Bar industry personnel who had been living happy and productive lives before the ban – raising families, paying off mortgages, and looking to bright futures – had seen those lives and futures laid waste at the single stroke of a pen.

This devastation of individual lives and an entire industry was simply covered up by the Clearway researchers as though it didn't exist when they presented their study to the media.

They could argue, like the slippery shyster throwing pennies to the bums, that technically their statement was accurate when they said "bar *and* restaurant employment" was not significantly harmed by smoking bans. The combined absolute employment figures indeed showed no statistically significant harm, since restaurants are rarely affected strongly enough to be forced into closure. Thus, when the two separate labor pools were added together, there was not enough harm done to the total combination to offer any conclusive and statistically significant evidence of actual harm – to that combination.

But, as David Kessler observed, there's a big difference between mere technical accuracy and actual truth. By lumping the relatively small number of employees in the pure bar service industry in with the vastly greater pool of employees in the general restaurant industry, the enormous harm to the bar employees was diluted and lost to sight.

The deception would have been exposed if the researchers had ever been taken to task for their lack of clarification by either the publishing medical journal or a responsible media.

In reality, this study actually showed absolutely nothing about the effects of the ban on bars. Any objective observer who looked at the basic data would almost be forced to conclude that the numbers were quite consciously manipulated to hide the pain and destruction to bar employees.

It is as though I was hired by the Ku Klux Klan to do a study showing that the closing of factories would not adversely affect the jobs of black workers in a town with a very sizeable white population and a very small black population. To get the half million dollar grant for that study, I submit a grant proposal promising results that would not only support these closings, but also help generate more closings in the future despite Liberals' concerns about black unemployment.

But then I gather the data and discover that while the white population only lost an insignificant 2% of its employment, the black population lost a very significant 11%. What I'd been hoping to find was that the employment losses in each group were either nonexistent or at least under a non-statistically-significant 6% so that I could say neither one was actually shown to be hurt by factory closings.

So what do I do? Simple: I take both sets of statistics, combine them, and show that *overall* there had only been a 5% drop in employment; not a statistically significant drop, even if it was quite real to the 5% thrown out on the streets. The much greater numbers of white factory workers (who only lost 2%) made it easy to hide the devastation dealt to the much smaller population of black factory workers (who lost 11%) when the two were averaged together. I am then able to "accurately" tell reporters that my research indicated that "white and black employment is not significantly harmed by closing factories." Of course, that statement, with my tacit encouragement and carefully worded public expansions on it, rapidly transmutes into news stories proclaiming that "Despite fears, black employment is not hurt by factory closings!"

Was it morally and scientifically correct to produce a study that simply lumped both sets of data together in this way? Was I being honest when I declared to the world that I had discovered that factory closures did not hurt the employment prospects of blacks and whites –

without giving the media the additional information about black employment that was in front of me? Was I being honest while presenting my results in terms that I knew would be misrepresented by the media as showing that blacks were not hurt? Or was I, in David Kessler's words, simply being "accurate" rather than "truthful"? This kind of hiding by combining seems to be exactly what Forster and Klein did in presenting their research.

As I dug a little deeper, I found the picture got even worse. I found the code number for the original grant proposal that funded the work. Now, in normal science, a research proposal states a hypothesis and promises to study that hypothesis objectively to determine if it is likely to be true. Antismoking science, however, is run under the standards of a different rulebook. Here is an excerpt from the original research grant proposal to Clearway's legal predecessor, the Minnesota Partnership Acting Against Tobacco (MPAAT), a very openly antismoking organization (Just note its name!) also funded by MSA tax dollars.

> **We believe that this research will provide public health officials and tobacco control advocates with information that can help shape adoption and implementation of CIA [Clean Indoor Air] policies, and prevent their repeal [and] contribute to MPAAT's overall mission by providing information that enables adoption and successful implementation of policies to protect employees and the general public from secondhand smoke exposure.[124]**

Suddenly, the reason why Klein felt it was "most appropriate" to lump bar employment in with restaurant employment becomes crystal clear: the researchers had promised their funders that they would produce research results that would support their "CIA policies." The *only* policies at that time which would qualify for MPAAT as CIA policies were total smoking bans – such alternatives as exhaust ventilation, air-filtration, or separated smoking areas were summarily dismissed.

So much for impartiality, so much for the scientific method, or double-blind controls, or testing an hypothesis, or even plain and simple basic scientific integrity. All of that seemed to be thrown out the window, lock, stock, and barrel. It had to be, because the goal was getting the grant. And once one has gotten the grant, the next goal is producing the promised results in order to guarantee another grant.

Producing results that showed bans were devastating to bar employment would certainly not have paved the way for future millions in grants from an organization that "couldn't recall" ever having funded such a contrary study. The researchers' work was bought and paid for by antismoking interests on the basis of a promise in the grant proposal that the science would produce only a certain kind of result — and those researchers seemed bound and determined to deliver the goods exactly as promised, facts and science and integrity be damned.

So now we've seen what it was that Dr. Klein meant – but didn't say – when she responded to my email by simply saying they'd felt lumping the two sets of figures together was the "most appropriate" form of research. Indeed, if one's goal was to procure future grants specifically designed to produce research promoting smoking bans, it was then clearly the "most appropriate" approach. In my earlier example of doing research for the Ku Klux Klan, I'd have guaranteed myself many further lucrative grants from the local Grand Dragon for my good work by following a similar path

Did they lie when they said their research showed no harm to bar *and* restaurant employment from smoking bans? Technically they did not. Just as in the case of my shyster lawyer, or my own fictional efforts in service of the white sheets, they can walk away secure in the knowledge that, by a trick of language, they can claim their research and statements were accurate. But just as assuredly, I'd say they were lying through their teeth in giving the impression to the media and the public that their work demonstrated no harm to bar employment.

The Klein Study: v. 2010

In 2010, a new chapter unfolded in the Forster/Klein story. The researchers, perhaps due to the criticism of their earlier study, produced a new piece of research that seemed to correct the main error of the earlier piece.[125]

The new study was titled, "Economic Effects of Clean Indoor Air Policies on Bars [sic] and Restaurant Employment in Minneapolis and St Paul," and at first I thought this was nothing more than a cheap trick to get two sets of headlines for the price of the same piece of research. The title was similar, except that this time the research was limited to just two cities rather than six. But then I looked a bit more closely and found I was wrong.

The first study had quietly lumped bar and restaurant employment together so the bar losses could not be seen. This time, the researchers kept the data for bars separate from restaurants, just as it had been given to them by the government. I was quite surprised to see that, even with the separation, the researchers had concluded that bar employment was not harmed by bans in the two largest cities they examined – so I decided to look a bit more closely.

Unfortunately, this second study was not as easy to dissect as the first one. You may remember my earlier discussion of how statistics can be so easily used to facilitate lies once they start going beyond the basics. The primary equation used by Forster and Klein in their second study is a bit too complex to reproduce here, but was described clearly in a single paragraph of their work:

> The processes of the Box–Jenkins ARIMA **(p,d,q)(P,D,Q)** model is described for a general seasonal ARIMA model for outcome variable y_t as [follows]: where **p** is the order of the auto-regressive process, **d** is the degree of nonseasonal differencing, **q** is the order of the moving-average process, **P** is the order of the seasonal auto-regressive process, **D** is the degree of seasonal differencing, **Q** is the order of the seasonal moving-average process, **s** is the seasonal span, phi_1 to phi_p are the seasonal auto-regressive parameters, phi_1 to phi_q are regular auto-regressive parameters, $theta_1$ to $theta_Q$ are the seasonal moving-average parameters, $theta_1$ to $theta_p$ are regular moving-average parameters, mu_t is a random (white-noise) error component, $alpha$ is a constant, and B is the backshift operator such that $B(z_t)$ equals z_{t-1}. To reduce hetereoscedasticity, we transformed each series taking the natural logarithms. ... The dependent and independent variables are log transformed, the exponentiated parameter estimate ... [etc.]

At first I felt somewhat defeated. With a half dozen Greek letters in the formal equation, a few Chi squareds, some autoregressive parameters with a tad of white noise and backshifting operators sprinkled in, and finally logarithmic transformations applied to exponentiated estimates of non-seasonal differencing, Klein had moved well beyond the depth where I felt confident enough to offer any hetereoscedastic criticism – or even a quiet burp.

My background experience with antismoking studies, particularly those produced by people who had seemed less than honest in the presentation of their earlier work, led me to believe this was all some sort of put up job. But I didn't know enough statistics to prove it, and when I contacted Elizabeth Klein to clarify some of the things that puzzled me, she was less than helpful – simply recommending that I read a textbook on statistics.

My annoyance with that response kept me from giving up. In *Dissecting Antismokers' Brains,* I had written about the experience of true believers in psychic phenomena confronted by a skilled skeptic who traveled around with them from one medium to another, cleverly exposing the fog machines, cold-air vents, levers and strings and pulleys, as well as the hammers-under-the-tables and ghostly-voice-makers. The skeptic exposed one after another after another, until finally he was confronted simply by a veiled psychic who claimed to hear dead people talking in her head. The true believers smiled triumphantly and declared, "See! THIS one is TRUE!" and the skeptic could find no way of disproving it despite knowing full well that it was just as fake as all the others.

I now found myself in the position of that befuddled skeptic and was initially quite discouraged, particularly given all the work I had previously put into describing the original Forster/Klein study for the opening part of this section. I couldn't honestly just ignore the new study, but I had no handle on how to show its trickery or weaknesses. But before I packed everything away and surrendered, the hard-headed Irishness I inherited from my father and the inquisitiveness I inherited from my mother asserted themselves and I took another look at the original data.

This time, I extended the data for the two cities in question to cover a period of three years before their smoking bans hit and three years afterward. What I found was quite interesting. Overall bar employment in Minneapolis and St. Paul had been in a steady and strong decline during the years previous to their bans; then that decline seemed to be temporarily interrupted to some degree during the bans' implementations, and then leveled out or resumed a less drastic loss afterward.

Superficially, the data still seemed to support the general idea that the bans had not been harmful and could, arguably, even have been helpful, but then I noticed something that Forster and Klein had clearly failed to address. Some unknown, very significant, and perhaps completely unrelated factor had been having a major impact on bar employment in the Twin Cities during most of the first decade of the twenty-first century. Whatever that factor was, it was completely unexplained and seemed to have been totally ignored by the Klein researchers. It was most certainly not a general economic decline experienced throughout the country, because similar drops during that time period did not appear in data from other areas. Without under-standing what that unknown but vastly overriding factor was, and without any knowledge of how it might have changed during the period under study, absolutely nothing could be said with any degree of confidence about the actual effects of the smoking ban on bar employment in these two cities. Two separate attempts to ask Elizabeth Klein about this were rebuffed in her answering emails, with no real help offered beyond that initial recommendation to go read a statistics textbook.

I still don't know the full story behind this second piece of research, but the clear reluctance of the researchers to examine or even acknowledge the importance of a third, and clearly quite important, unknown correlating variable leads me to believe that the results of the second Forster/Klein study should be taken with quite a bit more than a grain of salt. Like the psychic skeptic above, I may not be able to show just how the psychic is faking her act of channeling dead people, but I'm pretty damn sure that there's at least some sort of chicanery going on when I'm told that there's nothing of any importance beneath the locked trapdoor under the table. I might not be able to explain just why the sun rises after the rooster crows, but I'm not going to simply accept that the crowing wakes up the Great Sun-God... and I don't think you should, either.

Klein's Response

I have been very harsh in my judgment of Klein and Forster's research and its presentation, both here and in many Internet postings. After those postings and after several email exchanges with me about the preceding studies, Dr. Klein sent me an email in which she presented and strongly stated a defense of her work. While I personally did not feel that the defense was adequate, I thought it would be only fair to present it here in full as a counterpoint – despite its 600-word length.

I wrote her, saying

> I appreciated the care you put into [your email] in which you
> defended your research work. In my writings, I *do* try to achieve
> some degree of balance while still strongly arguing my points. I'm
> wondering if I may have your permission to present your side of the
> issue in my future writings by reprinting your email? I think it
> would be more effective and present your case more truly than if I
> simply tried to rephrase it.

Her response, in thirteen words, was a blanket refusal for me to
share her email. Unfortunately, it seems to be illegal to publicly share
email without permission, even if it is an official response from a public
figure funded by public funds about their public work. I was tempted
simply to leave it at that, but I think I should at least note that her two-
page epistle stated a robust defense of the researchers' reasons for their
research design and an assertion of their general integrity. Part of it
seemed sincere, part did not. Overall, the sense I got from a good part
of her email was almost one of desperation; the sort of "Not me!"
response that a little boy gives when he's caught next to the broken
cookie jar on the floor.

If Dr. Klein ever does give me permission to share her statement
of defense in her own words, I will be happy to consider it for any
future edition of *TobakkoNacht*. I truly do feel it's important to allow
people to respond when they are criticized as extensively as in this
particular case. My original conclusion to this section read as follows:

> So I leave it for my readers to judge. You've seen my analysis, you've
> seen the headlines and references to news stories about the original
> research, the original studies are available online, and Dr. Klein's
> defense of her work is presented above. Do you think it is an ade-
> quate defense in the face of how that work has been presented and
> used? Or is it not?

Due to her reluctance to share her own words here, I guess all I
can ask now is this: do you, after all you've read here, feel that any
defense she might have offered could really have been adequate? Or
not? And if it *was* adequate, do you think she would have been so
reluctant to share it publicly?

>>> Cleanup 1 <<<

Fundamentally, the Klein study illustrates what is wrong with nearly all the studies examined in these dissections. Science has taken a back-seat to politics. Even worse than that, science has become controlled by politics.

There was a similar scientific crisis within Stalinist Russia many decades ago. In 1928, a good Communist scientist by the name of Trofim Lysenko created a special farming technique that he claimed could make wheat grow early in the springtime. He promised that farmers could grow far more wheat if they followed his guidance.

Lysenko believed that his method of growing wheat would produce far more food than the newfangled genetic research that was taking place elsewhere in the world. To the great misfortune of the Soviet people, the hierarchy of the Communist Party believed in his methods and pretty much banned all agricultural research and practice contrary to Lysenko's techniques. As a result, tens of millions of people suffered and died early deaths from needless malnutrition and starvation throughout the Soviet Union in the mid-twentieth century.[126] Meanwhile, respectable Soviet agricultural scientists who questioned Lysenko's theories were ostracized by the official establishment, sent to labor camps, or even executed for daring to contradict the politically correct view.[127]

While no one in America today is being sent to labor camps or executed simply for disagreeing with the science as prescribed by antismoking researchers, there are certainly professors and researchers in academic institutions who have found themselves heavily criticized and cast out of the mainstream simply because they would not produce unscientific and politically motivated research. Unlike some of those whose work is examined on these *Slabs*, they refused to juggle their data into a "most appropriate" format to fit the result desired by the establishment and future grant funders. For their reward, they found themselves banned from essential communications channels and forced to defend their reputations against accusations and innuendos of being bought and paid for by the tobacco industry. To add insult to injury, they then find themselves in the position of having no one to turn to for further research support *other* than the tobacco industry, as they are quietly blacklisted by all the tobacco control grant givers. If such researchers are professors at schools whose boards then vote to bar *any* tobacco-connected research funding, they face a grim future indeed.

Professor Carl Phillips, whose work focused on Tobacco Harm Reduction at the University of Alberta; Dr. Kamal Chaouachi, a researcher specializing in studies of shisha (hookah) smoking; and Dr. Michael Siegel, a long-established Boston University researcher with a strong history of "approved" antismoking studies stretching back into the 1990s, have all found themselves targeted and ostracized for refusing to bend their science to politics. In the case of Dr. Phillips, the pressures were severe enough, both on him directly and through intimidation of students who dared to work on research projects in his area, that he eventually left his university position altogether and moved from Canada to set up a new career in Philadelphia.[128]

A perfect example of what can happen to even a lifelong career if a researcher refuses to bend and twiddle and frame his results to fit with the holy doctrine can be seen in the case of Professor James Enstrom of the University of California, Los Angeles (UCLA).[129] Enstrom had raised the ire of organized Antismokerdom back in 2003 when he and Dr. Geoffrey Kabat dared to publish a study (hereafter referred to as the E/K study) finding no real relationship between ETS exposure and either lung cancer or heart disease.[130] The E/K study used one of the largest databases for any such study ever undertaken, one created by the American Cancer Society itself. It wasn't perfect, but it basically suffered from the same negatives as many of the smaller studies from that time period. But since the result went the "wrong" way and since Enstrom and Kabat used a tobacco industry grant to finish their work, it was instantly met with a major firestorm of criticism in the *British Medical Journal*'s Rapid Response pages.[131] However, unlike the pointed and substantive criticisms later aimed at the antismoking-approved "Great Helena Heart Miracle" study, the 180 responses to the E/K study contributed more mud than light, with a number calling for actions against either the researchers themselves or against the editors of the *BMJ* itself for daring to publish such an abomination.

The calls weren't ignored, either. Although the official excuse seemed to focus on his more recent work examining diesel emissions,

Dr. Enstrom found himself officially notified in 2010 that he was being relieved of his thirty-four-year career at UCLA. Officially, they said that his research orientation didn't match with the university's "mission," but there was little doubt that his firing was largely payback for his insubordination in challenging the official antismoking ethos in his earlier work.[132, 133] It has taken some stiff fighting, and Professor Enstrom has found his position extended for the moment, but Lysenko's axe is still hanging over his neck.[134, 135]

The threat is not limited to North America either. In July, 2012, a retired GP, Brendan O'Reilly, was suspended from the BMA Welsh Council until 2014 "after he questioned the evidence behind the BMA's campaign to ban smoking in vehicles on BBC Radio."[136] Perhaps he should be grateful. Socrates had to sip hemlock for questioning authority.

The Kuneman/McFadden Economic Effects Study

To consider a case a bit closer to home, I only have to go to a study done several years ago by Missouri researcher David W. Kuneman with my collaboration. While we never sought formal journal publication for our research, our figures and analysis made a strong argument that smoking bans wreak economic devastation far beyond what had previously been conceived, and we supported that argument with simple figures drawn from fully accessible and verifiable public data.[137]

Neither our study's methodology nor its substance has ever been attacked, but its online publication by the Citizens Freedom Alliance earned it an entry on Stanton Glantz's* TobaccoScam website, a website intended to give people a general impression that omnipotent arms of Big Tobacco secretly direct the CIA, the NSA, MI5, CNN, Hollywood, Bollywood, and probably even the little toy-making elves at the North

---

* Stanton Glantz, usually referred to as "Dr." Glantz in the media when he bemoans the deadly threats of secondhand, thirdhand, and cinematic glimpses of smoke, is sometimes misidentified as a cardiologist, but actually has a doctorate only in mechanical engineering. A longstanding antismoking activist, he founded Americans for Nonsmokers' Rights and has brought tens of millions of dollars in grants to the University of California. Not surprisingly – considering those grants – he has managed to snag a Professor of Medicine position there, despite lacking a medical degree.

Pole. TobaccoScam made some rather questionable claims about our work and about Mr. Kuneman.[138]

They were correct in pointing out that our study was not peer-reviewed. Given our results, publication in a medical journal was unlikely, and since neither of us was familiar with the world of economics journals we simply never sought formal publication. But TobaccoScam then went on to file our study under the heading of "Fake Hospitality Results," – just in case anyone might have thought it was honest research worth taking seriously – and incorrectly stated that we "did not control for underlying economic trends and random variations in the business cycle." Actually, we had quite carefully corrected for economic trends by setting up comparison control states, and any concern about business cycles were subsumed in the extensive time frame we covered – almost a full decade of smoking ban activity.

Their negligence continued when they claimed our funding sources had not been revealed, thereby deliberately implying that we probably had Big Tobacco money behind us. Actually, we had explicitly and openly declared, in the proper section at the end of our study, that we had carried it out with our own resources: "DISCLO-SURES: The authors used their own time and funds to research and prepare this article."

Finally, TobaccoScam played up a rather ridiculous "tobacco industry link" by saying, "Kuneman is a retired research chemist who worked at Philip Morris." They left it simply at that, without noting that his research, twenty years in the past, had actually involved soda flavorings. In any event, we had clearly and openly disclosed this information in our competing interest section: "Kuneman, who smokes, worked for six years in the 1980s as a research chemist for 7-Up and still draws a small pension from that work. At the time of his employment, 7-Up was bought by Philip Morris. His current work and concern has no connection to that employment." Dr. Siegel strongly defended our work in his writings and pointed out that TobaccoScam even introduced the section containing their misleading presentation by saying, "Check out this section to see how the tobacco industry cooks the books behind the scenes."[139]

So why would TobaccoScam go out of its way to so unjustifiably and nastily attack our credibility? Could it be because the results Dave and I found were so incredibly damaging to their "smoking bans are cost-free" arguments? And could it be that they simply couldn't find any real grounds for attacking our work on the basis of its actual content and methodology? The motivation for this attack was quite clear. As you'll now see, our results were almost too damaging to be believed.

Mr. Kuneman and I used data taken directly from publicly available state economic records. We analyzed it in categories defined by the antismoking groups themselves so as to exclude accusations of bias or cherry-picking.* We used a Connecticut Office of Legislative Research (COLR) report that classified states as being either smoker-friendly or smoker-unfriendly in terms of bar and restaurant smoking restrictions. A state was classified as smoker-unfriendly if extensive statewide or local bans had been imposed to severely restrict smoking in bars and restaurants. Mr. Kuneman and I compared the four states with the highest levels of implemented smoking bans between 1990 and 1998 with the four that had shown the lowest levels.[140] The figures in Table 1 show clearly that the difference in economic growth rates between the two groups was astounding.

The smoker-friendly states had their general economies (as reflected through their Total Retail Trade figures), and most particularly their hospitality-based economies (Bar & Restaurant Trade), make great leaps during those eight years, jumping nearly 80% in some cases. Meanwhile, the four states that the COLR ranked as smoker-unfriendly, those with the strictest smoking bans, were a basket case by comparison. The hospitality segment of two of them actually showed raw dollar declines despite eight full years of national economic growth as reflected in the USA figures at the bottom.

We then went a step further and addressed the possibility that simple geography might have swayed our finding. We picked staunchly antismoking California as our subject, since it was closely adjacent to several states with weak or no bans. Again, we found the Free Choice states did far better than the state with the most onerous smoking bans in the entire country during those years.

---

* Cherry-picking occurs when researchers know that general data won't support their thesis, so they seek out special cases and limit their data to juicy "cherries" hidden among the brambles. This will be explored further in *Slab II*.

STATES WITH VERY HIGH BAN ACTIVITY

Year/State	Bar & Restaurant Trade		Total Retail Trade	
	1990	1998	1990	1998
CA	26.3	28.0	225	291
NY	13.1	15.8	124	148
MA	6.1	5.9	50.7	62.6
VT	0.46	0.44	4.5	6.0

STATES WITH VERY LOW BAN ACTIVITY

Year/State	Bar & Restaurant Trade		Total Retail Trade	
	1990	1998	1990	1998
NC	4.5	8.0	45.8	81.1
VA	4.4	6.9	47.5	73.6
MO	3.5	5.7	36.0	57.3
TX	11.4	18.4	120	190
USA	182	260	1807	2695

Source: *Statistical Abstracts of the United States*, 1992 and 2000, tables 1292, 1295. All figures in billions of dollars.

Table 1

Kuneman/McFadden: Multi-Year Economic Impact of Smoking Bans

Indeed, in looking at the data and thinking about how it would have played out in the long run if the study extended into the 2000s, it appears that, compared to the best Free Choice states, smoking bans may have actually cost California's overall economy an incredible one hundred billion dollars! Even for a state the size of California, and even spread over years, such a loss is nothing to sneeze at.

To be fair, I should emphasize the use of the word "may" in my analysis because there "may" have been other factors at play; factors that just happened to perfectly correlate with smoking bans or the lack of smoking bans by pure chance in every single state studied for the two factors. Unlikely, but possible. Possible... but *very* unlikely. And when we're talking about a hundred billion dollars lost to the economy of a single state, the far more likely and simpler possibility assumes commanding importance and Occam's approval.*

It might have been fun to sit down with Governor Schwarzenegger in his cigar tent and ask him what he thought California might have been like during his governorship if he hadn't needed to sit out in a cigar tent and if there had been an extra $100 billion in the state treasury. Maybe we could have invited California's antismoking guru "Dr." Stanton Glantz and California's antismoking Congressman Henry Waxman in to give their opinions as well – though we might have had to outfit them with gas masks. Something tells me that The Terminator might have had a few choice words, or perhaps something of more substance, to share with them.

In summation, despite their protests to the contrary, I feel the Klein studies provide an excellent example of how researchers lie with statistics and study design without "really" lying. In stark contrast, the Kuneman/McFadden study, although arguably just as biased in its motivation,† presented its material truthfully, rather than just accurately, in a form that was clear enough to understand even for those without a doctorate in Advanced Hetereoscedasticity.

Remember, that fancy-pants lawyer at his campaign rally wasn't lying either, and neither were most of the researchers behind most of

---

* The principle of Occam's Razor basically states that, if all else is equal, one should accept the simplest explanation as the most likely.
† Although without the half-million-dollar grant incentive.

the studies we'll be examining in this section. Technically, they almost all tell some distorted version of the "accurate" truth while knowing they can get away with adhering to that technicality.

As Bill Clinton might say, "It all depends on how you define the word 'is'."[141] My goal in examining these studies is to uncover the lies hidden behind the word "is" or "and," or any of a dozen other weasel words used by Antismokers stripping our freedoms away.

Beyond Bars And Restaurants

Bars and restaurants were not the only targets of economic concern with regard to smoking bans. In the last quarter of the twentieth century, politicians had looked to new horizons for income and set aside their moral qualms about gambling. Gambling had always been a lucrative vice and state governments decided it was time to take it from organized crime and let the State profit from the human weakness for a good bet.

In order to pacify the moralists, canny politicians initially directed that tax profits from gambling would be earmarked for things like childhood education; but eventually it simply became part of the core state budgetary process. Yes, the money from the lotteries, casinos, pull-tabs, and Keno machines might have gone to the children in a bookkeeping sense, but that simply freed up an equal amount of money from general taxes to be spent elsewhere. Taxpayers fell for the politicians' shell game in droves, and state after state voted to liberalize their gambling laws. The gold mine that Nevada and the mob had cornered for years quickly became an important source of base income for cities and states across the country.

Enter the smoking bans. Once bars had largely fallen to their dictates in many states, the Antismoking Crusaders looked hungrily at the last major bastions of indoor smoking: the casinos. The walls of that castle proved a bit harder to breach than the huts of the lowly bar owners. Politicians had known they were selling bars and small restaurants down the river when they imposed smoking bans, but they didn't care; State bottom lines weren't being hurt too obviously by the loss in small business tax revenues, and campaign-generous anti-smoking interests could always be relied on to produce showpieces like the Forster/Klein study to hide whatever damage there was.

But gambling revenues were a card of a different suit. Gambling receipts and taxes were unique, pure, and very carefully overseen and counted out to the penny by highly regulated government watchdogs and auditors specifically installed to guard against mob influence and creative book juggling that would impact state treasuries. Even states without casinos keep very careful track of their bingo, pull-tab, and Keno dollars. It was soon painfully obvious that, since smokers tend to be heavily over-represented in the gambling population in general, throwing smoking bans into the betting pool produced a very clear and undeniable result – instant losses of tens or even hundreds of millions of dollars that could not be covered up, mixed in with other funds, or claimed to be the result of bad weather or economic downturns.

Perhaps the clearest example of this, and notably not casino-specific, can be seen in the state of Minnesota. Minnesota had discovered that it could make an enormous amount of money off such things as pull-tab games, bingo, and other forms of charitable gambling, even without casinos. And, as usual in the case of governments caring for the most profitable sheep in their flock, very careful and very clear accounting was done regarding that money, with every penny watched over and accounted for by the Minnesota Gambling Board.[142]

In 2003 and 2004, the Minnesota Charitable Gambling Reports showed revenue streams on the order of $120 million per month from such sources. In 2005 and 2006, Minnesota was swept by partial smoking bans that discouraged gamblers in the affected areas from hanging around and spending their money in places where they felt unwelcome. Gambling revenues dropped instantly by roughly ten million dollars a month. Then, at the tail end of 2007, a full statewide ban came in, and again, an instant drop was took place, this time approaching fifteen million dollars a month!

A simple look at the raw numbers makes it abundantly clear that the "cost-free" smoking ban promised to Minnesotans actually cost the state over $300 million a year. With inflation and general trends factored in, the figures approach $400 million. When the worldwide economic meltdown hit at the end of 2008, it showed up as little more than a blip on the graph after the smoking ban disaster of the previous years. No matter how loudly Stanton Glantz declares to the media that "This whole economic argument is hogwash," it's hard to cover up figures as plain as the ones documented in that Charitable Gambling Report and illustrated in Figure 2.[143]

One might think that losses such as these, so blatant and so devastating to the pocketbooks of states and taxpayers alike, would have been enough to quash attempts to ban smoking in the casinos that generated such enormous sums of tax money – not to mention jobs. But, as we've already seen, even hard economic realities can become quite soft in the hands of skillful manipulators.

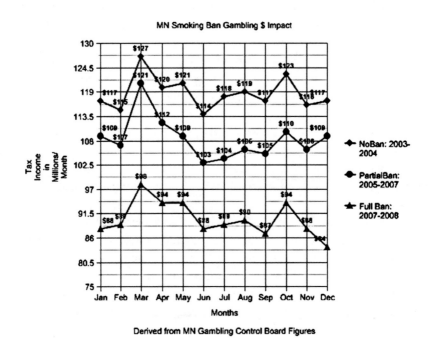

MN Smoking Ban Gambling $ Impact

Derived from MN Gambling Control Board Figures

Figure 2
Minnesota Revenues in Three Differing Ban Periods

In the last quarter of 2008, Atlantic City had been on schedule to move from a partial smoking ban to a full smoking ban. The casino moguls, who had been quiet as mice as they watched the bars and restaurants of their state being attacked by the Antismokers, finally stood up and spoke. Their earlier inaction had rested upon the sad misperception – as stated to me when I spoke to a New Jersey gaming association representative in 2005 – that they were simply "too big" for the State to ever touch. It was a bit late, but they'd learned their lesson and finally began to fight back – alone, but still powerful.

The casinos basically argued that the double punch of general economic woes and a total smoking ban together would result in the closure of up to a third of Atlantic City's casinos, plunging the city into an instant economic depression that it would never recover from, while at the same time dealing the state and its taxpayers a devastating blow. The state's earlier partial ban had already resulted in Atlantic City showing its first declines in its thirty-year gambling history with lost revenues being easily in the tens of millions of dollars. The casino chiefs asked for a temporary reprieve from a full ban to give the economy and their profits some time to recover. They asked for a one-year extension in which they would continue to ban smoking in all their bars and restaurants and on 75% of their casino floors, but could still allow Free Choice on the remaining 25% if they wished.

Antismoking forces were not pleased. They launched a campaign in full force, whipping up their astroturfed groups of senior casino workers to a frenzy and packing the council chambers with their brightly T-shirted supporters. Dealers booed and even screamed at Council members in front of TV cameras: "You gave us a DEATH SENTENCE!"[144] One news story before the Council vote attracted an anonymous Internet poster using the handle of "Tammy" who threatened to show up at the hearing "packing heat" with her friends if the full ban was tampered with.* Nonetheless, the Council voted five to four in favor of extending the partial ban for one more year before implementing a full one.[145] That implementation was supposed to come in November 2009. As of this writing in June 2013, I have not heard another peep from the casino workers who are now desperate to hold onto their jobs. They know what a full ban would do to them and their lives.

---

* "I will be there with my crew, packing heat, and ready to blow away the City Clowncil.[sic] The stakes are too high for my lungs, after my mastectomy from 30 years of smoking, to let these critters live another day!" –Posted by Tammy on Monday, October 27, 2008, 12:42 PM.

The experiences of Minnesota and Atlantic City are by no means unique. As the nation's economy staggered in 2008 and 2009, more numbers came in showing the true toll of previous smoking bans upon the gambling industry and taxpayers. States with total bans fared among the worst. A devastating study by Michael Pakko of the Federal Reserve showed that Illinois casinos posted a loss of 22% for the year and Colorado casinos were down 12%. Colorado would have shown a steeper decline except for the fact that its partial smoking ban had already impacted its casinos in 2007. Meanwhile, the Free Choice states of Indiana, Iowa, and Missouri showed 2008 growths of 2%, 5%, and 7% respectively, despite the general economic downturn.[146] Averaged together and accounting for the lost growth in addition to the actual raw losses, we're talking about roughly a 25% loss in real dollars that impacted taxpayers' pockets.

Even as late as August 2009, supposedly respectable anti-smoking extremists tried to deny reality. On August 14, 2009, Kathy Drea, vice president of advocacy for the American Lung Association in the upper Midwest, looked at Illinois' casino losses and said, "I would say the economy would be the number one issue. I think [the smoking ban] had very, very little to do with it."[147] She made that statement despite the fact that she must have known that nearby Indiana, Iowa, and Missouri had very different stories despite quite similar economic conditions. She simply ignored that data in making her public state-ments; after all, acknowledging it wouldn't have supported her goals.

She then went on to say that while there "might" have been a revenue decline, "it defies logic" to blame the smoking ban. How anyone could think that blaming declining gambling revenues on a smoking ban "defies logic" after seeing the figures from Minnesota, noting the history in Atlantic City, and reading the comparative figures from the Pakko study, defies logic in and of itself! Such economic facts haven't slowed down Stanton Glantz with his statements about eco-nomic fears being hogwash, nor altered his seeming beliefs that casino life under smoking bans was simply a field full of daisies and daffodils floating happily atop shimmering seas of wildly gambling nonsmokers who had heretofore cowered quietly in their homes. Such denial of crystal clear reality ranks up there with the 2011 statement by Deborah Arnott, head of the UK's ASH organization, claiming that, "it wasn't disparities in tax that led to the growth in smuggling." Right. People just smuggled cigarettes for the fun of it. Kind of like broccoli. Those broccoli smuggling gangs are wreaking havoc in the world today, taking over organized crime, infiltrating high schools, even sneaking sprigs of the evil green stuff into school lunches in primary schools![148]

The antismoking contingent of America's New Experiment With Puritanism gathered support from other quarters as well. Anita Bedell, executive director of Illinois Church Action on Alcohol and Addiction Problems, looked at the Illinois loss figures and said, "If they're down, it's probably a good thing. That means people are spending their money elsewhere." Like maybe giving it to Church Action centers?[149]

And Beyond Economics...

When one sees such denial being freely practiced by antismoking researchers, media representatives, and high-profile activists in the face of easily demonstrated contrary documentation, it's not surprising that the misrepresentation of studies dealing with the far foggier world of scientific and health issues is even more extreme.

It's almost funny to watch how, whenever Free Choice advocates succeed in publicly showing the falsity of the argument that smoking bans cause no economic harm, antismoking advocates immediately change horses and declare that the economics simply don't matter. They'll drop their failed argument, averring that the value of a human life outweighs the bulge of the pocketbook, and just go back to claiming that human lives are being seriously and immediately threatened by the lack of smoking bans despite any economic costs.*

The late 1990s and early 2000s saw increasingly strident claims about the unearthly deadliness of what Thomas Laprade of the Citizens Freedom Alliance famously calls "A little smoke from a handful of crushed leaves and some paper that is mixed with the air of a decently ventilated venue."[150] By way of contrast, a Director of California's Air Resources Board, Joan Denton, puts secondhand smoke in the same category as the most toxic automotive and industrial air pollutants.

---

* Of course, once that argument is successfully contradicted, they'll just bounce right back to the economic one as though it hadn't already been dealt with!

"Californians, especially parents, would not willingly fill their homes with motor-vehicle exhaust, and they should feel the same way about tobacco smoke."[151] Sir George Godber's 1975 conference had urged Antismokers to concentrate on creating and fanning fears in ordinary nonsmokers while inspiring guilt in smokers, and that's exactly what the Antismokers of the twenty-first century have done.

Prior to the appearance of this fear mongering, most nonsmokers generally hadn't thought too much one way or another about people smoking around them. Occasional contact with low or even moderate levels of smoke didn't really bother most people. It was a smell, a scent, an aroma, that was generally judged to be less offensive than body odor or cheap perfume unless a concentrated plume of smoke blew directly in one's face. The creation of fear changed all that.

Antismokers were faced with the problem of how to make that fear into an immediate threat rather than just a hint of a possible cancer seen off in the distance of forty years away. They needed an image as urgent and in-your-face as a screaming, drooling, chest-bursting heart attack exploding right there in the middle of a restaurant where some selfish smoker was "poisoning the air." While some small studies in the 1990s tried to bring such a concept to the public eye, none really succeeded in grabbing the major headlines until the year 2003.

On April 1, 2003, the University of California announced the amazing findings of a study titled, "Reduced Incidence of Admissions for Myocardial Infarction Associated with Public Smoking Ban: Before and After Study."* The preliminary data and findings were presented at the scientific sessions of the American College of Cardiology's 2003 meeting in Chicago.[152, 153] This study eventually became better known to the public as either "The Great Helena Heart Miracle" or "The Great Helena Heart Fraud," depending upon which side of the aisle you were on, and its guts will be laid out in *Slab II*.

---

* When I saw the news story claiming an instant 60% drop in heart attacks after a town's smoking ban and noticed the date, I simply assumed at first that I was reading an "April Fools" satire!

Slab II

The Great Helena Heart Miracle

Claims that smoking bans in public places have led to dramatic reductions in AMI incidence are not supported by the evidence. Scientifically invalid claims, though promulgated in the name of protecting public health, have adverse consequences.

- Michael Marlow

In early 2003, a press conference was called to announce the amazing results of a six-month-long smoking ban in a small community in the northwestern United States. Two doctors known for being quite active in antismoking circles[154, 155] and for making outrageous statements* in earlier efforts to ban smoking in the town of Helena, Montana,[156] claimed they just happened to notice an unusual drop in AMIs† when a ban finally was imposed, as well as a seeming bounce back to normal levels after it was rescinded. They enlisted the statistical and political advice of Stanton Glantz, and the three used the public stage of the American College of Cardiology's 52nd Annual Scientific Session to announce a study claiming that Helena's smoking ban had immediately slashed its heart attack rate by an incredible 60%! This was particularly notable since roughly 80% of Helena's workplaces had

---

* Robert Shepard: "...anything we can do to restrict [smoking] is justified."; Richard Sargent: "Working in a smoke-filled bar, that's the equivalent of smoking somewhere between 15 and 20 cigarettes per day."

† An AMI is an Acute Myocardial Infarction, a common type of heart attack. While there are technical differences between types of heart attacks when the term is used in the medical community, the press usually treats them all as being the same. The Helena study measured AMIs, but the news stories almost universally just spoke of "an amazing reduction in heart attacks."

already been covered by moderate to full bans of their own prior to the new total ban.[157] Even more telling was their claim that the heart attack rates rebounded to normal as soon as the smoking ban was lifted.[158]

The media went wild, fed by such things as the press-released proclamation by the CEO of the American Heart Association, M. Cass Wheeler, that "Banning smoking is the only logical response to the scientific evidence concerning the dangers of secondhand smoke."[159] Stanton Glantz chimed in with the note that "This is not the first study to find a link between long term exposure to secondhand smoke and heart attacks."[160] Richard Sargent, co-author of the study, added, "It is not alright to murder for profit. It's not right to poison people for profit, and that's their [bar owners'] argument. They have to be allowed to continue poisoning people even when we've demonstrated an immediate effect of it."[161] Cynthia Hallett, Director of Americans for Nonsmokers' Rights, added, "The bottom line is simple. Secondhand smoke kills."[162] Support even came from overseas as Vivian Nathanson of the British Medical Association took the opportunity to warn that "Second-hand smoke kills at least 1,000 people in the UK every year."[163]

A few months later, the Helena study was formally christened in a *Reason* editorial titled, "Miracle in Helena."[164] At one point a Googling of the Helena ban stories brought up over 50,000 hits from around the world. Here was seemingly incontrovertible proof of the urgent need for laws to protect innocent nonsmokers from an immediate and deadly threat. Incontrovertible, of course, only if the actual figures and study design supported the claimed miracle. Regrettably, despite gathering those headlines and uncountable TV news spots, the substance of the study was hidden from the general public and open professional scrutiny for a full year until the *British Medical Journal* (*BMJ*) finally agreed to publish it.

The study's publication saw the much-publicized "60% drop" claim melt down to 40% with no note or official attempt at explanation. The original graph of raw data simply disappeared altogether – also without note or explanation.[165] The peer-reviewed publication set off a new round of media fireworks, often introduced with the words, "And another new study," as though this was a second study supporting an earlier, vaguely remembered one from the year before. Meanwhile, the 60% drop claim was still staggering around like a brain-hungry zombie a decade later, being quoted as fact in May, 2013 by such bodies as Britain's National Health Service[166] and such sources as Wikipedia.[167]

The study's placement in the highly respected *BMJ* brought scientific prestige, but that prestige carried a stiff price tag. The *BMJ* allows for a more truly open peer-review process than most medical journals through their editor-reviewed online Rapid Response section. Such responses have to pass review by a journal editor but, if their facts are valid and their arguments reasonable, they have an equal shot at the microphone regardless of the academic credentials of the writers.

Peer-reviewed studies are considered to be the holy grail of researchers. If I conduct a study and simply publish it on an Internet website, it will be largely ignored since it has not passed the test of being peer-reviewed and approved by other recognized researchers and then published in an established and accredited journal. The study may be perfectly valid, quite significant, fully and easily verifiable and replicable, and extraordinarily important – but unless it has passed that initial publication test, it's likely to be overlooked by both researchers and the media. Well, that's not totally true; as we've just seen with Helena, a properly presented antismoking study can produce thousands of news stories and enormous political punch even a year before such review and publication – but for most ordinary researchers in most ordinary scientific fields, the rule generally holds.

What's sad is that, for the most part, the public and the press have no idea just how shallow and relatively meaningless even the peer-review of published studies can be. Researchers simply submit a study to an appropriate journal. If it's not total trash, the editors pass it on to volunteer reviewers for comments and recommendations.* Those reviewers can be friendly or hostile to the conclusions of the research, and are sometimes simply friends recommended by the study authors themselves. If the journal editors want to reject a study and want some justification for the rejection, they may deliberately pick a panel of hostile reviewers or simply get a single peer-review for *pro-forma* sake – as seemed to happen when Dave Kuneman and I submitted our counter-Helena study to the *BMJ*, *Circulation*, and *Tobacco Control*.*

---

* This particular action is what defines peer-reviewed research.
* Examined in *Cleanup 2* of *Studies on the Slab*.

In any event, the reviews come back to the journal with criticisms or suggestions for improvement, publication, or dumping, and the main journal editors themselves then decide whether the study should be accepted, rejected, or sent back for alterations and resubmission. The peer-review process generally serves as a gateway to help prevent truly shoddy research from being published, but it is by no means a holy *imprimatur* guaranteeing that published research is either valid or worthwhile.

The Helena study seemed to have more trouble than most during its peer-review period. It took almost a year of submissions and revisions before the study was published by the *BMJ*. But even after such revisions as reducing the headlined 60% drop to a more modest 40%, the responses that greeted Helena's publication took the form of a miniature tornado of pointed criticisms and questions of the sort that should properly have been dealt with during the pre-publication peer-review process itself. With Helena, that process was clearly a total failure as over a dozen significant, basic, and substantive criticisms and questions were published in the *BMJ*'s Rapid Response section within just the first ten days of the study's publication.[168]

In the normal world of respected and peer-reviewed academic research, the authors would have quickly appeared to answer those questions and either correct their errors or defend their work. The world of antismoking research is anything but normal, however. When the authors finally deigned to appear, after almost two months, they ignored virtually all those criticisms and questions and simply attacked responders as "following a well-established tobacco industry strategy."[169] No solid defense was offered to the claimed defects, no real answers were given to the questions asked, and no real scientific back-and-forth interchange whatsoever occurred regarding most of the concerns raised. Readers were just treated to mud-slinging and stonewalling.

Meanwhile, the researchers gloried in national media interviews and speaking engagements while touting their "miracle," and their promised study to show how post-ban AMIs continued to grow was forgotten, again without note or explanation. While it might be uncharitable to guess that preliminary follow-up research simply found AMIs randomly dropping again, it would certainly not be unreasonable to make such a guess. It's notable that although the full study was published a year after the initial presentation at the conference, the extra year of data available was simply excluded from the final analysis.

When I emailed Stanton Glantz several times in late 2012 with queries about the follow-up, peer-reviews, and the raw heart attack data, he responded with two terse one liners asserting that all relevant data was in the original study. I made a final vain attempt at communication, asking,

> **As for the data in the year following the close of the study itself, … obviously it was *not* in the study! It was gathered and analyzed AFTER the study. It is possible that you published an analysis of it and I simply missed it, in which case I'd appreciate the referral, or, if it was not published, I'd appreciate the data so that I could analyze it myself.**

…and I simply received no response at all.[170]

Could it be that the *BMJ*'s questions and criticisms were so baseless that they needed no response? When I published a Rapid Response 1,000 days after Helena's original presentation virtually none of the previous Responses had yet been adequately answered. Today, close to ten years later, they are still swinging naked in the breeze while the authors continue to make public appearances and claims supporting smoking bans with the "proof" of Helena.

Just as with the Klein economics study in *Slab I*, the story got even worse upon closer examination. Something that had always bothered me after Helena's publication was a vague memory of seeing the original graph data and having a feeling that it hadn't really backed up the colorful claims. Unfortunately, my attempts to verify those memories kept running into dead ends. The graph simply seemed to have been deliberately erased from the public record. Fortunately, the Internet's archival engine, the Wayback Machine, crawls through the net, caching literally billions of pages for future reference. Finding a particular page can be a little tricky, but fortunately I had saved the original URL reference in my files. Wayback retrieved the questionable graph and my suspicion was confirmed.[171]

Figure 3's resurrected graph is interesting on several levels. First of all, the jagged line drawn through that profusion of dots has an unusual quality. Note that the graphed line drawn through every pre-ban one year period shares precisely the same dips and peaks. Even the slight changing of angle as the first peak of each year descends is exactly the same for every year, regardless of what the fresh data actually showed. This wasn't merely a rolling average graph designed to smooth the spikes for clarity, it was a graph that slickly added together years of data, averaged it by month, and then just drew a line pretending that it looked quite regularly that way in each of the years – thereby hiding other peaks and valleys that would have made the ban months look far more normal.

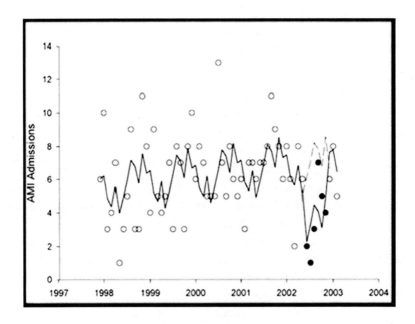

Figure 3

Original Press Conference Helena Graph

The justification for that funny line was the excuse of creating an "expected" pattern if no ban had taken place (the lighter, dotted line during the ban months), but the real purpose – to magnify the apparent impact of the ban and set the stage for the bounce-back claim – seems pretty clear.

Let's take a look at the raw data itself by removing the misleading lines while adding a little shading to show the lie to the bounce-back emphasized so strongly by the researchers and news stories. Free Choice advocate and blogger Dave Hitt's graphical talents produced Figure 4, allowing us to clearly examine that famous bounce-back and showing that the great bulk of it actually occurred *during* the smoking ban... not after it! [172]

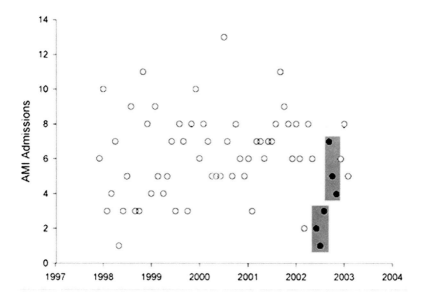

Figure 4

Hitt's Simplified Helena Graph, Emphasizing the Six Ban Months

Figure 4 does indeed show a clear drop in AMIs during the first three warm weather months of Helena's smoking ban in rather frigid Montana, a time when angry smokers and their friends likely drove to friendlier venues for their eating, drinking, partying, and heart attacks – outside Helena's jurisdiction and records.

The original presentation noted but passed over that out-of-town increase as simply being statistically insignificant rather than as offsetting the desired conclusion of a drop within Helena's limits. When looking at that drop, remember the scale of the graph. The inclusion or exclusion of just two or three heart attacks over several months would have strongly affected the outcome.

Now, move onward and look at the second half of the smoking ban period when heart attack rates returned to roughly pre-ban levels. According to stored National Weather Service temperature records for that part of Montana, the first three ban months enjoyed relatively nice weather: not a single day in that period with a minimum temperature below freezing. The second three months had forty-four days with sub-freezing temperatures. In that sort of weather most smokers and their friends likely just stayed in town for their partying, suffered their normal rate of heart attacks, and then went to the local hospital as usual – thereby showing back up on the graph.[173]

This bounce-back claim was an extremely important point, one emphasized to the media as absolute proof that immediate and drastic health effects occur upon implementing and rescinding smoking bans. A mere temporary reduction might have been too easily passed off as a coincidence, but the reinforcement of the claimed post-ban bounce back immediately resulting in new victims made the pro-ban position a lot more convincing.

Dave Hitt and I did our best to have the story revealed in Figure 4 properly publicized by the press. It would have dealt a killing blow to the national drive for smoking bans as the public discovered the lies behind antismoking tactics. Responsible press coverage might have spared tens of millions of people the economic and social ravages of their own Helena-copycat bans.

Sadly, such an exposure never happened. The news never made it beyond the little-seen online pages of the *BMJ*'s Rapid Response section and a few posts on Internet news boards and blogs. To this day, most folks familiar with Helena still believe heart attacks dropped by 60% during their ban and then jumped back up as soon as it ended.

Further analysis reveals that even the presentation of the initial drop was distorted. As we've seen, the line they drew through the data points was essentially meaningless. Unfortunately, drawing a line just connecting the sixty data points would produce such a disordered mess that the result would be indecipherable. However, there is a compromise – one that simplifies the data so the trends can be seen more clearly while also showing the random variation in that data.

Figure 5 takes those sixty values and averages groups of three. So, instead of sixty wildly different data points that set is reduced to twenty points representing quarterly averages. This averaging is both valid and useful in simplifying the data for meaningful analysis.

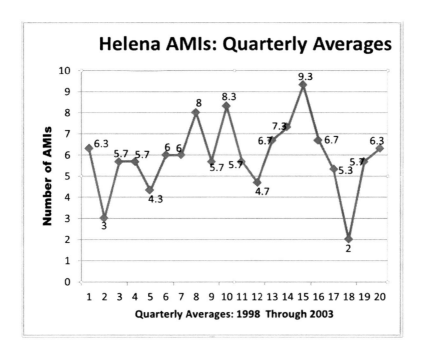

Figure 5

Helena Data Presented as Three Month Averages

At first glance, Figure 5 might seem to support the claims of the Helena authors. There certainly is a dip in the graph during the ban period that stands out. But note that the size of that dip is visually exaggerated by a completely random peak just prior to the ban. Obviously, the 2001 increase from 4.7 AMIs per month to 9.3 AMIs per month was not caused by a ban starting a year in the future; and the drop from that 9.3 down to 5.3 in the six months before the ban was certainly not caused by a ban that didn't even exist yet. But Figure 5 also shows something else quite interesting: a similar, slightly less intense dip back in 1998 – almost five years before Helena's ban.

The importance of that first dip again lies in Helena's low monthly numbers for AMIs. The two dips are so similar that a difference of just three heart attacks in the six-month period they represent would have totally changed the story presented. If there had been just one fewer in that 1998 period and two more in the first half of the 2002 ban period, then the Great Helena Heart Miracle of 2002 would have been about the same as the Non-Great Helena Non-Event of 1998. That fact alone should have raised a big, flaming, red flag and drawn Helena's conclusions into serious question during the *BMJ*'s peer-review process. Was it drawn into such question? Will we ever see what went on in those peer-reviews? I doubt it. Stanton Glantz ignored my requests for them altogether. Those peer-reviews are being treated like the secret tobacco industry documents that were kept under lock and key for so many years.

There is a final and very disturbing question. It is clear from the study itself that the authors could have separated AMI numbers for smokers and nons. The study very carefully describes its total sample: "38% of the patients with acute myocardial infarction in the study were current smokers, 29% were former smokers, and 33% had never smoked at admission." The researchers had the separate datasets right in front of them, just as Klein's researchers had. That division was vital if the researchers wanted to show that the highly emphasized protection for nonsmokers had played an important role in AMI reduction. Merely showing bans reduced smoking wouldn't have the same propaganda punch: folks still believed people had a right to smoke even if it killed them. But the Helena authors were ruled by the spirit of Sir

George Godber: a deadly threat to nonsmokers was required to create a widespread demand for government-imposed smoking bans.

The data could have shown only three results, that the ban:

(1) correlated with a decrease in nonsmokers' heart attacks;

(2) had no correlation with nonsmokers' heart attacks; or

(3) correlated with an increase in nonsmokers' heart attacks.

Given the historical background and media statements of the authors, it is obvious that even a very slight decrease in heart attacks among nonsmokers – even a single heart attack event – would have been noted with phrasing like this:

Although the strict test of statistical significance was not met, AMIs among nonsmokers were fewer than predicted absent a ban.

Did the authors make such a statement? No. So finding #1 clearly did not occur. Now, what about the second possibility? If there had been absolutely no change in nonsmokers' heart attacks they could have put such an observation in the best light by saying something like:

Unfortunately, the brevity of the ban prevented any statistically observable reduction in AMIs in the small nonsmoking population.

Did the authors make such a statement? No. So finding #2 is also highly unlikely. Thus, it becomes highly likely that the frequency of heart attacks among nonsmokers was neither reduced nor stayed the same during the ban. With that in mind, now consider what the authors might have done and said if nonsmokers had *more* heart attacks during the ban than predicted without the ban.

Obviously such a result would have been completely antithetical to everything the authors believed and wished to show. However, as long as the difference failed statistical significance, they could weasel their way out of it simply by emulating the Klein researchers: combine populations, confuse the results with the right words, and then, in an additional twist, bury the formal statement about it in part of the paper where it wouldn't be noticed. Did the authors do such a thing, use such words, and then bury it in such a location? Yes, they most definitely did, with the placement actually in the next-to-last paragraph in the ancillary "Relation To Other Studies" section, where they stated:

> **Researchers have predicted that smoke-free laws would be associated with a reduced incidence of acute myocardial infarction through a combination of reduced exposure to secondhand smoke and encouraging smokers to quit. ... While both of these effects are probably occurring, we do not have a large enough sample size to estimate their relative contribution to our results.**

Within the confines of the medical journals, the authors felt they had to give at least some lip service to the possibility that the reduction of AMIs might have only occurred among smokers or that AMIs might have actually increased among nonsmokers. By simply saying, "While both of these effects are probably occurring, we do not have a large enough sample…, " the impact of the ban on nonsmokers was inferred without actually being claimed. Word choices and phrasing for a point so central to the intention of one's research are not made carelessly. The fuzziness and low-profile placement both appear quite careful and deliberate as an attempt to reduce focus on the question.

Once the story made the newspapers, such care and constraints disappeared. In all the hundreds of articles citing the authors' statements about their research, I have not been able to find a single admission that their study had presented absolutely no evidence at all that heart attacks had been reduced among nonsmokers. Once they went to the media, they almost exclusively painted the picture of nonsmokers who'd been "saved" from breathing the "poison."

It is quite clear that the only message that the Helena researchers wanted to communicate was that secondhand smoke was immediately deadly to innocent bystanders. And indeed, that was the message that they communicated over, and over, and over again.

Unfortunately, the information and data needed to tell whether there actually was an increase in nonsmokers' heart attacks during the ban is not readily available. This question formed the third leg of my request in my emails to Stanton Glantz. And, as we've already seen, the only response I got was that the data was all "in the paper as it was published" – when clearly, it was not.

Glantz's claim reminded me of Richard Nixon's famous eighteen-minute gap in the White House tape recordings made public after the Watergate scandal.*[174] Unless some future researchers procure both the funding and the authority to examine the original data, and unless the Helena researchers willingly share the critical details of their subject selection criteria, there's simply no way for anyone to confirm the validity of Helena's research or for the standard scientific process of data checking and replication to be engaged.

In sum, as usual, we're simply asked to take things on faith. Smoking bans reduced heart attacks by 40% (or was it 60%?) in Helena; they had no deleterious effect on bar employment in Minnesota or anywhere else; they increased tax income from charitable gambling as nonsmokers flocked to smoke-banned gambling parlors, bingo halls, and pull-tab venues; they turned teetotaling, stay-at-home, smoke-phobic librarians into endlessly thirsty, high-tipping, devil-may-care party animals; they poured gold into the pockets of casino managers who were too dumb to count it; and they were probably responsible for the Virgin Birth as well.

We need to recognize and publicize the fact that many of the researchers behind such claims are little more than well-paid charlatans. Whether such advocate-researchers are guilty of intentional scientific fraud is an open question in my mind at the moment. I have no evidence of such fraud other than suppositions based upon the data and arguments I present in this book, and I certainly do not have the money to fight well-paid lawyers in attempts to get such evidence. Therefore, in the case of Helena and all the other studies I examine, I'll simply let readers form their own conclusions based on what they read here and in the cited sources.

That doesn't stop me from saying that I personally believe that the "Great Helena Heart Miracle" was as phony as a "Jabberin' Jesus's Jawbone Relic" in a fly-by-night sideshow tent.

---

* The Oval Office's carefully maintained tape recordings of Watergate conversations suffered a "mistaken" erasure by Nixon's secretary at a crucial point in the recording. Later investigation determined that the "mistake" was likely repeated nine separate times on one very specific section of the tape.

>>> Cleanup 2 <<<

Despite Helena's significant failings, the power of professional press releases fed to a sympathetic and headline-hungry media has inspired well over a dozen grant-seeking copycats. Most of them had similar findings since, as we'll see in a few pages, researchers doing preliminary work in areas that showed no drops in heart attacks could just move their focus to another location. After all, who would agree to fund research that showed smoking bans increased heart attacks? The traditional source for such contrarian funding, Big Tobacco, is now strictly off-limits to most respectable medical scientists. Even researchers daring to ask for their money might be refused; the industry has been badly stung by lawsuits and wants nothing that could come back to haunt them in courtrooms. A good part of Big Tobacco's recent court strategy rests upon claims that they have reformed and are now model citizens seeking to educate consumers about the harmfulness of smoking while offering them help in quitting. If you visit the Philip Morris or R.J. Reynolds websites nowadays, you might almost think you've stumbled upon antismoking strongholds by mistake.

Plus, no matter how valid a study might be, if it is funded in any way, shape, or form by the tobacco industry and shows anything even remotely positive about smoking, it will be immediately discounted as worthless by tobacco control researchers and the media, even if it manages to get published.[*]

So it's not surprising that the great majority of post-Helena studies on bans and heart attacks have concluded that the rates went down after those bans. Researchers and granting organizations want

---

[*] An amusing sidelight on this: Most antismoking research and activities in the United States nowadays *is* funded by Big Tobacco... but only after its money has been laundered through the Master Settlement Agreement for distribution to proper antismoking-oriented recipients.

the numbers to show heart attacks going down after bans, so that is what is found. Most of those findings still flunk the basic test of statistical significance even for mere correlation, but with the proper rhetorical padding that failing is usually overlooked by the media. Still, one might wonder how these studies manage to come up with those answers so consistently if they aren't at least partially true. The answer is threefold:

(1) cherry-picking;

(2) general heart attack reductions overall due to such things as statins, reduced primary smoking, and general dietary, exercise, and medical improvements,* plus a deliberate lack of reference to scientific control groups; and

(3) interpretation – as will be shown in graphic examples to follow, even the most amazing non-reductions can be "corrected" and then properly presented to the media.

The general reductions in the wider population pool are simply a matter of historical record, and the interpretational twists will be analyzed for each study, but the concepts of cherry-picking and control groups may not be so widely understood. Cherry-picking is the name for the practice mentioned a few paragraphs ago when I suggested that researchers would simply move on to a different town if their preliminary research indicated undesirable results in the town currently being analyzed.

Cherry-picking happens when researchers are more interested in arriving at a preferred result than in discovering the truth. It's a technique used either to please the people who are paying the most for the research or to support a researcher's own preconception about what the proper or most politically desirable answer should be. Simple

---

* What would have been a near-future heart attack fifteen years ago has today become a quick clinic visit to have a stent inserted through a tiny incision in one's leg.

human nature, when ruled by either greed or idealism, can easily overcome scientific disinterest. In terms of idealism, if a researcher truly believes that smoking bans reduce smoking and save lives, then that researcher would hesitate to produce research that might lessen support for bans. In terms of greed, if that researcher wants to pay off his mortgage, then he or she had better produce research that will attract future grants.

Imagine being a farmer going into the field to pick cherries. Some of the cherries are full, red, and ripe to the tongue. Some are past being ripe and have burst with rot. Some are too young, new, and green to be ready for consumption. Some are simply the victims of blight, worms, or pests, and are ruined. If I am a dishonest farmer trying to sell my field, I'll show someone a basket of only the very best cherries and claim they are representative of my entire field. However, if I am an honest farmer, I will pick a random selection of cherries from all areas of the field in order to show my potential buyer what it is actually worth.

I firmly believe that such dishonest cherry-picking, whether stemming from idealism or greed, is what we have seen repeatedly with these heart attack reduction studies. I believe that researchers have looked at heart attack rates in numerous small jurisdictions with smoking bans and then simply passed over those that did not show a reduction in heart attacks in favor of those that did show such a reduction.

Recently published research by Mathews *et al.* shows how easily such a thing could be done.[175] Figure 6, adapted from Mathews' original work, shows the incidence change in myocardial infarctions after smoking bans took effect in each of forty-three cities where the imposed bans represented a "meaningful increase in the level of restrictiveness" in smoking ban regulations. If you were a researcher with a pure-hearted antismoking bias or even just one who hungered after future millions in grants from "Tobacco Free Guppies," which city would you choose for a full study after doing a bit of preliminary analysis: Evanston or Flagstaff?

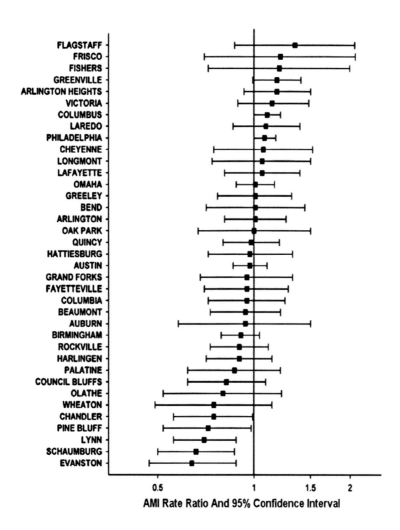

Figure 6

AMI Rates - 43 Cities With Strong Ban Changes
{Adapted from Mathews 2010}

Even using such cherry-picking, the effects being sought are either so rare or so small in size that the data often seems to have been jiggled or juggled in various ways. Most of the studies in this area of research show significant flaws in either design or interpretation, and the few that don't have flaws and still show significant post-ban heart attack reductions are likely just indicative of the luck of the draw – the few perfect cherries found in any randomly chosen basket. After all, if smoking bans actually had *no* effect, we'd expect to see heart attacks somewhat decrease in roughly half the towns with bans and somewhat increase in the other half.

So what did Mathews find? Did he find a strongly significant result in this large and varied study? Did he find massive Helena-type reductions to be the norm? No. He found literally no effect at all: an RR* of 0.99 with a 95% confidence interval range of 0.96 to 1.02 in the cities with the strongest ban changes - a result no more meaningful than claiming bias for a penny tossed a hundred times that produces heads on fifty-one of the tosses.

As for control groups, a proper study will also examine a second population group to help guard against the possibility that an observed effect is due to other variables than the one under examination. That second group should represent either a random population sample or a sample that is carefully chosen so as to be quite similar to the study group in every possible way *except* for the variable under analysis. In the case of smoking bans, researchers should ideally pick a non-ban control population that is geographically, racially, religiously, economically, and in all ways socially quite similar to the banned one under analysis. Almost all of the Helena-type studies that have been done tend to have skimped on having proper control groups, and in cases where such groups *do* exist, their data is not properly evaluated.[†]

I obviously can't analyze all of the headline-grabbing copycat studies here, so I'll present just a few of the more egregious examples

---

[*] RR stands for Relative Risk. An RR of 1.0 means that a condition has no correlation with a result. For example, if I studied the effect of temperature on a penny landing heads up, I'd find that a high temperature showed an RR of 1.0 – no frequency change of having the coin land heads up than if I'd tossed it in a lower temperature. However, trying to jog down a moving escalator after drinking a quart of good Irish whiskey might produce a very high RR for falling on one's hard Irish head.

[†] In the Helena study, the nearby control town showed heart attacks increasing while those in Helena decreased. The researchers ignored the contribution made by Helenians simply driving to that town for their carousing… thereby altering the database and eventual significance of results.

of seemingly deliberate bias that Dr. Siegel[176] and Chris Snowdon – the creator of the Scottish graph in the *Stratistics* section – have examined on their websites.[177] With their gracious permission, I have borrowed their graphic analyses and, where noted, some of their prose. Dr. Siegel's longstanding scientific record speaks for itself, and Mr. Snowdon's work, as evidenced in his highly respected book, *Velvet Glove, Iron Fist: A History of Antismoking*,[178] is, as always, both meticulous and detailed.

Iowa

In early 2011, Iowa researchers claimed that the state's smoking ban had brought about a "dramatic effect" in producing "a 24 percent decrease in Iowa hospital admissions for coronary heart disease."[179] Unfortunately for the researchers, the raw data for Iowa's actual heart attack death rates was available from the Iowa Department of Public Health and was put into graphic format by Dr. Michael Siegel.[180]

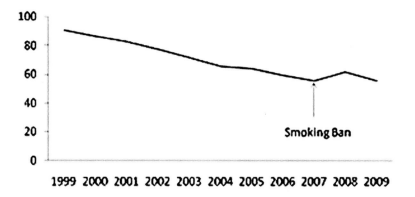

Figure 7

Trends in Iowa Heart Attack Rates

There's really not much need for detailed analysis here. The simple line graph in Figure 7 shows pretty clearly that there was no dramatic 24% drop in Iowa's heart attack rate following its ban. Indeed, if you wanted to combine this graph with the Scottish results examined earlier in *Stratistics* and some of the results in the next few pages, you might make a pretty strong argument that smoking bans *increase* heart attack deaths!

Not surprisingly, that simple reality never made it to any of the media broadcasts and stories.

North Carolina

As 2011 was drawing to a close, North Carolina was chafing under its recently imposed smoking ban and Ellen Hahn of the University of Kentucky was working hard with North Kentucky Action to spread local bans until the entire state could be conquered.[181] To help kill two birds with one stone, antismoking advocates needed a study showing what a great success Kentucky's tobacco-loving neighbor was enjoying with its ban. Thus was born a study claiming that "Emergency room visits by North Carolinians experiencing heart attacks have declined by 21 percent since (their) Smoke-Free Restaurants and Bars Law."* Just in case anyone had any doubts as to the cause of the decline, North Carolina's governor, Bev Purdue, offered this insight in the second paragraph of the study's press release: "Our goal was to protect workers and patrons from breathing secondhand smoke and we are seeing positive results."[182]

A generalized 21% drop in heart attacks due purely to protecting nonsmokers from the scourge of secondhand smoke in bars and restaurants would certainly be impressive, no? But, of course, even if the 21% figure was true, it would be utter nonsense to simply ascribe it

---

* Note the possibility for stratistical trickery here. Researchers can choose to measure five different things, looking for the one that will give them the desired answer. They can measure ER visits, total AMIs, deaths from AMIs, AMIs occurring in hospitals, or ER angina attack visits! Out of the five, at least one of them probably correlates with whether the governor's dog barks in the mornings – or anything else a researcher might want to show.

to protection from breathing secondhand smoke. After all, only 5% or less of the adult nonsmoking population actually worked full time in the affected venues, and passing patron exposures certainly couldn't make up the difference. As usual, the full story turns out to be quite a bit different than the public hype.

We'll again turn to Dr. Siegel, both because of the clarity of his presentation and because of his highly regarded record as a medical doctor who generally *supports* smoking bans. Dr. Siegel displays a quality that's quite rare among Antismokers: a dedication to scientific integrity. He strongly believes that phony heart miracles and exaggerated scare tactics work against public health goals in the long run and has laid his name and career on the line in speaking out against the scientific distortions so frequently utilized by other researchers in the antismoking community.*

In his analysis of the North Carolina study, Dr. Siegel was unfailingly polite and constrained. He noted that the researchers may simply have had their minds "clouded" by desires to see results supporting bans, but then he followed that up by noting that if such a level of mis-analysis ever came out of the tobacco industry they'd be blasted to hell and back for it.[183]

In his own words, starting with the boldfaced headline, **"Tobacco Control Science as Shoddy as It Gets: The North Carolina Smoking Ban – Heart Attack Report,"** Dr. Siegel wrote:

> **Believe it or not, the following data – which show a 21% increase in heart attacks among women during the first year following North Carolina's smoking ban – are the actual data behind the report out of the North Carolina Department of Health and Human Services which concludes that the smoking ban led to a 21% decline in heart attacks in the state.**

Siegel then offered the graph in Figure 8 to illustrate why his "Believe it or not" introduction was valid.

---

* I should emphasize again that Dr. Siegel actually supports mandated workplace smoking bans – a stand I personally believe is quite wrong. However, he has enough respect for scientific principles to recognize when studies supporting such claims are outrageously weak, and is willing to expose their false arguments – even if such exposure works against his general inclinations.

The graph is only for females in North Carolina, but to ignore that increase as though it didn't exist, while at the same time ignoring the fact that overall heart attack reductions actually slowed down after the smoking ban was implemented, displays nothing short of gross irresponsibility on the part of the original researchers.

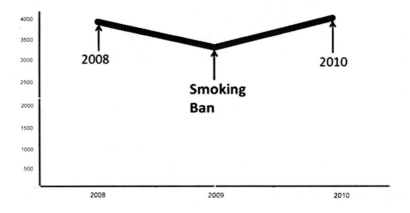

Figure 8

Annual Heart Attack Admissions Among Females – North Carolina

Scotland

Dr. Jill Pell made mega-headlines in 2008 with her claim that Scotland's smoking ban brought about a sudden 17% reduction in heart attacks. As we saw earlier in *Stratistics* (Figure 1), a simple bar graph of the data showed clearly that the drop – more like 7% than 17% – was simply a continuation of a decrease that had been going on for ten years. We also saw that in the second year of the Scottish ban, for the first time in over a decade, the number of heart attacks actually increased!

Christopher Snowdon's detailed analysis and criticism of Pell's study is so beautifully done that no further summary I could offer would really do it justice. I strongly recommend checking his website at http://www.velvetgloveironfist.com/pdfs/jillpell.pdf for more details and further insights on this remarkable, if regrettable, piece of work.*

Wales

The June 30, 2008 edition of the UK's *Daily Post* ran a story headlined "Fewer Heart Attacks in Wake of Smoking Ban." The story, as usual, leaned toward the spectacular, claiming "a big fall in treating heart attacks in the wake of the ban ... a fall of 13% between October-December 2006 and the same period of last year."[184]

I should note though, that to their credit, the *Post* and the article author Tom Bodden pointed out that Wales had seen a regular decline in heart attacks for several years prior to its April 2007 ban. However, the claimed 13% decline itself doesn't seem to be apparent in the figures from the National Health Service that formed the basis for the graph in Figure 9.[185]

Dr. Siegel, building upon an earlier Snowdon analysis, had this to say about the ballyhooed "big fall" in heart attacks among smoking-banned Welsh citizens:

> **There were 4,199 heart attack admissions in 2006 and 4,155 in 2007. Thus, there was essentially no change in heart attacks between these two years. In contrast, there was a 6.3% decline in ... admissions from 2005 to 2006 and a 10.3% decline in admissions from 2004 to 2005.**[186]

The fairly steady decline in heart attacks that had been going on since 2004 simply flat-lined in the post-ban years. An objective observer might even look at the trend line in Figure 9 and confidently say, "The nationwide smoking ban in Wales halted a longstanding decline in heart attacks. In the interest of public health, it should be rescinded immediately!"

---

* While visiting Siegel's and Snowdon's websites, readers are urged to check the analyses of other post-ban heart attack studies as well. If you have any doubts as to whether my criticisms here have been simply aimed at a few abnormally weak pieces of research, those doubts will be quickly dispelled by the many similar cases examined on just those two sites.

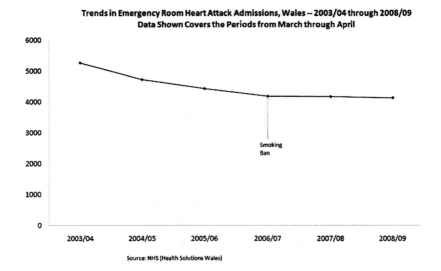

Figure 9

Trends in Welsh Heart Attack Rates

England

By 2010, the English smoking ban in pubs had been going on for what seemed like an eternity to both British tipplers and to surviving pub owners as they waited in vain for the promised review they believed would most assuredly reverse the disaster that had befallen them.* Her Majesty's Government, meanwhile, was desperately seeking to justify the ban despite the social and economic problems it had created. They

---

* In reality, when the promised review came due, the government tried to claim there was no real need for it since it was "obvious" that the ban was a "resounding success." They buttressed their claim by noting that the "overwhelming majority" of consultation feedbacks had supported the ban – while neglecting to mention that most of those supportive feedbacks had been filled out under the eyes of health workers at regional health offices.

were having a difficult time denying economic problems such the story of pub closures illustrated in Figure 10. That graph, created by Christopher Snowdon from figures supplied by the British Beer and Pub Association, is beautiful in its simplicity but damning in its content.[187] While it's true that, aside from an odd stasis in the late 1990s, the number of pubs in the UK had been declining for decades, the speed of the post-ban drop took on a new dimension. The five post-ban years saw almost as many pub closures overall as the previous twenty years combined: a 400% increase in closure rate! The drop becomes even more dramatic once one realizes that it's a drop in absolute numbers – i.e. a drop that one would expect to slow down, rather than speed up, as the number of pubs in the base decreased.

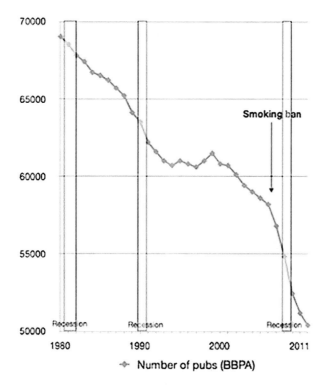

Figure 10

English Pub Closures 1980 – 2011

The economic disaster was impossible to hide, so to create and cement the impression that their ban was enough of a health success to justify the pain, British Antismokers needed to produce a "Helena" of their own – a study showing how wonderfully the ban was saving the lives of English citizens. To this end, Dr. Anna Gilmore, Director of the Tobacco Control Research Group, designed and carried out a study that showed there had been over a thousand fewer emergency hospital admissions for heart attacks in the first year after England's ban took effect. Dr. Gilmore praised the "important public health benefits" of the ban and declared that her study provided "further evidence of the benefits of smoke-free legislation," while the *Nursing Times* noted that, "Experts believe the findings clearly demonstrate the effectiveness of the smoking ban." Gilmore even stated quite confidently and formally in the study's final Conclusion, "This study adds to a growing body of evidence that smoke-free legislation leads to reductions in myocardial infarctions." The clarity of the drop was supposedly shown in the graph (Figure 11) that accompanied the study itself.

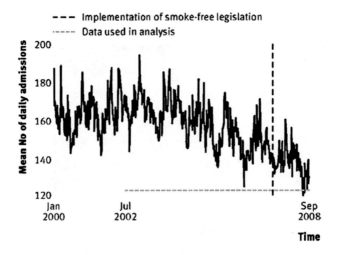

Figure 11

English Heart Attacks Pre- and Post-Ban

Hmm… ok… maybe the post-ban drop isn't quite so clear. Most of the rather blurry downward trend seems to have occurred before the ban. Fortunately, the intrepid Mr. Snowdon has once again come to the rescue and provided a graph of heart attack hospital admissions for roughly the same period that's a bit easier to see and understand. He has allowed me to reproduce it in Figure 12.[188]

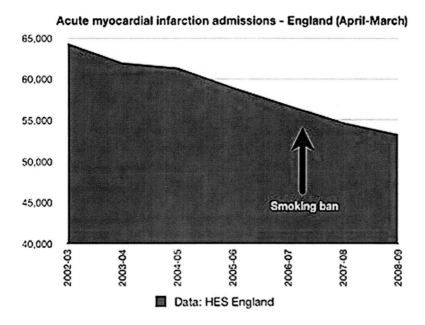

Figure 12

English Heart Attack Hospital Admissions

I regret having to admit it, but I'm afraid that even with Mr. Snowdon's expert graphical help, my own eyes still aren't quite up to the task of seeing the "swift and significant drop" that supposedly occurred after the ban came in.[189] It even kind of looks to me a bit as though the ongoing general decrease in heart attacks actually *slowed* after the ban. Maybe I'm just getting old. Maybe I should simply be comforted by knowing that "Experts believe the findings clearly demonstrate the effectiveness of the smoking ban."[190] Maybe their eyes are better than mine.

Or maybe this study, like all the others examined here, doesn't quite live up to the media hype with which it was presented to the public as part of a campaign to encourage acceptance of the British smoking ban in pubs. Look at the graphs and the evidence yourself: what do *your* eyes see?

For purposes of comparison, let's return to look at one more graph of ban effects on economics and how they are interpreted by the researchers quite differently than when they speak about the effects of bans on heart attack rates. A well-known British Free Choice blogger, Dick Puddlecote, was looking at Anna Gilmore's declarations about the supposedly impossible-to-deny drastic drop in heart attacks purportedly shown so clearly in Figure 12, when he thought it might be interesting to compare it to some assertions about ban economic effects made by one of Gilmore's colleagues, a Professor Linda Bauld at the Stirling Management School.

Professor Bauld's history includes high levels of activity with groups like ASH and the Scottish and British Tobacco Control Alliance, and has even extended to developing and mentoring a special website, Tobacco Tactics, devoted to discrediting people like Messrs. Snowdon, Puddlecote, myself, and any other researchers or advocates who dare to question the absolute goodness of smoking bans.[191] Her background might help to explain her interpretation of the data which produced the graph in Figure 13.[192] Professor Bauld believed the British smoking ban reflected what she called "international evidence" pointing to "a net positive effect on businesses" in general, with "no obvious effect" on hospitality businesses like pubs.[193]

Compare the supposedly "swift and significant drop" in heart attacks shown in Figure 12 to the supposedly "no obvious effect" on the pub closure situation shown in Figures 10 and 13. And when you're done, you're welcome to join me as I toddle off to the local optician for a new pair of spectacles. If anyone knows the addresses of Gilmore's or Bauld's opticians please send them to me... I think I may need their specialized lenses.

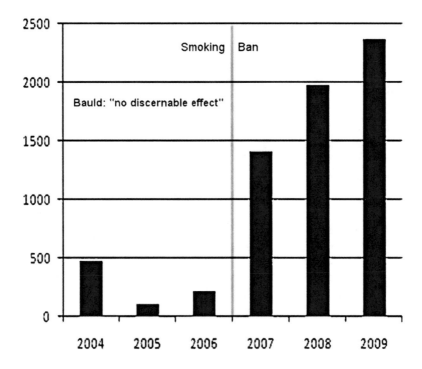

Figure 13

UK Annual Pub Closures

Why are these graphs and others like them so important? Simple: these bans have resulted in major and negative disruptions to many people's lives and livelihoods. Their implementation was justified and defended on the basis that they would be cost-free, that they would save people money in health care costs, that they would not harm the economics of pubs, and, most importantly, that they would bring about instant and drastic reductions in deaths as supposedly shown so clearly in those graphs.

None of those justifications has proven to be true. They have *all*, in various forms and degrees, been proven to be lies. And they have *all* resulted in significantly important harms to our social and cultural fabric as divisions have been driven between people and as social institutions have been weakened or destroyed.

Large Multi-Area Studies

You've now seen a decent sampling of the world of headlined post-ban instant heart attack reduction studies as of June 2013.* But this story has another side that also needs examination. Despite a stunning lack of coverage by the media, there have been five post-ban heart attack studies done on scales much larger, more diverse, better designed, and better controlled than Helena and its copycats, and all five have concluded that the instant drops supposedly found by the smaller, cherry-picked, Helena-type studies were most certainly not representative of larger scale long-term reality. We will now examine the Kuneman/McFadden study,[194] the RAND/Stanford study,[195] the Marlow 50 States study,[196] the Mathews 74 Cities study,[197] and the Rodu Six States study.[198] I will examine the first three in some detail myself, then turn the floor over to Dr. Siegel for his comments on the last two.

The Kuneman/McFadden Heart Attack Study

In 2005 and in 2009 there were two similar studies conducted on much larger and more varied population bases than previously examined. The 2005 study was carried out by Missouri researcher David W. Kuneman with my collaboration (hereafter the DK/McF study)[199] and the 2009 study was the product of a prestigious group of academic researchers drawn from the National Bureau of Economic Research, the RAND Corporation, and Stanford University (hereafter the RAND/Stanford study).[200]

---

* Actually, as I was finishing up this section (February, 2013) Dr. Siegel had just headlined a new one from Spain on his blog. The Spanish researchers claimed an 11% post-ban decrease, but when Siegel checked the actual figures he found a slight increase instead. Par for the course.

The DK/McF study has never been published in the medical journals, despite three submissions by the authors. The story of those submissions and the negative responses to them is best appreciated in an article I wrote for the American Council on Science and Health (ACSH) shortly after its third rejection. Titled, "A Study Delayed..." it is available at http://www.acsh.org/factsfears/newsID.990/news_detail.asp/

The statistical methodology we employed may not have been as sophisticated as that of its successor in 2009, but the study's scope and results were similar. Mr. Kuneman and I examined changes in heart attack rates over substantial periods of time in states with widespread smoking bans and compared those changes to ones occurring in states with virtually no smoking bans. Our study, done by two totally unfunded researchers working on their own initiative and without obligation to or dependence upon any granting organizations or interests, found no real effect on heart attack rates from imposed smoking bans. As Boston's Dr. Siegel described it in his analysis,

> Using data on emergency room admissions for acute myocardial infarction (heart attacks) from the Agency for Healthcare Research and Quality's HCUP database, McFadden and Kuneman examined trends in four states that implemented statewide bar and/or restaurant smoking bans - California, Oregon, Florida, and New York - and five states with neither a statewide smoking ban nor widespread local smoking bans - Arizona, New Jersey, South Carolina, and Iowa. They also examined trends in heart attack admissions for the United States as a whole. ... [showing] that when one examines population-based data for an entire state, one does not find any evidence of a dramatic decline in heart attacks immediately following the implementation of smoking bans. This casts serious doubt on the conclusion of the Helena, Pueblo, Piedmont, and Bowling Green studies.[201]

That finding was expanded and vindicated by respected organizations and researchers four years, and many bans, later. If the original DK/McF study had been properly published by the *British Medical Journal* in 2005 as a corrective to its 2004 publication of the Helena study, it's quite possible that the disastrous bans that have plagued the United Kingdom, Europe, Canada, Australia, parts of Asia, and many US cities and states would never have materialized. Legislators supported those bans largely because they believed the claims that the health of the workers in bars, restaurants, and other businesses was being immediately and severely threatened by the slightest presence of smoke in the air. Such claims have been made at every hearing and in every campaign for every smoking ban that I have ever participated in or followed in the years after 2004, and the legislators believed those claims because they were presented as being undisputed... since the medical journals refused to publish any studies that disputed them!

It is my strong belief that the publishers of the original Helena study, the chief editors of the *British Medical Journal,* were profoundly derelict in their duties in this regard and should be taken to task and held fully accountable for that dereliction and its results. When presented with valid research seriously challenging the conclusions of an earlier publication that had far-reaching social, political, and economic ramifications, they should have gone out of their way to re-check the validity of those earlier findings and then worked with the new researchers to bring the corrective study up to whatever standards were required for prompt publication (if, in truth, they felt that the study did not meet such standards initially). Instead, the *BMJ*'s rejection letter for the DK/McF study simply stated that, *"Our main problem with the paper was that we did not think it added enough, for general readers, to what is already known about smoking and health."*[202]

That statement was made despite the 50,000 hits that Google produced on the Helena study, clearly showing the wide interest of "general readers," and despite the fact that our results were diametrically opposed to the results of that study, thereby showing that it most certainly added to what was "already known." The *BMJ*'s rejection letter offered no option for revision and resubmission, an option frequently given to authors of papers supporting the antismoking position, and upon an appeal it was simply passed to a colleague of the main editor, who, unsurprisingly, upheld the initial decision that its results were neither new nor interesting. Given the enormous public-policy impact of its publication of the Helena study, I believe that, overall, the *BMJ* was deeply and inexcusably negligent in its handling of our submission, and, as a result, significantly damaged general respect for the process of peer-review as a whole, while contributing to significant societal harm. The entire review process seemed unusual and questionable."[203]

I've already noted the lack of response by the Helena authors to the criticisms and questions about their work in the official online Rapid Response pages of the journal, but it has also been interesting that none of those authors have ever publicly mentioned our contrary work.

Note that I said publicly. Because what's even more interesting than their lack of public response is the fact that Drs. Shepard and Sargent evidently made a *private* response addressing our criticisms and our study. It was a response sent out to an international anti-smoking private email list generally kept hidden from prying public eyes. But in their zeal to discredit the DK/McF work, a summary of one of their emails made it to an antismoking-maintained website known as "The Tobacco Industry Tracking Database" – a website where files are kept on anyone engaging in "badthink"; i.e., ideas thought to be favorable to the tobacco industry or smokers' rights.[†] I only happened to find it while defending myself from what I felt was a libelous accusation made in an Internet forum. The accusation was based upon the description of my nefarious activities – writing letters to the editor, commenting on news stories, and other such offenses – in that database.[204]

The heading of the entry identifies it as a summary of a letter written after our attempts at publication, and the body of it is of particular interest since it notes the primary content of the email as consisting of a "response and talking points" to our study. Look below and examine the database entry yourself and ask: What did that response and those talking points consist of? Were there scientific criticisms? Were there notes of methodological weaknesses? Was our

---

[*] Highlighted by the unusually long period of time our study was held, the assignment of only a single peer-reviewer, use of a "mini-hanging" committee instead of a normal one, a suggestion that we simply present our original research as a Rapid Response (which would have rendered it ineligible for future publication), the silly nature of the primary rejection basis (our findings were clearly *not* "already known"), and the formal appeal process consisting simply of the opinion of the chief editor's subordinate.

[†] Actually, this is the same database that contained the criticisms about Dave Kuneman's 1980s' employment at 7-Up and our supposed non-disclosure of our funding.

data devalued or our data selection methodology criticized? Here is how it appeared in the pages of that database.

> **Sargent, R.P., Shepard, R.; "[Email with talking points re: the Helena, Montana heart study and the self-published study criticizing it by Kuneman and McFadden]" [n.s.], [March 2007].**
>
> This email was written in response to a criticism of the authors' study of the effect of a smokefree ordinance on heart attacks in Helena, Montana. The email contained a response and talking points to a study published in the web by Kuneman and McFadden. Sargent and Shepard pointed out that the Kuneman and McFadden critique was self-published without peer-review on the web, and not published by a journal. In addition, three subsequent, peer-reviewed studies on this health effect of smokefree ordinances had the same results as the Helena study.[205]

So the answer to the above questions is a simple "No." The only criticism they were able to mount against this challenge to their work was noting that the *BMJ* and other journals had refused to publish it. It should also be noted that their comment on peer-review was obviously incorrect since, in the course of the submissions, the work had been fully peer-reviewed by at least five different peer-reviewers – although four of them hid their identities behind a cloak of anonymity.

GlobaLink Interlude

So how secret is the email list that the Helena researchers posted their criticisms to? Well, of the twenty-seven references trying to tie me to the tobacco industry on their search page, this is the *only* reference where this so-called "Tobacco Industry Tracking Database" resorted to the n.s. (no source) notation – not even allowing the email group name to reach the public. After discussions with several tobacco control insiders, I feel confident, however, in saying that this emailing was to the private listserv email group within an organization known as GlobaLink.[206] GlobaLink's history includes banning and ejecting Dr. Michael Siegel from its ranks once he began criticizing antismoking excesses publicly[207] and ostracizing Dr. Kamal Chaouachi for daring to question the deadliness of shisha smoking.* GlobaLink basically serves as a communication and coordination hub for professional antismoking activists around the world. This ensures that they are attacking their targets effectively and in coordination across national boundaries while sharing, to maximum effectiveness, the billions of dollars available for their efforts.

It is a conspiracy theorist's dream, tied in closely with both the United Nations' World Health Organization and with Big Pharma, the masters of the international medical nicotine trade. Its members and associates hail from a hundred or more countries and institutions of learning and hold positions at many major universities. Its main focus is to eliminate smoking and tobacco products from the world, one step at a time. If someone in their membership begins to show signs of deviant thinking, like Dr. Siegel or Dr. Chaouachi, they are banished. Improper questioning of central doctrine is not tolerated within the group and the risk of leaks cannot be accepted.

Quietly, by coordinating across national boundaries, they can make it seem like there's a worldwide popular demand for their *cause du jour* while pressuring recalcitrant nations with warnings that they'll be "left behind" and perceived as underdeveloped and backward if they don't step into line. If pure psychological pressure isn't sufficient to ensure conformity the GlobaLink heavies can appeal to the UN's WHO Framework Convention on Tobacco Control. The FCTC is a feel-good treaty that almost every nation in the world has signed, probably without much thought to its ultimate consequences. They might well have been slower to put their pens to parchment if they'd fully understood the very sharp teeth built into it. It can control UN aid money if a country starts balking at things like mandated smoking bans, putting bloody body pictures on cigarette packs, levying the first worldwide consumer product taxes, or whatever else the new internationally-based antismoking government demands.

---

* Dr. Chaouachi, previously mentioned in the section *Lysenko's Axe*, is a researcher who has been quite vocal in international conferences defending the hookah café culture by speaking out against secondhand smoke claims made in regard to hookahs and shisha. Shisha is the more general name for Middle-Eastern-type water pipe smoking.

As those who've read *Dissecting Antismokers' Brains* are aware, I've always strongly rejected conspiracy theory explanations in the area of antismoking activities. There are just too many natural confluences of motivations and convenient influxes of vast sums of money from sources like the MSA to make dependence on mysterious conspiracies necessary. That being said, I have to admit that GlobaLink and what it represents comes pretty close to being as much of a real-life, highly coordinated, worldwide and well-hidden true conspiracy as any I've ever run across in all my years of political and social activism.

Perhaps the scariest thing is that it's relatively new. Most of what I have seen and written about in the antismoking movement over the years came about independently of such coordination. Now that such coordination *does* exist, things are only likely to get worse – far worse – and spread into other areas of behavior control "for our own good" through the deadly conjoining of well-meaning idealists with the self-serving greedy, the fanatical moralists, and simple neurotics who've gotten swept up in the perfect storm.

The RAND/Stanford study: Kuneman/McFadden Confirmed

To return to consideration of the large post-ban heart attack studies, after the Kuneman/McFadden study was rejected by the formal world of medical journals, it might have forever remained little more than an obscure Internet oddity except for the fact that in 2009 – four years after that rejection – its results were strongly corroborated and extended by a far better funded and far more statistically sophisticated study done by a far more prestigious set of researchers. On April 3, 2009, the previously mentioned RAND/Stanford study, "Changes in U.S. Hospitalization and Mortality Rates Following Smoking Bans," was announced to the world by Kanaka Shetty and his associates.[208] This study, similar in scope and style to the DK/McF study, was far larger than any of the Helena copycats that claimed heart attack reductions after bans. It covered an eight-year period and included over 200,000 admissions for acute myocardial infarction in addition to examining over two million general heart attack deaths across the United States – over a thousand times as much data as was covered in Helena.

Not unexpectedly, unlike the numerous smaller pro-ban studies that had repeatedly garnered front-page media attention during the preceding five years, the announcement of the RAND/Stanford study merited little coverage despite the prestige of its originators and researchers. Tobacco.org lists it in their bibliography of studies with nothing more than the rather unimpressive notation that

> **(The RAND/Stanford study) examined databases, including the Nationwide Inpatient Sample, to compare short-term changes in myocardial infarction or other disease-related hospital admissions and mortality in regions with clean indoor air laws with control regions. The researchers found that clean indoor air laws were not associated with statistically significant short term declines.**[209]

The *New York Times* somehow seemed to entirely miss noticing this study covering between 200,000 and 2,000,000 heart attacks, despite the fact that it had previously given prominent Section A editorial and news coverage to Helena's study of just a half-dozen or so heart attacks per month.[210, 211]

The *Washington Post* had given the Helena study prime page A-1 headline coverage in 2004, including a statement by a CDC official, Terry Pechacek, indicating that similar studies would soon be coming from New York, California, and Delaware.[212] Somehow, that wealth of quickly arriving corroborative studies never seemed to materialize, but it was an impressive promise at the time and made good press – kind of like politicians' campaign promises made with the knowledge they'll never be implemented, but meanwhile imprinting an image in the public mind to help achieve their short-term political goals.

Despite its earlier intense coverage of Helena, the *Post* did not take much more notice of the thousandfold-larger RAND/Stanford study than the *Times* had. And the medical journals that scrambled all over each other to publish the smaller, cherry-picked studies supportive of bans seemed to find no more room for this comprehensive and should-have-been-definitive work than they had found for the earlier DK/McF research.

Fortunately, professional journals in other fields tend to be a bit more objective, and in the winter of 2011, the RAND/Stanford study passed peer-review and found a formal home in the *Journal of Policy Analysis and Management*.[213] Perhaps not surprisingly, even upon such publication, the *New York Times*, the *Washington Post*, and the rest of the major media ignored the news of the publication just as thoroughly as they had ignored the initial announcement.

Michael Marlow's 50 States Study

In January of 2012, economist Michael Marlow, professor of economics at California Polytechnic State University, applied some of the rigorous statistical tools of economics to the examination of heart attack incidence after smoking bans.[214]   He recognized the deficiencies of earlier small studies in examining time periods that were too short to be truly meaningful, and which likely also suffered from self-selection sampling bias – a more polite term for what I have previously referred to as cherry-picking. Dr. Marlow used panel data analysis and found that bans did not exert a significant effect on AMI incidence over the period 2005-09 throughout the entirety of the United States.

Using this massive database, covering a full five-year period, and controlling for numerous other variables related to heart attacks such as race, income, exercise, and diet, no significant effect of bans on heart attack rates was found. However, as with the earlier two studies, Marlow's findings found no welcoming home in the media. Unless one actually scoured the subject literature, it is unlikely they'd ever know of the these three large, notable, and quite newsworthy studies – despite the full peer-reviewed journal publication of two of them.

The Mathews 74 Cities and Rodu Six States Studies

The other two large multi-area studies, the 2010 Mathews 74 Cities study* and the 2012 Rodu Six States study, offered similar results with negligible or nonexistent reductions in heart attacks after smoking bans. I've already discussed the Mathews study briefly, but overall for these two studies I can offer no better analyses than those offered once again by Dr. Siegel. He has graciously given me permission to quote extensively from his blog articles on them.[215]

---

* The Mathews study provided the original graph we adapted in Figure 6 while discussing cherry-picking.

[Mathews' study] examined rates of heart attacks among persons ages 65 and older in 74 cities across the United States which adopted strong smoking bans during the period 2000-2008. The researchers compared the heart attack rate in each city during the year before and after the smoking ban was implemented.

When Mathews included all 74 cities that enacted smoking bans during the study period (regardless of strength of the ordinance), he found an overall decline in heart attack rates of just 3%. However, when the analysis was restricted to the 43 cities whose newly enacted ordinances represented a significant increase in protection from secondhand smoke, Mathews reports that there was absolutely no change in the heart attack rates across the sample of cities. A figure shows that heart attacks decreased in some cities and increased in others. All told, heart attack rates decreased by an insignificant 1% among these 43 cities.

Based on these findings, I pointed out that the study fails to support the conclusion – being widely disseminated by anti-smoking groups – that smoking bans result in dramatic, immediate reductions in hospital admissions for acute myocardial infarction.

Not surprisingly, the antismoking community took a different view of the Mathews study than Dr. Siegel. They emphasized the 3% decline in heart attacks across all seventy-four cities, ignoring the fact that for thirty-one of those cities, the imposed bans represented no truly significant change in the lives of their inhabitants. They also ignored something else that was rather important – the fact that heart attacks had been declining across the entire country during this period regardless of smoking ban implementations.

The importance of this second factor is seen in Siegel's figures for national annual declines during the bulk of the study period.

2002-2003: 4.5% decline
2003-2004: 8.0% decline
2004-2005: 7.0% decline
2005-2006: 4.5% decline

As you can see, the 3% decline among the seventy-four cities that had banned smoking was actually *less* than the decline experienced countrywide. The 1% decline among those forty-three cities with significantly stronger newly imposed smoking bans showed those cities actually faring far worse than the average! An objective observer, if they were to conclude anything from this data, would be forced to conclude that smoking bans *increase* rather than *decrease* heart attacks.

As Dr. Siegel points out in another analysis, Rodu's study of six states arrived at a similar conclusion.[216]

> The authors examined age-adjusted ... heart attack mortality during the 3 years before implementation of the smoking ban and during the first year after the smoking ban was implemented. These trends were also compared with those in the 44 other states without smoking bans.
>
> The results were that in four of the six states (California, Utah, Delaware, and South Dakota), the smoking bans were not associated with any significant short-term decline in heart attack mortality. In one of these states – South Dakota – there was an 8.9% increase in heart attack mortality during the first year of the smoking ban which was significantly different from the expected decline of 7.2%.
>
> In two of the states – Florida and New York – there were declines in heart attack mortality during the first smoking ban year that were significantly greater than previous trends [in those states]. However, these declines were not significantly different from the declines during the same year observed in the other 44 states.

Siegel goes on to note that Dr. Rodu's paper formally concludes that "The major finding of this study is that state-wide smoke-free laws resulted in little or no measurable immediate effect on AMI [acute myocardial infarction] death rates." With the conclusion so boldly stated, at least Dr. Rodu did not have to suffer the indignity of having his work perverted to supposedly support an opposing position.

Summation

Each of the five preceding large-scale studies covered populations much larger than the hyper-headlined mini-studies that followed the Helena pattern of making incredible heart attack reduction claims in order to fuel support for new smoking bans. All five of them can be classed as respectable and honest science. But of the five, only one was accepted for publication in a medical journal, and even that one – Dr. Rodu's – was published in the *Journal of Community Health* rather than in one of the mainstream, high-profile journals that were so eager to publish the multitude of ban-supporting micro-studies.

This brings us to the close of our dissection of The Great Helena Heart Miracle, its copycats, and its extensions. The lesson this section has taught about the integrity of medical science in the area of the effect of smoking bans on people's health is a sad one, but an important one. We have seen how statistics and data can be twisted, and we've seen how reality, no matter how clearly displayed and presented even in the supposedly pure world of science, can be ignored.

With that as a base, some of the craziness ahead will be easier to understand.

Slab III

Instant Heart Attacks And Killer Kornflakes

They're GREEEEEAAAATTTT!

- Tony The Tiger

As whimsical as the title of this section sounds, it actually reflects both the basis and the refutation of the deadly serious, but profoundly incorrect, claim that ordinary people need to fear for their lives just from being in a room for a few minutes with someone smoking. One researcher warned against even walking into a room with smokers.[217]

The first prominent study to promote such fears was carried out by a Dr. Ryo Otsuka in 2001. Otsuka supposedly showed that simply being near a smoker for thirty minutes could kill you.[218] The hype and fraud flashed around the world with the same roar that later greeted the Helena study, but if you actually examine the research instead of just reading the headlines, you'll discover that what Otsuka actually found was nothing more than a small, temporary, barely measurable change in blood vessel activity in an unusually insulated population subgroup exposed to abnormally high levels of smoke in a confined and stressful situation for a full thirty minutes straight.

The smoke level used by Otsuka, 6 parts per million (ppm) of carbon monoxide, was four times the 1.5 ppm levels the US Federal Aviation Administration had found in its 1992 analysis of smoke levels in the middle of the smoking sections of pressurized airplanes.[219] This was not simply a case of a physical reaction to just being near a smoker in a social situation; it was a physical and psychological reaction to being metaphorically locked into a veritable gas chamber.

Otsuka's study took nonsmokers who religiously avoided smoke in the course of their daily domestic, occupational, and social lives, forced them to sign papers acknowledging the potentially dangerous health conditions, and then stuck them in a smoke-choked room for thirty minutes. The actual result? A small circulatory system change similar to what might normally be found simply after eating a typical meal. The most amazing thing is that there were no heart attacks just from the emotional stress!

There was no control. Even a high school science project would have had a sham control model and protocol signing with subjects exposed to harmless but irritating odors and fog. The control study results would probably have been identical. If high school students had submitted such a poorly controlled and constructed study as a science project and drawn similarly extreme conclusions, they would probably have been flunked.

Why wasn't even a simple control set up? Could it be simply that the results would have negated the point of the study and that maybe the antismoking grant money would have dried up? Perhaps... I honestly can't think of any other reason. It seems that Otsuka's study didn't show a physical reaction to smoke, but instead simply showed a physical reaction to fear and stress – conditions promoted more by Antismokers than by smoke. Even if the reactions *had* been a direct result of the physical components of the smoke, the concentrations used by Otsuka bore no resemblance at all to actual conditions that would commonly be encountered in today's world.

Otsuka was at fault for deliberately using extreme experimental conditions without reasonable controls. The media was at fault in not reporting the reality of those conditions or the likely reaction of extreme nonsmokers forced to remain in such conditions for a full half hour. And Antismokers are at fault for using this study to convince people that simply being near smokers for a few minutes causes heart attacks.

Otsuka's study, just like the Helena study, inspired numerous copycats using conditions that were sometimes even more extreme in their efforts to imply supposedly deadly results. One such study that stands out was done in 2007 by Dr. Danilo Giannini of the Angiology University of Pisa, Italy. Dr. Giannini's research was published in the journal *Angiology* with the intimidating title of "The Effects of Acute

Passive Smoke Exposure on Endothelium-Dependent Brachial Artery Dilation in Healthy Individuals."[220]

Dr. Giannini exposed nonsmokers to secondhand smoke at carbon monoxide levels of 35 ppm – levels over 2,000% as smoky as in the middle of old airplane smoking sections – and then compared the subjects' Flow Mediated Dilation (FMD) to that of non-exposed subjects. That's clearly not the situation most of us think of when we hear the phrase "secondhand smoke exposure." Under such conditions, no one would have been surprised if the nonsmoking subjects had simply exploded, but all that Giannini observed was a moderate and temporary change in FMD. With findings similar to Ryo Otsuka's in style and effect, Dr. Giannini claimed his findings could be related in some sense to similar changes that might, if maintained steadily over many decades, predispose some individuals to heart attacks.

The fact that his smoke concentrations were at such outrageous levels, the fact that even at those levels nothing approaching an actual heart attack was induced, and the fact that he ignored the need to set up a control group, were all simply lost on the press and ignored by politicians when the study was made public. Instead, as usual, it was held up as "yet another study" supporting the idea that just being in a room with a smoker, even briefly, was a deadly peril comparable to sharing a cage with a rabid tiger on crack cocaine.

Are the Otsuka- and Giannini-type studies actually meaningful outside the propaganda sphere of antismoking research? Did they show changes that normal people should worry about? If we ignore the lack of proper controls and the emotional stress experienced by the subjects in those experiments, were the findings themselves something that might indicate a risk in the usual sense of the word? And do such studies justify headlines like that of CBS News in warning that even "A Few Whiffs of Smoke May Harm Your Heart"?[221]

The relative justification of the use of the word "risk" in such matters might be better understood if we compare its use in a situation less emotionally threatening than being locked in a smoke-filled chamber. How about comparing it to the "risk" of eating a nice, healthy, breakfast staple – a bowl of cornflakes and milk?

As Dr. Siegel summarized it in an article on his website,

A new study[222] published in the June 19 issue of the *Journal of the American College of Cardiology* found that the hyperglycemic state associated with eating Corn Flakes results in measurable decrements in endothelial dysfunction, as assessed by endothelium-dependent flow-mediated dilation ... Based on the same type of information in this study regarding Corn Flakes, anti-smoking researchers and organizations – including the CDC – have concluded that 30 minutes of exposure to secondhand smoke causes heart attacks. ... Of note, you don't see the cardiovascular disease researchers (in the Corn Flake study) jumping to this startling and sensational conclusion (nor) any warning in their article that eating Corn Flakes might trigger a heart attack, or that we need to ban Corn Flakes to prevent heart attacks.[223]

Siegel goes on to say that "Both ASH and ANR are disseminating lies to the public in making these statements, which exaggerate, extrapolate, and distort the science in a way that turns a potentially important piece of information into a lie."[224] What Dr. Siegel neglected to do in his analysis, however, was to compare it directly to the type of study carried out by Dr. Giannini. Dr. Giannini used secondhand smoke exposure levels almost 2,000% higher than what most would consider a hearty exposure. Imagine the extent of endothelial dysfunction reaction if a breakfaster was forced to consume 2,000% of a similarly hearty meal – about twenty extra-large bowls of cornflakes with a gallon of milk and a quart or so of strawberries sprinkled on top!

It's clear that on the basis of the likely results, Killer Kornflakes Consumption beats out ETS Exposure by a country mile in terms of posing a deadly threat to our children. Cornflakes should not be sold to children. They should not be sold to child abusers posing as responsible parents who feed them to innocent children. They should not be displayed openly in stores. They should not be associated with brightly smiling cartoon characters and healthy sports figures. Boxes should be a drab muddy green with 80% of the remaining packaging devoted to full-color pictures of putrescent bloated corpses, fat-oozing arteries, and bloody bursting hearts! The base $4 price of a box of flakes should be slapped with a tobacco-style tax of 400% to make a carton of Tony The Tiger ring in at a nice, child-safe $20. And finally, any film portraying cornflakes or their paraphernalia (bowls, spoons, milk, etc.) should be slammed not just with an R rating but with a full-blown X.

Giannini's research may soon bring headlines blaring the warning, "A Few Flakes of Corn May Harm Your Heart." Unless someone's been lying to us about the deadly threat of casual ETS exposure.

>>> Cleanup 3 <<<

These "instant heart attack" studies were meant to overcome people's general lack of overwhelming worry about vague possibilities of inconsistently theorized slightly increased health risks forty years after they might have already been run over by a drunk trolley driver on crack who'd just gotten hit by lightning during a force five tornado in the midst of an earthquake. By broadcasting the idea to nonsmokers and legislators that ordinary, innocent people might just happen to walk into a restaurant, breathe a wisp of smoke, and keel over dead, they were able to pack a far more powerful emotional punch than by simply preaching about lifelong exposure studies resulting in theoretical increased percentages.

As a side benefit, these studies gave some support to the Helena-type claims of drastic heart attack reductions in the immediate wake of bans. Of course, scientifically, the theoretical justification was nonsense once it was actually examined and compared to real world experiences. As we've seen, the researchers measuring these bodily effects generally stuck nonsmokers into environments many times smokier than anything they were accustomed to, and then deliberately neglected to set up proper scientific control procedures. And then, just to top it all off, even *with* such biasing of the experimental setups, they were still unable to show anything at a clinical level that a doctor would consider an imminent heart attack threat in any normal sense of the words. Still, these studies addressed an urgent publicity need, and Antismokers have never been too picky about such little details as the difficulty of jamming twenty chain-smokers into a child's bedroom or 100,000 of them into a neighborhood bar![225]

Up to this point, we've examined studies and claims that lay at least somewhat within the far borders of the realms of what could perhaps loosely be argued to be in the wide general ball park of semi-reasonable results and pseudo-science from outer space. While a person might be surprised at bars doing well after a ban, or at the size and speed of the drop in Helena's heart attacks, or the "instant" nature of dangers from what they are led to believe are ordinary, passing encounters with smoke, still these are things that might seem believable to most people when stated as fact by "responsible authorities."

We will now wander off this path of at least semi-arguable rationality and enter a whole new and very strange world of scientific analysis.

We are in the midst of the biggest public health crisis in the history of the world, and nobody even gets it. Nobody understands how important this is. ... How much tax will it take to reduce consumption?

- Robert H. Lustig, MD,
{speaking of the dangers
of Sugar and soda pop.}

**Welcome To Krazyville in Slab IV, folks!
Hang on tight. It's quite a ride!**

Slab IV

Thirdhand Smoke (THS)

Curiouser and Curiouser...

-    **Alice**

In January, 2009, a news story hit the papers and the airwaves that was so outright silly that I initially thought it was simply a well done satire put out by an Internet joke website like The Onion or repeated from a Saturday Night Live skit.* It was not. It was being disseminated as serious news by such generally respected sources as the *New York Times* and *Scientific American*.

A team led by a Dr. Jonathan Winickoff from Boston's Massachusetts General Hospital for Children had published a study in the *Journal of The American Academy of Pediatrics* titled "Beliefs About the Health Effects of 'Thirdhand' Smoke and Home Smoking Bans."[226] That study was presented almost universally by the media as a revelation that even parents who only smoked outside the house were exposing their children to a deadly and toxic risk. The *New York Times* listed several such toxins, and highlighted the danger to innocent children by noting that just a few thousandths of a single gram of one of them, Polonium 210, had been successfully used as an assassination weapon against a rogue KGB agent a few years earlier.[227]   Even *Scientific American* was hoodwinked into running an article on the study where they bemoaned the fact that "only" 65% of nonsmokers

---

* This was my second time to be caught like this, with the first instance being the April 1, 2003 release of the Helena story. In the world of antismoking research, it is becoming progressively harder to differentiate between reality and satire.

agreed with the statement, "Breathing air in a room today where people smoked yesterday can harm the health of infants and children." The *SA* article then went on to note that "smokers emit toxins" from their clothing and hair, and even quoted the ubiquitous Stanton Glantz as saying, "The level of toxicity in cigarette smoke is just astronomical when compared to other environmental toxins such as particles found in automobile exhaust."[228]

In reality, as any responsible journalist who'd even glanced at the actual study would have known, the "new research" consisted of nothing more than examining a sample of random opinions given by ordinary householders responding to a telephone survey.* Individuals were called at home and simply asked whether they agreed or disagreed with the statement noted above, "Breathing air in a room today where people smoked yesterday can harm the health of infants and children." Of course, the assumption in most people's minds while taking the survey was that the questioner was asking about leftover traces of secondhand smoke in the air of a closed room rather than asking about the then unheard of concept of "thirdhand smoke." Even in terms of thinking about day-old secondhand smoke, the great majority completely dismissed the idea that there could possibly be any significant harm even to young and innocent infants.

Of the dozens of news sources I examined upon publication of the study, only four presented the details of the study without showing a strong antismoking bias. Two of these were small regional UK publications, the *Tyrone Times* and *Bedford Today,* and both specifically took Britain's Daily Telegraph to task for failing to note that the research was simply based on an opinion survey![229, 230] Articles presented by *Global Health Law* and, surprisingly, by the normally antismoking British National Health Service also both portrayed the study accurately as being simply a random survey padded out with comments and opinions expressed by the researchers in hopes of making their work more publishable, headline-worthy, and effective as a behavior modification tool.[231, 232]

---

* Amazingly, not a single major mainstream news story seemed to even pick up on what was clearly presented in the title of the study: "Beliefs about the effects..." Reporters almost universally took those "beliefs" as being actual substantive scientific findings!

Some independent commentators, most prominently Jacob Sullum, the editor of *Reason,* strongly questioned the tone of the stories about the study,[233] and several respected doctors and researchers, such as the previously cited Michael Siegel[234] and Geoffrey Kabat, senior epidemiologist at the Albert Einstein College of Medicine and author of *HYPING HEALTH RISKS: Environmental Hazards in Daily Life and the Science of Epidemiology,*[235] pointed out the dangers of creating and promoting irrational fears purely on the basis of an opinion survey. But they were, notably, the exceptions. Most of the popular news sources accessible to the general public, ran stories that were far closer to the yellow journalism of the *Times* and *Scientific American.* They focused on unrealistic doses and scary chemical names, generally giving the impression that new scientific research had actually proven a deadly threat from something brand new called thirdhand smoke.

In general the mainstream media simply repeated what they were told. And what they were told religiously followed the standard advocacy guideline that even the shoddiest research should be tolerated if it promotes a good cause, such as a belief in global warming or a fear of smoking. Twenty years ago, Dr. Alvan Feinstein, Yale Professor of Medicine and Epidemiology, criticized this attitude strongly. He cited the example of a leading public health authority* who had made the following observation about the data used to condemn secondhand smoke: "Yes, it's rotten science, but it's in a worthy cause. It will help us get rid of cigarettes and become a smoke-free society."[236]

My own perception of Winickoff's work came both from my introduction to it via the *New York Times* and my subsequent research into the content of the study itself. I was infuriated by the idea that a news source as respected as the *Times* could present news of a simple opinion survey in so biased a fashion as to have parents whisking their babies out of grandparents' arms for fear they'd die like assassinated KGB spies, and I set out to show just how outlandish the comparison actually was. Despite my previous publication failures in the *Times,* I first tried a simple letter to the editor. Perhaps the article had been written hastily from a press release and if the problem was pointed out, the *Times'* editors might actually be anxious to run a corrective item.

---

* He refrained from naming the source, for obvious reasons.

Dear Editor, Of Polonium and Politics...

Roni Rabin's Jan. 3 article, "A New Cigarette Hazard: 'Third-Hand Smoke,'" confused science with opinion survey. The study did not "focus on the risks posed to infants and children." It focused on people's opinions of such risks: a very big difference. The article then extended statements by the study authors in a way designed to raise unreasonable fears among parents.

In two separate places in the story, including the spotlight ending paragraph, the article highlighted the threat to children from radioactive polonium in "third-hand smoke." For a child to absorb the featured "dose that was used to murder former Russian spy Alexander V. Litvinenko," that child would have to have not just crawled upon but actually have thoroughly licked clean a full ten square feet of smokers' flooring each and every day for almost three trillion years: 300 times the age of the known universe.

The actual "threat" posed to a child from normal exposure is less than the "threat" posed by an occasional beam of sunshine sneaking through a well-curtained window. To even use the language of "threat" or "danger" in this regard is wildly misleading.

We're not talking about mere editorial license here. We're talking about a degree of yellow journalism that makes the sinking of the Lusitania look like a picnic.*[237] When an article in the Times repeats such fear-mongering without concern, there is something very seriously wrong. It's no wonder we've seen cries for smoking bans when such fear is fed to the public by supposedly responsible authorities.

Alas, this letter fared no better than any of my dozens of past and future attempts with the *Times*. Despite being solidly supported by the reference material shown below, neither the *Times* nor any of several other newspapers that had published stories about thirdhand smoke were willing to publish any corrective analysis.

---

* The term "yellow journalism" was created in early twentieth century America when newspapers fell largely into two categories: the "responsible organs" that reported the news accurately while their owners' editors' editorial opinions stayed on the editorial pages, and the cheap tabloids printed on unbleached, yellowed paper, screeching opinionated headlines designed just to rile the masses and sell copies. Yellow journalism hit its low point in 1915 when it played up the tragedy of a German U-Boat sinking the British passenger cruiser Lusitania. That reportage threw us into World War One. Sadly, we've seen its resurgence all too often when the media deals with the smoking issue.

Now, some might think the *Times* may simply have had doubts about the accuracy of my statements. I was aware my figures might seem extreme, so I added a reference statement with my formal letters:

From *http://acsa.net/HealthAlert/RadioBacco.html* we see that "the intake of polonium by a typical smoker is about 0.72 picocuries/pack." Litvinenko's exposure is judged to have been around 5 millicuries.

A 15/day smoker gets about 0.5 picocuries/day. A nonsmoker working with smokers might get, at most, about 1% of that, 5 femtocuries/day.

A millicurie is a thousand microcuries, a million nanocuries, a billion picocuries, or a trillion femtocuries. It would take that nonsmoker a trillion days to absorb the dose that killed the Russian.

Of course, that's secondhand smoke. The article referred to "third-hand smoke" absorption by a child from smoke-exposed surfaces. An estimate for the amount remaining on the 10,000 square feet of walls, ceilings, furniture, drapes and floors in a reasonably ventilated 2,000-square-foot home can't be quantified with any exactitude, but it seems unlikely that it would be more than 1%. After all, if you're in a room with open windows or vents, does the smoke sink down to the floor like a rug or get sucked/blown outside?

So, let's assume that 1% actually does remain and spreads out over those 10,000 square feet. With 15 cigarettes having been smoked while the child was at pre-school hours earlier and the house then thoroughly aired out, we'd have 1% of 5 femtocuries spread on that surface.

Let us suppose you don't watch your child very carefully and let us further suppose your child deeply loves licking an entire ten square feet of floor sparkly clean every day while you are watching Oprah on TV. That child will then have ingested 1/1,000th of those 5 femtocuries into his system: 5 "attocuries."

So, how long would it take such a child to get the 5 millicurie "killing dose" of the Russian corpse featured by the *Times*?

In 1,000 days, our child would have licked up 5 femtocuries.
In one million days, 5 picocuries.
In one billion days, 5 nanocuries.
In one trillion days, 5 microcuries.

It would take one *quadrillion* days (2.74 trillion years, almost 30 billion centuries) for that child to absorb 5 millicuries. The universe is only about 10 billion years old, so the child would have to lick floors for 274 cycles of our universe to match our radioactive Russian.

Of course, there's also the whole annoying fact that the half-life of polonium is only 138 days, so we'd have to completely ignore the basic laws of physics in order to justify the story's thesis. And since the child would normally excrete most of that polonium, we'd have to cruelly refuse to change his diaper until the end of that period... not a very pleasant task, but maybe the third-hand smoke researchers would volunteer.

Even if someone wanted to quibble with my estimates, changing 1% to 10%, or 10 square feet to 100, or 15 cigarettes to 150... or even ALL THREE in attacking my argument... we'd *still* be talking three billion years of exposure along with a suspension of the basic laws of biology and physics that normally govern our universe.

Now do you see why I feel this comparison was so misleading and needs to be corrected? Other elements in "third-hand smoke" might be somewhat more concentrated, but still nothing that wouldn't demand thousands of years of assiduous tongue-licking.

As noted, the corrective letter was not published by the *Times*, despite being well backed up and properly referenced.

While variations of this communication were also submitted to other guilty papers, I felt that the misfeasance by the *Times* itself, both because of the prominence with which it played up the KGB reference and because of the responsibility a media vehicle takes on when it succeeds in prominently positioning itself as an accurate source of information, deserved an extra follow-up. After a week had gone by with no response, I dutifully followed up by sending the following to the main editor and publishers of the *Times.*

Dear Messrs. Sulzberger, Heller, and Rosenthal,

On Jan 2, 2009, the *Times* ran an article ("A New Cigarette Hazard: 'Third-Hand Smoke' " by Roni Rabin) about researchers who examined the results of an opinion poll and concluded that parents who were so afraid of secondhand smoke that they also believed something called "third-hand smoke" could be harmful were the parents most likely to ban smoking in their homes.

HOWEVER... the article did not present the information in such a form that a reader would understand that. Instead, the article quite clearly portrayed the research as indicating that there was danger, extreme danger, to children from the "toxic stew" that might be brought into a home on a smoker's clothing or that might lie hidden invisibly on surfaces ready to attack innocent children at a later time.

The article went out of its way to emphasize one particularly scary element in tobacco smoke, "polonium-210, the highly radioactive carcinogen that was used to murder former Russian spy Alexander V. Litvinenko in 2006." The article made no mention at all to worried parents that their children, even under the most extreme "thirdhand smoke" conditions, would have to be exposed for literally three trillion years in order to pick up such a deadly dose. To some extent, the writer can offer the excuse that this was the sort of approach she picked up from interviewing the researchers, and to some extent, that excuse may be valid. I am quite sure that the researchers did indeed emphasize such things rather than merely talk about the implications of a survey on people's ill-founded beliefs. But that is no real excuse for the article's publication.

The *Times* engaged in the rankest and most base form of yellow journalism for the sake of headlines imaginable, driving fear, not just upon a nation, as in the case of the Lusitania, but as a wedge into the very substance of our social relationships and our families. And then, after such deplorable behavior, when the problem was pointed out to them in at least a half dozen literate letters to the editor that I am aware of (as a highly visible activist in this area, people often cc me on letters they send to papers, and this article created a bucket load), the *Times* refused to print even a single one of them or to run any sort of corrective article that might ameliorate the damage it had caused.

The *Times* article will almost certainly result in the destruction of families as one spouse wrongly blames the other for deaths or illnesses of their children, and the *Times* will be morally responsible for that destruction. Perhaps someday it will be held legally responsible as well; you must remember that the First Amendment does NOT cover falsely yelling FIRE! in a crowded movie theater... which is exactly what this article did.

I will send you my own letter to the editor, along with the supporting documentation originally accompanying it, so that you can see that my analysis of the *Times'* irresponsibility is not overstated. This case is being widely discussed on the Internet on various newspaper websites and fairly high profile blogs. I can assure you the matter will not be laid to rest at any point soon.

I hope you will intervene and see to it that some corrective action is taken.

The communication was ignored and ultimately the only people to be aware of this mockery of science were those reading postings about it on the Internet. The great majority of the population who would have any thoughts at all about the concept of thirdhand smoke today would simply "remember" that it had been verified and proven to be very dangerous... perhaps even more dangerous than smoking itself! New Jersey GASP's website enthusiastically headlined, "New Studies Indicate That Thirdhand Smoke May Be More Dangerous Than Secondhand Smoke!"[238] And, as we'll see later on in *Slab VII*, one noted researcher – who seems to enjoy chopping up unborn baby rats' lungs – has actually made a serious research claim along those lines, noting that he was the first to show that "exposure to the constituents of thirdhand smoke is as damaging and, in some cases, more damaging than secondhand smoke or firsthand smoke."[239]

Thirdhand smoke is "more damaging" than secondhand smoke or even smoking itself?? This is the sort of thinking that the *Times'* uncritical reporting of the initial Winickoff study has led to.

But could it be possible that this was all just an innocent misunderstanding? Perhaps just simple, scientific illiteracy on the part of the *Times'* editors, reporters, and publishers resulting in an unfortunate, but unintentional, misinterpretation of a report on a survey? Or perhaps simply a case of sloppy reading of the study abstract (which is all that most reporters seem to generally read)? Free Choice advocates have sometimes been accused of exaggerating the extent to which such sloppy reporting is a creature of intent, rather than mere happenstance. In this case, though, the study clearly noted at its very start that "There is no safe level of exposure to tobacco smoke" and "Children are uniquely susceptible to thirdhand smoke exposure," thereby plainly setting the stage for later misdirection. Even a tabloid reporter operating under a deadline would have been hard pressed to miss the statement of intent in the study's formal Conclusion:

> **This study demonstrated that beliefs about the health effects of thirdhand smoke are independently associated with home smoking bans.** *Emphasizing that thirdhand smoke harms the health of children may be an important element in encouraging home smoking bans.*[240] **(Emphasis added.)**

The tale of "Thirdhand Smoke," a term that previously had been used only by nightclub comedians poking fun at antismoking zealots, was spread far and wide as real science by a press operating under the rule of printing news "for a good cause" rather than under the rule of accurately reporting health information to the public. This study and the stories about it were designed to create further pressure on smokers by pitting parent against parent in pitched battles over purely imaginary threats to their beloved and innocent children. There is no way of knowing how many homes and families may have been literally torn asunder under the stress of spousal battles in which each side believed passionately in the correctness of its position, all in the name of pressuring smokers to quit. The "unintended" collateral damages are considered inconsequential by the activists in the antismoking movement, but the *New York Times* and other media outlets that aided and abetted this abomination should be deeply and utterly ashamed.

>>> Cleanup 4 <<<

While Winickoff's study made the biggest splash and was the first one to give a formal, popular name to the concept of tobacco smoke left on surfaces, it does not stand alone. Several years earlier, the previously mentioned advocacy journal *Tobacco Control* published a study by a Dr. Georg Matt, a lecturer in psychology at San Diego University.[241] In that study, Matt tried to argue that infants were threatened by microscopic traces of nicotine in the air – as well as nicotine they'd absorb from eating dust off the floor – even if parents constrained their smoking only to outside the house. Matt found that the level of nicotine in the air of such smokers' homes was more than double the level found in nonsmokers' homes: 0.22 micrograms per cubic meter as opposed to just 0.09 micrograms per cubic meter. Matt's research was character-ized in headlines even two years later as showing that "Babies 'suffer third hand smoke'."[242]

This is, once again, simply an example of the Commander Almost Zero Fallacy where double or triple virtually zero is *still* vir-tually zero. When talking about fractions of micrograms dispersed throughout entire cubic meters, we're talking pretty much about zeroes. It's like talking about the risk of being hit by lightning in your second floor bedroom on a clear day being double the risk of being hit in your first floor living room.

As Christopher Snowdon pointed out in an article on this study, "the legal limit of workplace exposure in the United States is 500 micrograms per cubic meter, some 2,500 times more than was found in the smokers' households."[243] To put that into a clearer perspective, wouldn't you be concerned if you were told that the level of some poison in your home was 50% higher than a legal limit for that poison? Of course you would be. But if you take the numbers that are being used to frighten parents about thirdhand smoke and compared them to the legal safe limits for workplaces, you would find that the workplace limits were not just 50%, or even 500%, higher – *but a full 250,000% higher!* The levels measured in the homes were almost unimaginably inconsequential, but by preying upon scientific ignorance, campaigners were able to magnify the importance of those levels to such an extent that we now find parents who actually worry about their offspring even briefly visiting the home of a smoker.

So, Winickoff was not the first to devise such theories, but he was the first one that the media was foolish enough to take seriously. Winickoff's fifteen minutes of fame was followed just a year later, in the *Proceedings of the National Academy of Sciences of the United States*, by a study under the direction of Dr. Mohamad Sleiman of the Lawrence Berkeley National Laboratory.[244]

Dr. Sleiman decided that rather than measure non-carcinogenic nicotine or incredibly sub-nanoscopic amounts of polonium, he would check for two substances called NNA and NNK, "tobacco specific nitrosamines" (TSNAs). These TSNAs are known to sometimes cause cancer when administered in high doses to rats, and the National Toxicology's 12th Report on Carcinogens reported that the levels of NNK and NNN (another TSNA) absorbed by a man who regularly chewed snuff for thirty or forty years might be similar to the levels found to produce cancer in hairless mutant rats.[245]

Compared to snuff, tobacco smoke itself has much lower levels of these chemicals and they're generally not thought of as being a significant contributor to total cancer risk, even in dedicated smokers. There's clearly not enough of these chemicals in tobacco smoke to form even the most absurdly remote risk from secondhand or thirdhand smoke absorption. However, what Dr. Sleiman realized was that the traces of non-carcinogenic nicotine left on household or automotive surfaces might transform into carcinogenic NNK in the presence of high levels of nitrous acid. He observed that low levels of nitrous acid are common in home environments and went on to test for the formation of the nasty NNK in the presence of what he called "high but reasonable" levels of nitrous acid found in homes and cars.

His research actually looked somewhat credible at first, but then my eye caught something that seemed out of place. Dr. Sleiman had used a level of 60 parts per billion of nitrous acid in his experiments. Knowing the way antismoking researchers work to please their funders and garner publication, I wondered if that truly was a "high but reasonable" estimate of what would normally be found in homes. The question was quite important because, without those levels of nitrous acid, the NNK simply wouldn't be formed at levels that were even measurable, much less harmful! Upon checking several sources, I discovered that the average level of nitrous acid in the air of homes was between 3 ppb (parts per billion) and 5 ppb.[246, 247]

The good doctor hadn't used normal exposures of nitrous acid, or even double, triple, or quadruple normal exposures in order to get his scary results. He'd actually exaggerated the average exposure of about four parts per billion by roughly 1,500% in order to get any measurable levels at all of this toxin so he could produce a paper warning parents of the "danger" their children were in.

Actually, if you lived in a home with nitrous acid levels at the study level of 60 parts per billion in the air I think the least of your worries would be micrograms of nicotine on the floor transforming into picograms of NNK and then being licked up for billions of years by chronically underfed and eternally constipated children accumulating it in their bodies.[*]

Some might think that this last thirdhand smoke study was about as outrageous as things could get. But the summer of 2010 saw one even nastier. We'll look at that and then take a quick peek into an even weirder future before moving on.

Terrifyingly Toxic T-Shirts

The study in question is an example of "Science By Press Release." As you may remember from our discussion of the Helena Study, that's the term for heralding scientific results to the media before they've been properly confirmed, reviewed, or published – and sometimes even before they exist. In this case, Dr. Timo Hammer and his colleagues at the Hohenstein Institute in Germany put out a press release titled, "When Baby Smokes Too!," in which they raised the scary specter of TTT – Terrifyingly Toxic T-Shirts![†248] Basically the study – at least as it was presented in the press release – claimed that someone being anywhere near a smoker outdoors who then returned inside to hold a

---

[*] In fairness, however, we should note that even 60 ppb is pretty close to zero!
[†] My term, not Hammer's.

baby, could be causing "major damage" to the child's skin and nerve cells as their clothing transferred "neurotoxic poison" which would poison, mutate, and even outright destroy those young, innocent cells. The story called to mind horror movies about epidemics of flesh-eating bacteria, but with no bacteria needed.

Details of the future study are still a bit sketchy even a year after the initial press release, although Dr. Hammer was very gracious in answering several questions I put to him in emails. Somehow, they plan to expose shirts or swatches of shirt material to "average" amounts of outdoor smoke such as might be encountered by being near various numbers of cigarettes on an average open air balcony on what I guess would be an ideally average breezy day. I guess the subjects in question (assuming they find anyone suicidal enough to actually stand several feet from a burning cigarette in the open air) would have to be on a rotating disk so that an ideally average amount of smoke might blow back toward and be evenly filtered from the air as it passes by the shirt.

This last "filtering" part is a little tricky since Antismokers generally claim that smoke cannot be effectively filtered from the air even after being forced through ultrafine, high technology HEPA air filter mesh specifically designed to attract and capture such smoke, but, as we've already seen, a little trickiness isn't a big problem in anti-smoking research. Still, that's the theory behind the experiment: that *some* amount of smoke containing a measurable amount of this "neurotoxic poison" will indeed be filtered out of the air, and then deposited on the shirt, and then brought inside, and then potentially be transferred onto and into the skin cells of the defenseless child crying to be picked up and cuddled by Deadly Daddy or Malignant Mommy.

While it's impossible to make detailed criticisms of the research before the research has actually been published, the model experiments detailed in the press release were a far cry from the reality of smoking on a balcony and then holding a baby. It seems quite fair to attempt some criticisms of the results as shared in the press release and picked up and published around the world. One Crusader who took advantage of the opportunity to terrorize parents was, not surprisingly, John Banzhaf, the earlier-mentioned cruise-ship-dancer-cum-lawyer who'd originally founded Action on Smoking and Health. ASH put out a press release on Hohenstein's press release (a press release squared? A secondhand story on a thirdhand study?) and headlined it, "Tobacco Smoke Residue Causes Massive Damage in Babies' Skin." Yes, you read that right, "massive damage." According to the lead paragraph in ASH's press release,

> **Parents who do not smoke in the presence of their children, including even those who smoke only outdoors, nevertheless put their children at serious risk of "massive damage" to both skin and nerve cells, since a neurotoxin in thirdhand tobacco smoke penetrates the child's skin....[249]**

It is not until and unless you reach the fourth paragraph in ASH's press release that you would discover that this "neurotoxin" is simply the same nicotine that we all know is in tobacco, eggplants, tomatoes, and potatoes. That wouldn't have sounded nearly as scary in a leading paragraph, eh?

Of course, nowhere in the Hohenstein press release or in ASH's reprise of it is any real mention made of the amount of nicotine the children will be exposed to. Absent that information, I decided to conduct a thought-model experiment similar to the one I'd formulated regarding the unfortunate KGB spy assassinated by Polonium 210. The amount of nicotine produced by a cigarette is astronomically greater than the amount of Polonium 210, but it's still interesting to see the amounts actually being talked about.

Say you smoke a cigarette in a room about 5 meters x 3 meters x 3 meters – the size of a small dormitory room – with fairly normal ventilation. While it may vary a bit from brand to brand, your cigarette will put out about a milligram of nicotine. After your five minute break, probably about 99% of the smoke would either be inside of your body or floating in the air or blown out a door or window. It's likely that about 1%, at the most, would have been "deposited" anywhere to become thirdhand smoke. With no furnishings, we'd have about 200 square meters of room surface and perhaps about 2 square meters or so of body surface. So 1% of that 1% (2 out of 200) would end up sticking to you. Let's say your T-shirt takes up a good third of your body surface. We're now talking about 1/3 of 1% of 1% of that 1 milligram of nicotine being on your shirt. That's .01 x .01 x .33 x .001 grams (.000000033 grams or 33 nanograms) on your shirt.

But the study wasn't talking about being in a small room. It was talking about being outside on a balcony with normal, balcony-type breezes. How much smoke would you expect to stick around and glom onto your shirt out there compared to what you'd get in a small, smoky room? A reasonable estimate would probably be about a thousandth as much ending up sticking to your shirt from your outdoor exposure as you would have had from indoor exposure. We are now talking about (roughly) 33 picograms of nicotine on your shirt.

Now you go inside and scoop up little Donnie Dumpling in your arms. For modesty's sake, let's allow him some clothes. Now you're holding your sweet one in front of you. The largest area of your shirt surface that he can rub himself against if he's very active is perhaps about 1/3 of it, an area containing about ten picograms of nicotine. Your tiny totty has some hand, arm, leg, and face area exposed, but most of him – say 90% – is covered in clothing, so only about 10% of his bare skin will come in contact with your nasty shirt... exposing him to about one picogram of nicotine.

How much of that transfers to his skin? In terms of transference properties, nicotine probably acts in a manner fairly similar to any somewhat-greasy dirt-like substance. If you're wearing a shirt covered with greasy dirt, some would rub off on the baby in ten minutes or so. We want to get you cleaned up, so we take away the dirty baby and hand you a nice clean one. We could probably do this at least ten times and *still* have some dirt getting rubbed off onto the new babies and *still* have you sitting there in a substantially dirty shirt. It seems fair to say that in ten minutes, only 1/10th, at most, of that picogram will be rubbed onto the child's skin: about 100 *femtograms*. Of that 100 femtograms, it's likely that at least 90% will soon get rubbed or washed away, leaving at most 10% to get "absorbed" by those hungry little baby cells that supposedly gobble up nicotine like candy.

That's ten femtograms. Is that a "risk"? Let's think about it in perspective. We'd have very few qualms about feeding our little one a teaspoon or two of *BabyGoober's Eggplant Mush* for lunch each after-noon, true? Eggplant happens to have about 1 milligram of nicotine per 10 kilograms of product. Two teaspoons of the delicious *Mush* would weigh about 10 grams and thus have about .001 milligram, or one microgram, of that deadly neurotoxic poison. (By the way, don't think you can get away with feeding your toothless toddler *BabyGoober's*

Terrific Termaters instead. Tomatoes are also a nice nicotine source for paranoids to worry about.)

For the moment, we'll stick with the eggplant and the one microgram of nicotine. Say your tender one loved eggplant so much that you fed him two teaspoons of it every day from weaning until he was about three years old: 1,000 days. That would be one milligram of nicotine. How long would you have to cuddle little Donnie before his endangered skin would absorb that same amount of nicotine? How many cigarette outings on that balcony would you have to take to transfer 1 milligram through his skin at a rate of 10 femtograms per guilty smoke break? Basically, we need to know how many femtograms there are in a milligram and then divide the result by ten, so, just as we did for thirdhand smoke in general, we'll run the numbers.

1 milligram = 1,000 micrograms
1 milligram = 1,000,000 nanograms
1 milligram = 1,000,000,000 picograms
1 milligram = 1,000,000,000,000 femtograms

Dividing that one trillion by ten lets us see that it would take roughly 100,000,000,000 (one hundred billion) smoke breaks to give your child the same "healthy" amount of nicotine they'd get from those smushed eggplants. If you smoked ten per day, it would take ten billion days. That's a bit over 30 million years of cuddling – over a hundred times as long as *Homo Sapiens* has walked the face of the earth – and you could still rest assured that you haven't "poisoned" them with anything more than the "deadly neurotoxins" they would get in a healthy diet that included a reasonable daily serving or two of pureed tomato or eggplant. This hardly seems to justify the headlined titles of the press releases – "When Baby Smokes Too!" and "Tobacco Smoke Residue Causes Massive Damage in Babies' Skin" – but fear of this "risk" will still likely cause a sad number of parents to tell a loving Grandmom or Grandpop that they're not welcome to touch their blessed grandchildren for fear of causing neurotoxic poisoning and massive skin and nerve damage.

The existence of such neurotic fears is not imaginary. On one Internet board, an Antismoker unwittingly gave credence to the sad problem when she complained about "a grandmother who took her grandchild to the movies" but who insisted on sitting several seats away from the child because of concerns about the smoke on her clothing. The reason the tale made it to the net was because the grand-mother ended up sitting next to the complaining Antismoker, who then worried about her *own* poisoning from "the stench" of the smoker!

I cannot think of any other field of scientific endeavor where this kind of reckless disregard for human feelings on the basis of absolutely ridiculous pseudo-scientific extrapolations would be tolerated. The social and personal damage caused by playing on people's fears and abusing their love for children in this way should be considered a matter of criminal liability, with substantial penalties.

Knowing that establishing the concept of the deadliness of thirdhand smoke in the public mind will go a long way toward moving concerns about secondhand smoke into the realm of indisputable fact by comparison, the antismoking lobby wants to play up this new "risk" as much as possible. However, how can they get researchers to jeopardize their serious academic reputations and devote their time and energy to creating convincing studies about something so ob-viously bogus? The sad and simple answer: money.

The uncharitably inclined might wonder about the rationale for the funding of this T-shirt study. The director of the Institute behind the study, Professor Hofer, noted, "We are currently looking at textile coatings which could neutralize the toxins from the cigarette smoke and so reduce the danger of 'third-hand smoke'." The press release goes on to say, "by working with skilled partners in industry, it is possible that they may achieve this goal quite soon. Especially for developing baby-friendly clothing to be worn next to the skin."[250]

Interesting. The "goal" of reducing the "danger" of thirdhand smoke. I wonder how much profit will eventually be made by selling hyper-expensive nico-neurotoxin-proof, Surgeon-General's-Approved-Hazmat-Clothing to grandparents who want permission to hug their grandchildren? And how toxic might the chemicals in those unusual garments prove to be themselves?

Before ending our look at thirdhand smoke, I'd like to point out that the Hohenstein Institute and its concern about Terrifyingly Toxic T-Shirts by no means stands alone in the thirdhand smoke loony bin. One afternoon while wandering the wilds of the Internet, I came across a very serious post from a young lady named Vicki. Vicki professed to be quite seriously worried about the awful dangers of yet another variation of this new threat. As with my first glances at the Helena study and Winickoff's blathering about thirdhand smoke, at first I thought it must be a joke, but when I checked, the placement of it within the context of the discussion after Winickoff's *Scientific American* article made it quite clear that it was serious.

> **I'm a true believer in third-hand smoke, and I'm not even a child. At work I had to share a telephone with a smoker. I developed breathing problems, had a swelling in my mouth, and also had a "suspicious" breast biopsy. I started cleaning the phone off with "Wet Ones" wipes. My swelling went away, plus my breathing problems went away. Still have to be tested again to see if my biopsy is benign, but I'm pretty optimistic.[251]**

Thanks to the Antismokers, we are now living in a world where innocent people have been made fearful of "catching breast cancer" because they've used a telephone previously used by a smoker. It would be hard to find any examples of more extreme, hate-based disease paranoia even in the "coloured only" accommodations of the 1950s' southern US or in the heart of pre-*Kristallnacht* Nazi Germany.

Slab V

Outdoor Tobacco Smoke (OTS)

For years, walking through Times Square and other plaza areas around Manhattan I've battled second-hand smoke. I've actually crossed the street to avoid people walking and smoking in front of me. Why didn't you include sidewalks in this ban, mayor?[252]

- Randi Kaye, CNN Anchor

Outdoor Tobacco Smoke: The Beginnings

From the mid-1980s through the early 2000s, Antismoking Crusaders focused on exaggerating the harms of indoor exposure to ETS in order to facilitate the passage of bans in hospitals, airplanes, theaters, restaurants, malls, bars, charity bingos, casinos, and even strip clubs – basically any setting they could get away with defining as a workplace so they could appeal to people's sense of fairness in wanting to "protect the unfortunate workers."

To digress for a moment, it's amazing how the antismoking movement has managed to change even the meaning of such basic words as "employee" and "employer" in pursuit of their goals in this area. They have twisted the definitions so that volunteers at a private club are now marked as employees in legal passages such as this: "[An employee is any person] who performs services for an Employer, with or without compensation.... 'Employer' means a person (who) utilizes the services of one or more Employees at an Enclosed Place of Employment." That quote is from the legislation titled "Public Peace

Morals and Welfare, Smoking In Public Places" in the City of Shawnee, Kansas."[253] Note the wonderful circularity with regard to employee, employer, and place of employment, and the Orwellian redefinition of the base word itself.

OK, maybe you think I just pulled an oddball law from a small town in Kansas filled with smoke-chasing tornados and little dogs named Toto and made a big thing about it.

Nope.

This is pretty much standard boilerplate from antismoking lobby groups that has been inserted into modern smoking ban laws almost everywhere. You can find it in ban laws from Shawnee, Kansas,[254] to Somerville, Massachusetts;[255] from the entire State of Georgia[256] to the entire State of Ohio[257]; and in many places in between.[258] If you Google the following as shown:

"performs services for an Employer with or without compensation"

… you'll get over 8,000 hits. Now, if you add the condition of

-smoking

… to the end of the query – thereby asking Google to only return pages that have that oddball definition of "Employee" in references that do *not* concern smoking – you will find that the 8,000 hits have shrunk to just *one* hit.[†] This rather unique way of legally redefining the words "Employee" and "Employer" for political purposes has evidently been used solely for the benefit of the Great Antismoking Crusade! Orwell would be proud.

This linguistic digression has a point: it nicely illustrates the extreme social contortions that have been needed just to bring about

---

[*] Shawnee's special attention to smoking is interesting. In the introduction to their ban law, it is noted that Shawnee recognizes four types of crime: felonies, misdemeanors, traffic infractions, and tobacco infractions. Smokers got a whole new category of criminal law written just for themselves. I guess they should feel honored!

[†] As of January 30, 2013.

effective indoor smoking bans. It has not been until the last ten years or so – basically since the turn of the century – that even the craziest Antismokers called for laws controlling exposure to outdoor tobacco smoke (OTS) in venues like restaurant patios. The concept that the brief exposures one might encounter for an instant or two while passing by smokers in a park or near a doorway or happening to catch an occasional diluted whiff of smoke while sunning at the beach could possibly be harmful in any normal sense of the word was simply too absurd for even the deeply neurotic to consider. But never let it be said that the fear of appearing ridiculous ever slowed down a true Antismoker for long!

Enter Mr. James Repace, a man famed in legend for showing up at his EPA job wearing a full gas mask to protect himself from the "outgassing" of his office's building materials, and who is even better known for claiming that it would take winds of tornado intensity (i.e., 300 miles per hour) to blow even the faintest suggestion of smoke out of a window.[259] In recent years, Mr. Repace has made quite a handsome living for himself as an "International Secondhand Smoke Consultant" using his credentials as a "Health Physicist."[260] Of course, with only a 1968 Master's degree in physics and no degree at all in "health," his credentials may seem a bit thin, but he *did* work for two or three years in the early 1960s as some sort of x-ray technician/instructor/researcher at a New York hospital.[261, 262]

While I don't actually have the details on how this particular study about traces of OTS came about, I'm guessing that Mr. Repace wanted to go on a cruise but also thought about how nice it would be if he didn't have to actually pay for it. How could such a thing be arranged? Simple – he could take along his little air testing kit and maybe get a grant for measuring the deadly threat posed by smoke in the cruise ship's casinos!*

Unfortunately for Mr. Repace, the ship's casinos evidently had pretty efficient filtration and ventilation systems. They just weren't smoky enough to be truly scary. While measurements in landlubber

---

* Of course, to be fair, Mr. Repace (and, I would guess, at least one assistant) could be said to have paid for it with their work. Remember that air-sniffing in eleven restaurants (as opposed to the five or so locations here) was billed at $70,000 in Minnesota.

smoking bars commonly gave readings up in the area of 150 to 400 nanograms per cubic meter, the readings in the well-designed smoking cruise ship casinos were giving readings of only 11 nanograms per cubic meter. To make things even worse, some dastardly folks in the cigar bar decided to have a birthday party and when the birthday cake candles were lit for five minutes or so, the readings shot up by 400% over the measurements taken while people were simply puffing blithely away on their smokes.[263]

Whatever grant hopes he may have had seemed about to go up in smoke – so to speak. Ever enterprising, it seems he began measuring the OTS in one of the outside cigar/cigarette bars elsewhere on the ship and made an amazing discovery. He found that there was, on average, almost as much smoke (about 9 nanograms per cubic meter) in the air by the smokers at the outside bar as there was inside the well-ventilated smoking casinos!

This was, once more, the clever use of the concept of taking very small numbers, repeating them until they sounded important and scientific and official, and then simply ignoring the fact that any possible exposure was so small to begin with that no sane person would ever worry about it. Mr. Repace seems to be rather fond of the Commander Almost Zero Fallacy that we explored a bit earlier in the *Stratistics* section.

While I don't know all the details of how it ultimately turned out, I expect he may well have succeeded quite handsomely in either writing off the cruise as a business expense or perhaps even getting it paid for completely under a grant. I can only envy such skills.

Mr. Repace's contribution went far beyond entertaining his fellow passengers on a nice cruise, however. He had made the concept of worrying about outdoor smoke acceptable, and the path he blazed was soon followed by others. One of the more sophisticated studies in the area has already been lampooned in the *Satirical Smoke* section under the title of "Fun With Stan And Jamie," and, as noted in the introduction to that satire, actual experimental conditions were truly not much different than as outlined there. Neil Klepeis published "Real-Time Measurement of Outdoor Tobacco Smoke Particles" in 2007.[264]

Klepeis employed five different types of sophisticated measuring devices, and measured smoke in both natural surroundings and in carefully defined experimental conditions, both indoors and out. As noted in the satire presented earlier, he set up such situations as one where he surrounded a little "sniffer" device with a ring of cigarettes spaced around it at a distance of a foot or so. What he found was that in such a situation, the smoke concentrations in the middle of such a small circle could be comparable, at least sometimes – occasionally – to the smoke concentrations in an indoor bar or restaurant where smoking was permitted. He also found that as soon as you moved a few feet away from a smoker and had even a slight breeze blowing the smoke away from you, not even the most sophisticated scientific technology of the twenty-first century could detect any perceptible levels of offending pollution.*

The Klepeis study is probably best appreciated simply by reading a little more detail about its arrangements – which, you'll note, really aren't much different than their portrayal in the earlier satire.

> The E1 experiments consisted of six outdoor patio experiments on a single day in which a cluster of single PAS, NEPH, and GRIMM monitors were surrounded by five burning cigarettes at distances of 2, 4, or 6 ft., ... in concentric pentagonal arrangements ... The design of the indoor experiments was nearly identical ... except that only distances of 0.25 and 0.5 m [10 to 20 inches] from the burning cigarette were monitored, and ... for one of the two living room experiments, a small fan was introduced to explore the effect of controlled air directionality.[265]

It should be noted that in Klepeis' actual experiments, no one's hair was set ablaze nor did any innocent people (or Antismokers) get blown up. Just standard, boring antismoking research with lots of expensive equipment and many, many micro-, nano-, and pico- measurements. Did Klepeis examine any health effects from such exposures to smoke? No, of course not. They don't exist. All his research did was use various types of very sophisticated, important-sounding, and ultrasensitive instrumentation to detect even the smallest traces of smoke in the air. Is there any background research in the literature that would indicate a threat from the levels and durations of exposures that Klepeis recorded?

---

* Unfortunately, it seems that James Repace was absent during this phase of the Klepeis research. It is rumored that he may have been busy, sitting outside a smoking bar in Kansas waiting for a tornado to demonstrate his theories.

No. If there are, I have never been able to find any, nor have any of the hundreds of antismoking advocates who I've challenged over the years on Internet discussion boards.

We'll take two final quotes from Klepeis before moving onward. The first is simply amusing for its terminology. At times during the experiments, "peak concentrations" were measured when wisps of smoke passed by the little sniffers. The term "wisps of smoke" doesn't sound all that impressive though, so the researchers named and described the phenomenon thusly: "Microplumes are defined as thin concentrated streams of smoke, or some other air pollutant, that follow complex trajectories during periods of release." So the next time a wisp of smoke wafts your way outdoors, you can sound suitably educated as you complain to the smokers about their microplumes. If they hit you, I would like to make it clear that I am not in any way legally or morally responsible for your injuries unless I burst into laughter.

And the closing quote, although one of the most important in the entire study, was obviously not given much prominence in the press: "OTS disappears almost instantly when tobacco sources are extinguished."

Unfortunately for those leading the Great American Antismoking Crusade, despite some pleasing sound bites about "measured outdoor concentrations approaching those found in indoor smoking situations," the Klepeis study was just a bit too technical to really be scary to most people. While news stories about it tried to make it sound like a real threat had been precisely quantified, the imagery – with all the different measurements and instruments and nanograms – was simply too sophisticated and complicated to play well in Peoria. It really didn't have the impact that Antismokers felt was needed to frighten college students into acquiescing to bans throughout the great outdoors of their campuses. Something more intimate, more gutsy, more poisonous, toxic, and deadly sounding was clearly needed as Smoke-Free Campuses geared up for its big end-of-the-decade push in 2009. As noted earlier in the *Stratistics* section, researchers at the University of Georgia came up with just what the doctor ordered as that campus was considering its own outdoor smoking ban.

One of the most useful aspects of the OTS scare-mongering has been the push it gave to banning smoking on university campuses. Back in the 1970s and '80s, it had been common for high schools, and even some middle schools, to establish smoking areas for their students in order to control the problem of smoke-filled bathrooms and to reduce the fire threat from hidden smoking and hastily disposed-of butts; but by the 2000s, the zero-tolerance-for-drugs atmosphere, the encouragement of the idea that a "snitch mentality" was socially acceptable, and the deliberate positioning of nicotine as "more addictive than heroin" had succeeded in almost totally eliminating such areas. College campuses were a different story, however. While smoking had been largely banned in classrooms, offices, and even a good number of dormitories by use of the indoor secondhand smoke arguments, college students still puffed freely in the open air of campuses, asserting their rights as free adult citizens enjoying a legal pleasure in the great American outdoors.

The specter of possible health effects from OTS offered a wedge toward changing that. In 2005, only about 200 of the nation's campuses lived under outdoor smoking bans, but in the next five years another 300 followed suit. Antismokers took advantage of this growth in an effort to make it sound as though campus bans were becoming universal. Of course, they never mentioned that the total number of US school campuses numbered over 5,000. Fortunately, university student populations were often well-read enough to see through the smoke-screen and proved quite resistant to the call of the bandwagon effect.

Enter the University of Georgia's push for a smoking ban in 2009/2010. The idea was not very popular on this southern, tobacco-country campus and the efforts toward a ban were running into a brick wall. Something was needed to convince the nonsmoking student population that its health was endangered by the common, but highly diluted and usually momentary, encounters with tobacco smoke on campus walkways and outside entrances to campus buildings.

University researchers showed themselves ready and able to take on the challenge, and in November of 2009 they published a study showing that exposure to tobacco smoke outdoors could result in an increase in salivary cotinine levels – a fairly reliable indicator of exposure to tobacco smoke.[266] How much of an increase and in what sorts of situations? Anything at all like the exposures one might get walking around a campus or through a doorway with some smokers nearby?

Well, first of all, the exposures were based on six full hours of sitting around in the smoking areas outside of bars on busy weekend nights in June of 2007. Most folks would agree that such a situation represents a fairly major and quite unusual exposure to "Outdoor Tobacco Smoke" (OTS). Nevertheless, in the news articles splashed around newspapers and on campuses where smoking bans were being pushed, the claim was simply made that "exposed" students – without any details as to the exposure – showed a 162% increase in a certain smoke-induced chemical in their blood (cotinine). That represented an actual increase of less than 2/10ths of a nanogram per milliliter – .2 ng/mL – over the blood levels of the unexposed control group.

Remember the importance of the Commander Almost Zero Fallacy: increasing something that's virtually zero by 162% still leaves you with … virtually zero. It is absurd to try to scare people with fractions of nanograms of exposure to just about anything. According to the classic Benowitz/Jacob Study of 1994,[267] the average smoker in the course of a day's smoking will have a measured blood level increase of roughly 300 ng/mL of cotinine. That's about 2,000 times the increase observed in these volunteers after a full six hours of sitting around in smoke pits outside of college bars on busy Friday nights. Of course, that information never made it to the news media or to the students. All that was hyped was the scary sounding "162% increase from outdoor exposure to tobacco smoke," an exposure that obviously has nothing in common with the actual normal student exposure to the half minute or so of exposure that an average nonsmoking student might encounter while spending a day on a smoking campus. It also sounds considerably more impressive when it's not noted that smokers themselves are walking around every day, quite normally and happily, with levels that are 2,000 times higher than even the most chronic non-smoking smoke-pit-inhabitants.

By very selectively presenting the actual scientific findings of the University of Georgia study (which, if it had been reasonably and fairly interpreted, would have strongly indicated how completely nonsensical it was to worry about such things) and by ignoring the context of such previous findings as the Benowitz/Jacob Study, the Smoke-Free people succeeded in convincing many innocent students that their health and their lives were at risk.

Once again, think about Commander Almost Zero. At a half-minute of exposure per day compared to six hours, and considering the 2,000 times figure noted above, a student with an average daily half minute of exposure to weekend smoke-pit clouds of smoke would have to walk around a smoking campus for (2,000 x 720) one and a half million days to absorb the amount of cotinine-producing nicotine that an average smoker gets in a single day of smoking. To get that one day's worth of smoking from such exposure, a student would have to work on their degree for about 4,000 years – with no weekends, no holidays, and most definitely no summer vacations.

Another way of analyzing the exposure would be to consider how many student-years of campus life it would take, on average, to produce a single lung cancer. This is quite a relevant question since a common complaint heard from student antismoking activists is, "Why should I have to risk lung cancer from just walking to class?"

For this model, let's assume that all the smokers on a particular campus are rather inconsiderate and congregate within ten feet or so of doorways to smoke between classes. If our innocent, nonsmoking, cancer-fearing student has to pass through such "clouds" of smokers ten times a day, every day during their school year, just how serious would their risk be? Can we find a way to take at least a reasonable guess?

Yes, we can. Walking through such a cluster of smokers might take about three seconds.* Walking through such doorway clusters of smokers ten times a day while attending classes in five different buildings would give a student a total of about thirty seconds of diluted outdoor exposure. As we saw earlier in *Stratistics,* even if we accept on faith the EPA's risk evaluation arising from the fairly smoky confines of smoking workplaces in the 1950s and '60s, it would take an exposure of about 40,000 worker years to produce, on average, a single lung cancer. But a "worker-year" consists of about fifty weeks, with each week giving forty hours of exposure. That's 2,000 hours per year, i.e., 120,000 minutes of exposure in a single worker-year. For our single average lung cancer, we would thus need to multiply those 40,000

---

* A reasonable estimate based upon our earlier research into average walking speeds for the young and fairly fit.

worker-years x 120,000 minutes/year = 4.8 billion minutes. For the purpose of the current calculation for our students, that would be 9.6 billion half-minutes of exposure over the course of 9.6 billion days.

But the smoke that our student would be walking through outdoors is generally not really of the same smoke density as you'd find indoors in those old smoking workplaces. With breezes blowing in an outdoor environment, it's likely that, even around doorways, it would seem reasonable to say the average exposure would be about 10% what it was in those old smoky offices. So, to be equivalent to that worker exposure, the students would have to work on their degrees for 96 billion days.

To actually get a single lung cancer from such exposure, our poor perpetual grad student would have to hit the books on the average for roughly 260 million years. You may remember the earlier calculations in *Stratistics* concerning passing by smokers on the sidewalk. They were presented a little differently, and also assumed a single exposure per day at a lower intensity since they did not involve really walking through the middle of smokers; but with those factors taken into account, the end results of such exposure were roughly consistent with these: several billion years of daily exposure to single incidents.

If our grad student only walked through one crowd of smokers a day, and then finally graduated after only five or ten billion years of coursework, he'd have lucked out on his fifty percent chance of surviving the Deadly Doom Of Smoke and finally be allowed to take a night off to go see his girlfriend. Now that would take dedication! And a very patient girlfriend.[268]

It's also worth remembering: even *that* extraordinarily low level of risk would only hold true if we accepted the EPA figures – with their juggling of statistical goalposts and stretching of confidence intervals – as actual fact. If we used normal statistical standards we might see those billions of years turning into billions of centuries.

>>> Cleanup 5 <<<

The incredibly paranoid concern about actual health effects from brushes with wisps of outdoor smoke simply hasn't been around long enough to spawn a lot of copycat studies, but the push by antismoking groups to further denormalize smoking by getting it banned at beaches and in parks has made for a lot of media focus, and thereby built up public fears even without multiple studies in the headlines.

You may remember that in the original Commander Almost Zero subsection of *Stratistics*, I spoke about the student who worried about outdoor smoke forming "thirdhand outdoor smoke residue" on campus benches. I offered computations there showing that even if a student assiduously licked an entire bench clean every single day, it would still take him or her literally three quadrillion years to honorably suffer a KGB spy kind of death. But what if the student just sat on the bench normally and happened to rest his or her hands on it while doing so, and then, later on, happened to rub a finger along their lips or eyes? Well, in a case like that, the dosage would be roughly a million times less (figuring in the exposed finger area, the amount of transference, the degradation by prior transference elsewhere, etc. – quite a rough estimate obviously, but good enough for this purpose). In such a scenario it would take that student three *sextillion* years to die. That's 3,000,000,000,000,000,000,000 years, or three hundred billion times as long as our universe has existed.

I realize that you may be thinking that I've gone off the deep end myself at this point. But please, remember: Don't shoot the messenger. As the following tale makes clear, I have barely even touched on the full extent of the insanity that has been encouraged in the building of antipathy toward smokers.

Shortly after this chapter was originally written in 2011, an Internet poster provided the perfect capstone to its pyramid of insanity. Once again (When will I ever learn?) I was initially sure that this posting had to be satirical, despite the dead seriousness of its tone, but then I realized that it had not been posted to a forum dealing with smoking, but actually sent in as a pleading query to a forum dealing with supernatural phenomena. The writer claimed to be having a "problem with smelling cigarette smoke, but no one smokes at home," and then went on to note that no one else could smell it despite the fact that she herself was forced to use an inhaler because the smoke was giving her asthma attacks even after she covered her head with a shirt.

It was at that point that we moved into realms even more surreal than SHS, or THS, or OTS*, as the writer outlined her conviction that the offending smoke odor was being sent to her by a smoking "spirit" who kept appearing to her despite the fact that she had prayed and "commanded it to leave in the name of Jesus."[269] As noted, the context and the tone of the posting left no room for thinking of it as being anything other than a deadly serious and heartfelt plea for help in dealing with her "smoke problem."

This innocent Internet poster may be the harbinger of a whole new realm of antismoking paranoia: fear of attack by what can only be called ODS: Other Dimensional Smoke![†]

UPDATE: March 17, 2013: "A cavalry officer who died in the days of the British Raj has been declared a saint in India with worshippers claiming he smokes cigarettes left as offerings at his tomb.... Believers tend his grave in Musa Bagh cemetery. One devotee said: "We leave him food and cigarettes as tributes. The cigarettes glow like somebody is inhaling them. We know Shah Baba smoked so there can be only one answer...he is smoking the cigarettes."

(No, this is not a satire either: it's a full-blown news story, courtesy of the *Daily and Sunday Express* in the UK.)[270]

---

* Secondhand smoke, thirdhand smoke, and outdoor tobacco smoke
† If studies of ODS become a big thing and get massive government funding, I just want it to be on record that I should get first crack at the grant treasure trove for creating the term!

Slab VI

Karz With Kidz

Can I drive daddy? Please?
No son, you're not old enough.
Please Daddy? PLEEEEASE??
OK. Here, sit in my lap.
You steer. I'll do the brake and gas.

- Unknown

In the 1990s, the perpetual search to find even more ways to denormalize smoking while also making it more difficult and less enjoyable was still fairly hidden from the general public. The idea of social engineering – treating humans like lab rats whose behavior should be adjusted to proper norms by the societal equivalent of electric shocks – was simply unacceptable back then. Smoking bans were mainly justified by appeals to public health concerns. Advocates defended bans by saying such things as "This is not social engineering. It is an example of the accepted practice in our society of protecting the public from significant health threats,"[271] or, "This is not social engineering any more than a stop sign at a roadway intersection is interference in how we drive. It is all about the greater common good."[272]

At those times when the threat from casual exposures to traces of smoke were questioned, Antismokers simply fell back upon the stratagem of defending the lives of the poor enslaved workers "who should not be forced to choose between their health and a paycheck."* They

---

* Again, a simple Googling of variations of that phrase will bring up thousands of examples of its use!

would never utter a word nor even a hint about social engineering or behavior modification. It wasn't until 2009 that we began seeing high-profile antismoking advocates like Mayor Bloomberg openly making such statements as "This city is not walking away from our commit-ment to make it as difficult and as expensive to smoke as we possibly can."[273]

This need to hide their real motivations while still gobbling up freedoms slice-by-slice led to increasingly inventive uses of the "save the children" argument. One which swept the media by storm in the early 2000s was a push for laws focusing on smoking in cars.

Previous efforts to reduce smoking by banning it while driv-ing had failed, despite the plausibility of smoking being a risk factor for accidents. After all, most folks have seen smokers take their eyes off the road for a second to find, or light, or extinguish a cigarette, and many have even seen or heard of instances of cigarettes being dropped in the laps of drivers or of times when the glowing "cherry" fell off the end while someone was driving.* So, it seemed natural to expect that studies would just confirm the danger and that antismoking advocates would have yet another arena in which smoking could be banned and smokers' freedoms limited.

Unfortunately for the Antismokers, their plans didn't work out as desired. No matter how many dollars were spent in the effort, not even the most dedicated statistics juggler could confirm that smoking was a significant cause of auto accidents when compared to other vari-ables. Drinking, cell phones, snacking, kids, radio adjusting, even simply having passengers in a car all soundly beat out smoking as accident-causing driving distractions.[274]

The most definitive study in this area is probably the one that was done for the American Automobile Association in 2001.[275] In the resulting report, Dr. Stutts *et al.* offered a descriptive analysis of five years' worth of National Accident Sampling System data as well as specific information on accidents in North Carolina. They found that smoking contributed to 0.9% of crashes, while auditory entertainment

---

* There have been numerous complaints that the new, legally mandated, "Fire Safe Cigarettes" are particularly prone to creating accidents with these falling "cherry blossoms."

(radio, CD, etc.) and other vehicle occupants, respectively, contributed 11.4% and 10.9%. In other words, radios and fellow passengers caused twenty-five times as many accidents as smoking did. "Stratistically," smoking was 2,500% safer in terms of accidents!

Eating, drinking, and cell phone use were all ranked below 10% as distractors, although all of them were still far more dangerous than smoking.* And since a good number of accidents are not due to distracted driving, the meaningful percentages came in even lower.

A possible explanation for the low ranking of smoking is simply smokers' awareness of the potential for accidents while fiddling with their smokes, thus encouraging extra caution during such times.

It looked like this was one holy sanctum where smokers were safe. But no, "Never Say Die!" shouted those seeking to extinguish the burning butts. "What about the *CHILLLLDRENNNN?*" Sound bites and heart-wrenching Photo-Shopped pictures were created around the concept of helpless, gasping, choking tykes trapped in rolling gas chambers. Babies were presented as being smothered in clouds of smoke not just twice as dense as at home, but in clouds that were 10, 23, or even 72 times as dense and deadly![†] Studies competed with each other in producing ever more sensationalist statements about little Morky or Mindy being forced to breathe in atmospheres deadlier than the EPA's "Dangerous Level," as murderous parents toked on their death sticks.

Just as with all the other studies we've dissected so far, these claims weren't worth the paper they were printed on. To see why, let's look at a few general statements first and then return to examine two of the major and most often-referenced efforts in this area: Offermann 2002[276] and Klepeis* 2008.[277]

---

* It should be noted, however, that cell phones simply weren't all that common in 2001. The study's 1.5% figure might well be over 10% in 2013. Meanwhile, the smoking/accident figure would likely have decreased due to passenger sensitivity about smoke and generally lower population smoking rates.

† Yes, as we'll see in a few pages, one study actually claimed that smokers' car nicotine levels are literally 72 times as high as in "smoke-free" cars, a classic Commander Almost Zero Fallacy.

* Yes, Klepeis of the elaborate outdoor smoking experiments again; but this time moving his magic into our cars.

Most reasonable people hearing the phrase "a parent smoking in a car with a child" would picture someone driving along a street or highway with the windows at least moderately opened to allow the smoke to blow out of the car. Smokers tend to open their windows while smoking, even when driving alone. The number of parents who would sit in a parked car with the windows rolled up tight and the air vents sealed while continuously smoking up a fog next to their baby blue eyes is probably smaller than the number of card-carrying Ku Klux Klan members who voted for President Obama.

And yet, imaginary parents puffing away in such sealed-up gas chambers provide precisely the kind of scenarios for some of the scary figures quoted in news stories about smoking in cars. Figures such as the 23x one noted a few paragraphs ago have been presented by government bodies and used in official testimonies, only to later be tracked down and found to be simply misapplied numbers plucked out of the air by advocacy groups.[278]

Another trick favored and featured by antismoking researchers in this area is similar to the one used in the outdoor smoke studies: focusing on the momentary conditions of what they call "peak concentrations" (i.e., the "microplumes" mentioned a little earlier), while deliberately confusing those exposures with ones that last continuously over 24-hour or 365-day EPA guideline periods.

Think back to the last time you were in a car or a social situation with a smoker sitting right next to you. Occasionally, the air will waft the wrong way and, for a moment, a concentrated plume of smoke will blow right into your face (or into a researcher's "sniffer monitor") from the burning tip of the cigarette. It doesn't happen often in a moving car with the windows cracked even moderately open, but even then, such moments occasionally exist.

That is what is meant when researchers cite figures for peak concentrations. Such figures are completely meaningless when compared with the EPA outdoor air standards for contaminants inhaled and exhaled with every breath, for 24 hours a day / 365 days a year, but that is exactly the comparison Antismokers make when presenting these "smoking in cars pollution studies" to the public. For individual tiny discrete moments, the air quality in a particular few cubic centimeters of space in these cars could indeed be far worse than the EPA's level for 24-hour constant and inescapable exposure. Actually, if that were all the air one had to breathe, it's unlikely even the hardiest adult would survive for a single hour. But in terms of a moment of exposure, it's kind of like having a cup of coffee at 160 degrees* and taking a tiny little sip from it – you'll enjoy it and your health won't be damaged at all. But if I immersed you in a cannibal's kettle at 160 degrees for 24 hours, you'd be soup. Heck, you'd be deader than a hard-boiled egg in 24 minutes!! That's why you should ignore the "peak readings" in stories about studies like these: they're nothing but a propaganda tool used to frighten innocent people.

The EPA itself – even though it doesn't issue press releases warning about it – is actually quite aware of the danger of this sort of misuse of their data and cautions against it in their official documents. In their guidelines for the proper scientific interpretation and public use of their data, they explicitly warn against taking data for any period of less than 24 hours and applying the 24-hour standards to such findings.[279] While Antismokers will speak of those guidelines when referring to findings covering periods of a few hours, minutes, or even seconds, the EPA's strict rule of application calls for observations shorter than 24 hours to be averaged out over full 24-hour periods with unmeasured periods set to a pollution level of zero for meaningful comparisons to their health standards.

Such a strict application is clearly not reasonable in extreme situations – e.g., with our cannibal kettle or cup of java, or in a garage with a very high carbon monoxide reading for twenty minutes – but the EPA's warning is clearly meant to prevent precisely the kind of wanton abuse that is so often employed in antismoking arguments regarding briefer exposures outdoors, in cars, or even during an eight-hour workday. That warning has been consistently, blatantly, and deliberately ignored by antismoking advocates in their quest to terrify the nonsmoking public and increase support for smoking bans.

Another gambit used to gull the gullible is to compare situations that are so dissimilar that the comparisons are totally pointless, a subset of our Commander Almost Zero Fallacy. On August 25, 2009, a new study was featured in *Tobacco Control* titled "Secondhand tobacco

---

* Actually somewhat cooler than the service industry standard of about 180 degrees Fahrenheit.

smoke concentrations in motor vehicles."[280] The news release about it warned that "After 1 to 3 cigarettes, airborne concentrations of nicotine were 72 times higher in cars with smoking compared to smoke-free cars."[281] That sounds pretty impressive, even downright scary, until one stops to remember what's being compared here: cars *with no one smoking in them* – and therefore with no inherent source of nicotine at all – versus cars where people were actively smoking. It's similar to the previously analyzed OTS study where students in the middle of smoke pits were found to have 162% more exposure to nicotine than students who weren't around smokers at all.

It's sort of like saying tomatoes have 162,000% more nicotine in them than apples. That could be quite true, since apples have pretty much no inherent nicotine in them at all, but it says nothing at all about the danger or safety of tomatoes – which are quite safe to eat despite having quite measurable quantities of that highly addictive and deadly neurotoxic poison in every luscious bite. And, as we'll see in more detail later on, some popular brands of baby shampoo have 87,000 times the concentration of formaldehyde as the smoke-filled air of your corner pub – but that doesn't mean that the mum shampooing little Edgar's auburn curls should be locked up for child abuse.

Such figures and comparisons are absolutely meaningless in terms of any real measure of concentration and beyond absolutely meaningless in terms of any effect on someone's health and well-being. It's like claiming that suburbs with swimming pools are more dangerous to live in than suburbs without swimming pools because, on average, there would be more deadly chlorine gas in the air of the pool-loving 'burbs.

One of the researchers in the *Tobacco Control* study above, Miranda Jones, gave a bit of insight into the motives for presenting smoking-in-cars study findings in this way when she observed in the media release that "Fifty-three percent of the smokers surveyed said that *being unable to smoke in the car would help them to quit smoking altogether.*" [Emphasis added.] With that statement, Ms. Jones nicely demonstrated the true motivation for these studies: the promotion of the social engineering goal of a reduction in smoking. The health threat to children is just a boogeyman created to support the larger behavior modification program.

For the glory of pushing a smoking ban in cars, even the specter of thirdhand smoke was brought back into play as the colorful concept was invoked by one of Jones's co-authors, Patrick Breysse. Repeating the sacred mantra of the Surgeon General, he wisely intoned, "There is no known safe level of exposure to secondhand smoke. ... exposure to hazardous components of secondhand smoke can occur long after smoking has stopped." Breysse then went on to note that "air nicotine concentrations in motor vehicles were ... even higher than concentrations measured in restaurants and bars that allow smoking."[282]

That last observation might seem meaningful until one checks the reference that was used. A little investigation into that reference – a study of smoke in venues in European cities – uncovers the fact that a median level of about 70 micrograms per cubic meter was actually listed.[283] Since the Jones/Breysse Study states "Median (IQR) air nicotine concentrations in smokers' vehicles were 9.6 micrograms per cubic meter," the authors seem to be declaring that the observed 9.6 in vehicles is a larger number than the observed 70 in European cafes. Math like that might not be too out of tune for antismoking research nowadays, but it's nonetheless annoying for any reader who managed to pass third grade arithmetic and who happens to be sharp enough to check the references.*

The fact that the Jones research was funded by the Flight Attendant Medical Research Institute (FAMRI) may have had something to do with the presentation of its findings. FAMRI's funding comes from a lawsuit against the tobacco industry and its mission statement specifically declares that they are dedicated to sponsoring "research to combat the diseases caused by exposure to tobacco smoke...."[284] Clearly, producing studies that will cause smokers to stop smoking in their cars and thereby encourage them to quit altogether is in tune with "combat(ing) the diseases caused by exposure to tobacco smoke" – even though the diseases in question are those affecting the smokers themselves.

----

* Sloppy referencing is all too common in antismoking-oriented studies. At one point when I pointed out such a problem, I received, instead of an honest admission and an apology, simply an excuse that it was probably the fault of a lazy grad student.

Given the FAMRI funding source, it would have been far more surprising if their research had concluded, "Oh, it's *fine* to smoke in a car with your kids. The smoke all gets blown away out the windows!" than if they simply concluded – as they did – that 9 is a larger number than 70. After all, the first conclusion would cost them their careers while the second would never even be noticed unless some cur-mudgeonly obsessive Free Choice advocate actually tracked down and read the silly little references in their research.

You may remember that at the start of this section I promised to look at two major studies, one by Offermann in 2002, and one by Klepeis in 2007. Many of the general limitations in these two studies have already been dealt with here, but there are still a few points to add. Offermann measured particulates in cars under several condi-tions, but gave, at least in my opinion, enormously undue emphasis to the unlikely "I'm gonna smoke up a storm with all the windows sealed up tight so I can suffocate Little Boozums In Da Bassinet!" scenario. Indeed, with the windows rolled up tight and the ventilation carefully shut off for maximum suffocation jollies, the PM 2.5 level headed up into the 2,000 ppm range for about three minutes.

Of course Offermann couldn't totally ignore the fact that some parents might occasionally crack open a window while smoking so he also ran one version with the driver's window opened three inches. In that scenario the air in the car headed up toward the 100 ppm range for five minutes. If that were done twice a day, every day with your little one by your side, they'd be getting, on average, an extra 1 ppm of particulate matter added on to the 50 ppm or so they'd be likely to be getting if they drove around a city with you all day anyway!

Offermann's conclusion however, ends up being: "Indoor concentrations of ETS can be especially significant in automobiles due to the small indoor air volume," while referring specifically to the windows rolled up scenario and confusing the minutes of exposure to 24 hours a day of exposure. Why would he do that? Impossible to say, although applying the same sort of lens we applied to Jones/Breysse might offer some insight: Offermann's study was funded by the "Tobacco Free Project, San Francisco Department of Public Health, paid for by Proposition 99, the 1988 Tobacco Tax Initiative."

Do you think the funding just *might* have played a role in how the study was structured and presented?

As for Klepeis 2008, this time the lead authorship position was traded among the three researchers over to Ott. The measurements were a bit more detailed than Offermann's research several years previously, but the conclusions weren't that much different. As in the case of Offermann, this study emphasized the "smoke 'em up wid all da winders closed" scenario and pointed out that if you took your little one out every day and smoked a couple of cigarettes like that during your drives *and* made sure that you continued to keep those windows sealed up tight and the fans/ac off for at least an hour or two while doing so, then you *might* move the general air pollution average up by as much as 42 ppm for the day, actually up into the "moderate" EPA readings!

Having already been familiar with Klepeis, Ott, and Switzer's work from the previous year in their outdoor study, I wasn't surprised at the results. Nor was I too surprised when I saw who paid them for it: FAMRI. Yep, the same folks who paid for the Jones/Breysse study. All three of these "car air pollution" studies were paid for by smokers with money either extorted through the MSA and laundered by the federal and state governments or pulled from Big Tobacco in the settlement with the flight attendants. All three were dependent upon funding sources clearly and openly dedicated to doing anything and everything possible to reduce, and, if possible, eliminate tobacco from the face of the earth.

With that knowledge as background, perhaps it becomes a little clearer why every news story you see about this topic seems to start from the same conclusion… i.e., that *something* must be done!

>>> Cleanup 6 <<<

The "no smoking in cars" movement never had a particular high-profile study to claim as a flagship. Instead, as we've seen, there have just been lots of little studies and odd numbers thrown about and a surprising amount of success in having legislation passed despite the lack of a showpiece to hang a hat on. Probably their success can be attributed to their concentration on the welfare of children while promoting the imagery of rolling gas chambers and abusive parents.

However, there have been other concerns about kids and cars, particularly once one moves into the area of child drivers, i.e., teenagers with drivers' licenses. Teenage driving has always been known to be more risky than driving in general, despite requirements for car instruction, skill testing, and an absolutely zero tolerance for anything to do with drinking and driving by teens. Recently, there's been a whole new concern added in: the tendency of teens to text while driving.

In November of 2010, Dr. Fernando Wilson published a study titled "Trends in Fatalities From Distracted Driving in the United States, 1999 to 2008" in the *American Journal of Public Health* in which he used accident and cell phone use statistics to estimate the excess numbers of vehicle deaths caused by drivers engaging in text messaging while driving.[285] Using a database showing an excess of 6,000 fatalities for every 24,000,000 text messages, his work might have seemed impressive to a statistician but, unfortunately, the most important message – that it truly is outrageously dangerous to text while driving – probably sailed right over the heads of most teenage drivers. I wrote and asked him for a copy of his research, which he nicely provided, and while reading his study, it occurred to me that he might actually be able to make some good use of the antismoking scare propaganda that's currently being aimed at teenagers!

Using the EPA lung cancer figures as a basis for comparison, and knowing that the power of MTV and other such messages to teens has convinced a great number of them that their lives are being imminently threatened if someone lights up in a car with them, I decided to compare Dr. Wilson's figures for texting to the EPA figures for ETS exposure fatalities to see if perhaps this unreasonable and exaggerated fear could actually be put to use in a positive form.

Using Dr. Wilson's figure of 6,000 fatalities per 24,000,000 text messages and the EPA's 19% increase in lung cancer after a lifetime of daily, intense (1950s/1960s levels) workplace smoke exposure (one excess case per 40,000 worker-years), I was able to calculate that a passenger in a vehicle where the driver engaged in one texting episode per hour would experience 13,000 times the excess risk incurred by a passenger in a vehicle where someone smoked one cigarette per hour. Conversely, driving with a driver who engaged in a single texting episode carried the same excess risk as riding in a car for years while a fellow passenger smoked 13,000 cigarettes.* True, the risk of fatality from texting involves possible immediate death while the ETS risk involves possible postponed death decades in the future, but that realization only serves to make texting seem even more risky and more serious by comparison.

Communicating that enormous difference in relative risk to teenagers who turn pale at scenting the merest whiff of smoke could quite conceivably save thousands of young lives from being snuffed out in car accidents. I shared this idea with Dr. Wilson in the course of several emails, but never received any further responses from him. Whether he's afraid of going up against antismoking interests who might interpret such a comparison as weakening their own efforts or whether he simply didn't feel the comparison was valid, I can't really say; but I believe my figures, and my comparison based on those figures, were quite valid. If Dr. Wilson's basic research on the risk of a driver texting was correct, then it would seem that, indeed, it is literally over 10,000 times as dangerous to share a car with a texting driver as to share one with smoking passengers.

---

* Note: I am not comparing the risks of texting while driving to the risks of smoking while driving or to the risks of driving itself here—that's a whole different area of research. My comparison is simply one of the excess danger theorized from ETS exposure in a car compared to the excess danger theorized from being a passenger while a driver is texting.

Slab VII

The Giblets

A census taker once tried to test me. I ate his liver with some fava beans and a nice chianti.

- Hannibal Lecter

The Medical Examiner looks around his morgue. The cadavers have all been dissected, their mysteries unraveled, their entrails carefully picked apart, and the secret guilt of their lies and indiscretions exposed. Even the minor organs, neatly set aside during the main dissection of each cadaver, have been weighed, measured, and properly disposed of.

At last, he can go home and rela... but wait, his erstwhile assistant Igor – who bears a startling resemblance to a well-known Californian Antismoker reputed to have an embarrassingly microscopic aspect to a certain part of his intimate anatomy – hobbles into the laboratory.

With clear strain on his deformed back, he carries a heavy bucket filled almost to the brim with small but gory objects. He sets it down with a groan and looks up with a glint of fear in his eyes. "Master! Master! You cannot leave yet!"

The Examiner growls, "Igor, what is it? I have plans to go home this evening and watch Dexter. *His sister is SO cute, even if she does pollute the purity of the show with her cigarette smoke."*

Igor winces – he's always secretly liked that sister. "Master, you cannot leave until you've examined the Giblets; the little pieces strewn around antismoking crime scenes without enough substance to be body-bagged on their own."

The Examiner glares at his flimsy excuse for an assistant, but there's no way around it. The repast is not finished till the dessert plates are empty and the cordial is quaffed. "OK, Igor, dump the whole mess here on Slab VII."

Igor carefully upends the bucket and watches its macabre contents slide out onto the dissecting slab. "Master, if you work quickly, you may still get home in time to watch Dexter dismember his victims during family hour."*

The Examiner sighs, nods sad acquiescence, and returns to his instruments....

Antismokers are endlessly creative. They have more money than God with which to produce studies, and they have countless thousands of graduate students eager to please their mentors with the "right" sort of results produced by "properly" designed research in the eternal quest for the golden ring of a doctoral degree. If their results come out "wrong," it will be much harder to get that degree or be published – and publication is the lifeblood of academia. Remember the story of the studies involving post-ban heart attacks? The meaningless micro-studies were published by the dozen because their conclusions supported the goal of ban promotion. Meanwhile, far larger and better designed studies went begging when it came to publication in medical journals simply because they produced the "wrong" answers.

There are far too many of these abortions to look at them all in the detail with which some of the preceding ones have been examined. But the little ones – the ones that simply squeeze into the interstices of the evening news every week of the year – cannot simply be ignored. These studies are what form the mortar in the construction of the background groupthink that is so necessary for the true denormalization of smokers to take root and grow. In the following pages, you will find examinations of some of those smaller studies that have reached beyond the medical journals to sway the thoughts and feelings

---

* Actually, it seems the producers have recently caved under pressure, forcing Dexter's sister to give up her naughty habit. To make amends for past sinfulness, the program—starring a serial-killing cop as a hero—now inserts bits of antismoking dialogue into its script (e.g., upon finding an ash in one of the many pools of blood featured on this child-friendly show, Dexter's partner exclaims at the unusual evidence, "Hey, it's 2010! No one smokes anymore!") Meanwhile, onscreen smoking is limited to a raspy-voiced sleazoid motel owner and a gang of pedophiles living under a bridge. No, I am not kidding. NBC's *Hannibal* learned the *Dexter* lesson well: nary a single whiff of smoke at all to disturb the children among its mounds of impaled and mutilated corpses and its weekly, but delicate and refined, cannibalistic feasting and recipe lessons.

of millions as the merest breath of tobacco smoke has been portrayed as deadlier than wartime gas attacks. These studies may not have spawned as many copycats or extensions as some of the previous headliners, but they still captured their own fifteen minutes of fame in the news and they still stuck in people's minds – just as they were intended to do.

As with that earlier examined research, the designs and the presentations made about the studies I'll analyze here contain fundamental flaws that are so numerous that they convey more confusion and baseless fear than knowledge or true understanding. There is obviously no way in a single volume to examine all the studies that Antismokers claim support their positions, but the fact that I could so easily find so many, so vulnerable, so execrable examples of truly bad science should make it clear that their "mountains of studies" are riddled with far too many weaknesses to be accepted as solid.*

Antismokers combine these flawed studies to produce the impression of massive and unified results that justify the promotion of smoking bans and the denormalization of smokers. They also display simple contempt for the honest development and communication of meaningful science and sound advice to both legislators and the concerned public. They are an insult to scientific methodology and a disgrace to the concept of a democratic society in which citizens depend upon truthful information in making informed choices.

Both *TobakkoNacht* and *Brains* emphasize the message that Antismokers lie. They may slip some bits of truth in amongst the lies, and occasionally the lies will seem small things – padding for the packages to hide the paucity of their contents – but the main product, the conclusion that smokers are a deadly threat to others and that widespread government mandated smoking bans are harmless, justified, necessary and inevitable ... *that* is a lie. And that is the message that I hope every reader of this book will walk away with and remember as they continue to see stories of yet "another new study" in the news.

---

* I could, of course, still be accused of cherry-picking, but as you will see later in the *Slings And Arrows* section, I have dealt with this possibility quite thoroughly on the Internet over the years by simply offering my opponents their own free choice of studies for analysis. The end results are the same.

At this point, we'll turn the analysis of these mini-abominations over to the poor tired fella who's just cleared off the main *corpora delecti* and invite you to enjoy *The Giblets*. Chianti served upon request.

Of Mice And Men

In 2007, a study was published in the *Journal of Periodontology* showing that subjects with periodontitis (a potentially serious gum disease) who were exposed to secondhand smoke were supposedly more likely to develop bone damage, the number one cause of tooth loss. According to the president of the American Academy of Periodontology, "This study really drives home the fact that even if you don't smoke the effects of secondhand smoke can be devastating."[286] That statement was followed up by a caution from DMD Dr. Kenneth Mogell, "... secondhand smoke has effects well beyond what we might have thought!"[287] And finally, as usual, a web page titled, "Secondhand Smoke Harms Children's Health," dragged the children in with a warning that "Periodontal disease [is] a leading cause of tooth loss.... No amount of secondhand smoke is safe for children. If you smoke, ... quit. It's important for your health and the health of your children."[288,289]

Sounds pretty serious, right? Kind of makes you think that parents who smoke will end up raising herds of toothless young geezers who'll be laughed at in school. And that's exactly the image that was meant to be conveyed. But what the headlines and quotes artfully hide is that the study was done using highly concentrated clouds of smoke and that the "subjects" studied were actually thirty-six specially bred Wistar *rats!* Dr. Nogueira-Filo *et al.* published their research with the intimidating title of "Low- and High-Yield Cigarette Smoke Potentiates Bone Loss During Ligature Induced Periodontitis"[290] Most newspaper readers and TV news viewers would never know the crucial facts: not only did the researchers study rats' teeth instead of children's teeth, but the rats were exposed to levels of smoke far beyond anything ever experienced by any child on the face of the earth.

Basically, the rats were locked in a chamber measuring eight-tenths of a single cubic foot while the smoke from ten cigarettes was pumped through it. Eight minutes later, the half-suffocated animals would be dragged out, revived if necessary, and set aside to wait until their next visit to smoking hell. They got these treatments three times a day for a month and were then mercifully put out of their misery and dissected to determine whether any symptoms of damage to their gums could be detected. Oh, and to load the dice even more, the rats were initially given ligatures (wounds) around which the gums were valiantly trying to heal despite the repeated tortures of the antismoking scientists.

How much smoke were they exposed to? Was it anything like what little Johnny and Janie might suffer while living with a smoking parent? Well, ten cigarettes in an eight-tenths cubic foot chamber would create the same concentration as a thousand cigarettes burned in a standard phone booth of about eighty cubic feet.* The experiment basically modeled a situation in which you would take your child to a dentist for a particularly nasty dental procedure, one that required deep stitches in his or her gums, and then brought that child home to be locked in a phone booth three times a day while you blew the smoke from 1,000 cigarettes at a time through that booth. The study showed that if you did that for a whole month, their gums might not have healed quite as well as if you hadn't done that – although that's only necessarily true if your child is a specially bred Wistar rat.

To bring it closer to the real world, say you lived in a two story home offering 1,000 square feet per floor and ten-foot-high ceilings. That home would have roughly 20,000 cubic feet of air space; the equivalent of 250 phone booths. So, to duplicate the conditions of the experiment, you'd need to sit down and smoke roughly three quarters of a million cigarettes a day while your little ones tried to watch the Teletubbies through the haze. And then, when you brought your sweet and somewhat desiccated little loves back to the dentist a month later, you'd find that maybe they weren't doing quite as well as you'd expected.†

But what about exposures at a far lower level? Say, 750 instead of 750,000 cigarettes a day? Or just 75? Or maybe only 7 cigarettes smoked several years ago by a maiden aunt on your front porch? Well,

---

* Phone booths are usually only around 70 cubic feet in volume, but for mathematical simplicity we'll assume we had some help from Dr. Who in building this one.
† That's assuming that anyone in your home or the surrounding neighborhood actually survived the 750,000 cigarettes per day regimen in the first place.

according to the claims of the newest research headlined in 2011, even that maiden aunt of times past could be a heartless killer of children yet unborn.

The research, headed by Dr. Virender Rehan of UCLA, a principal investigator at Los Angeles BioMed, was published in the July, 2011 issue of the *American Journal of Physiology*,[291] and it was, as usual for this sort of stuff, headlined all over the world. With "Thirdhand Smoke" abbreviated as THS here, some of the typical headlines were "THS Hurts Infant Lungs,"[292] "Unborn babies at risk from THS"[293] "THS Dangerous to Unborn Babies' Lungs,"[294] and "THS Affects Infant's Lungs."[295] Quotes from those stories included such notes as:

> **Prenatal exposure to toxic components of a newly recognized category of tobacco smoke ... can have as serious or an even more negative impact on an infant's lung development as postnatal or childhood exposure to smoke ... long after smokers have finished their cigarettes ... [THS is] a stealth toxin because it lingers on the surfaces in the homes, hotel rooms, casinos and cars used by smokers... babies [are] especially vulnerable to the effects of thirdhand smoke ... The dangers of thirdhand smoke span the globe ... more damaging than secondhand smoke or firsthand smoke ... pregnant women should avoid homes and other places where thirdhand smoke is likely to be found to protect their unborn children against the potential damage these toxins can cause to the developing infants' lungs.**

A scary picture. An invisible stealth toxin. More "negative impact" than secondhand smoke. More damaging than firsthand smoke. Babies especially vulnerable. A danger that spans the globe! Several stories emphasized the concept that even touching a surface in a home where smokers might have smoked a long time ago could lead to a lifetime of respiratory pain and suffering for innocent children not even born yet.

None of the stories went into any detail at all about the actual research other than occasionally mentioning a few of the scary chemical names of the "stealth toxins" left behind by smokers. In order to find out more, I had to request a copy of the study itself from the researchers. Given all that I've seen, I should not have been surprised by what I found. Nonetheless, I was.

The study didn't examine mothers touching surfaces in homes where someone smoked in the past. It didn't examine mothers being hugged by smokers. It didn't even examine mothers being touched by someone who might have once walked through a room where George Washington might have smoked a pipe before sleeping and leaving one of his ubiquitous signs.

The study once again simply examined rats. More specifically, it examined baby rats. More specifically than that, it examined tiny unborn baby rats who were bloodily ripped out of their mommy rats' guts and then torn wide open so that their innocent little unborn rat lungs could be yanked out, thrown on a slab, chopped brutally into teenie-weenie one millimeter cubes, and then soaked with concentrated solutions of chemicals that can just barely be detected at nanogram levels in nitrous-acid filled rooms where people have smoked heavily. Some isolated cells in those little bits of tortured fetal rats' lungs were then found to have undergone changes that could be related in some vaguely arguable way to abnormalities in human lungs that might sometimes correlate with conditions that were nebulously correlated in some way to asthma.

None of that information was given in the news stories. Almost none of it was provided in the study abstract. The little that was provided in the abstract would have been quickly overlooked by most reporters after they were hit with the following opening line: "The underlying mechanisms and effector molecules involved in mediating in utero smoke exposure-induced effects on the developing lung..."

If any reporters *did* manage to stay awake after that, rather than simply heading straight to the press release with all its juicy quotes (and no hint of rodents), they might have noticed the one mention of the word "rat" in the following excerpt: "Fetal rat lung explants were exposed to nicotine, 1-(N-methyl-N-nitrosamino)-1-(3-pyridinyl)-4-butanal (NNA), or 4-(methylnitrosamino)-1-(3-pyridyl)-1-butanone (NNK)." Even that explicit mention would have been blasted out of almost anyone's consciousness once they hit the phrase "breakdown of alveolar epithelialmesenchymal cross-talk, reflecting lipofibroblast-to-myofibroblast transdifferentiation."[*]

No one, anywhere in the world, reading any news stories that I was able to find in English, would have had the slightest clue that this study had done anything other than observe the horrible effects of

---

[*] No, I did not make any of that up.

thirdhand smoke exposure on human children who had suffered from their mothers' unwise visits to those George Washington tourist traps.

In case there is any doubt about how this research was misrepresented to the public, let me present a quote from ModernPregnancyTips.com,[296] a source that you would certainly expect to be concerned about presenting such information accurately. It's also a source that you would expect to be responsible about correcting unreasonable fears that might plague mothers-to-be. In the story on their website, though, not only did they quote the concerns of the original THS creator, Jonathan Winickoff, in warning about the danger of even "touching [the] toxic substances [on] contaminated surfaces," but they then compounded the fear by explaining the study's findings as follows:

> The researchers on the study looked at the way that these tobacco toxins affected the normal lung development in infants. They found that exposure during the prenatal period caused significant disruption in the normal lung tissue growth, which can lead to serious respiratory ailments later in life...

Note the use of the word "infants." Do you see *anything* there that even hints that they simply chopped up fetal rat lungs and poured chemicals on them? To make the irresponsibility stand out even more strongly, my attempts at adding corrective material for their readers – material that may have actually *saved* some pregnancies by relieving the emotional stress on expectant moms who read the article – were simply censored into oblivion by Modern Pregnancy Tips.* Did that censorship result in the death of any unborn children from the unjustified stress it surely caused some pregnant women? No one will ever know.

The final nail in the coffin that showed how the researchers wanted their research to be perceived can be seen in an article in *Science Daily,* where they state "[Dr. Rehan] said this is the first study to show

---

* If you happen to be near your computer you might want to compare http://tinyurl.com/iCytePage to http://tinyurl.com/CensoredPage – you'll see where the corrective posting was removed by the webmeister. The "iCyte" page copy is a dated capture of the original, saved at the iCyte website, as will be explained in more detail later.

(that) the exposure to the constituents of thirdhand smoke is as damaging and, in some cases, more damaging than secondhand smoke or firsthand smoke."[297]

So thirdhand smoke is now claimed to be more damaging (at least "in some cases") than *firsthand* smoke??? By the time we get to fifthhand smoke, thermonuclear weapons will have been rendered obsolete! As a statement to the media by a professional, and supposedly responsible, scientific researcher, such wording is simply unforgiveable. "The constituents of" may be an important qualifier to scientists, but as a media statement to the general public, the message was clear: a deadly threat from invisible traces left behind by smokers can be more dangerous than actually smoking.

When you look at the reality of the findings of the study compared to the ultimate social effects that this sort of misleading presentation will have on untold thousands, or even millions, of families, it is hard to avoid the feeling that the researchers engaged in outright criminal conduct roughly equivalent to screaming *FIRE!* in the middle of a crowded movie theater after seeing Humphrey Bogart take a puff in Casablanca.

Boys Versus Girls

Moving back to at least some semblance of human reality, 2011 also saw headlines about a new study showing that exposure to plain old secondhand smoke causes high blood pressure.

The catch here – and the point that made it especially newsprintworthy while plucking on our heartstrings and leading us to condemn heartless parents who smoke at home instead of at work or at the bar – was that the smoke was supposedly causing high blood pressure in children as young as eight years old.

Again, we saw the news flashed in lurid headlines in the *Independent, The Guardian, US News & World Report, UK Mirror, UPI, MedIndia,* and tabloids everywhere. "Passive Smoking Blood Pressure Risk,"[298] "Passive Smoking Raises Blood Pressure in Boys,"[299] "Secondhand Smoke Could Lead to Hypertension Among Boys,"[300] and "Secondhand Smoke Can Raise Boys' Blood Pressure and Cause Heart Disease."[301]

This study is another example of "Science by Press Release," or "Science by Conference Presentation."[302] Just as with the initial presentation of the Helena Study, the research had been neither peer-reviewed nor published. In this case, it had simply been presented at a meeting of the Pediatric Academic Societies. Dr. Jill Baumgartner of the University of Minnesota's Institute on the Environment conducted this study of 6,400 children aged eight to seventeen and found that boys regularly exposed to secondhand smoke at home had an average systolic blood pressure reading 1.6 mmHg higher than those who had no such exposure. Virtually every headline on every news story I have seen focused on this finding. However, virtually every headline on every news story also ignored the fact that the study had a stronger and actually more significant finding: exposure to secondhand smoke at home seemed to *lower* the blood pressure in girls by 1.8 mmHg.

Was that a meaningful drop? Should parents be arrested for child abuse if they refuse to blow healthy tobacco smoke into the faces of their little girls? Of course not. But the decrease observed in the girls was still larger than the increase observed in the boys and it was totally ignored in the headlines and in most of the news stories.[*]

In this case, the main researcher, Dr. Baumgartner, was responsible enough to note the girls' findings separately to the media, even though she tried to downplay it by indicating that the lower figures for girls might simply have resulted from them having some sort of special, sex-specific "protection" from the smoke exposure effects. That statement was still misleading, though, as "protection" would result in an effect in the same bad direction, but less. It would not produce an effect in the opposite and healthy direction that was actually greater.[†303]

Dr. Michael Siegel was not as forgiving in his own analysis of the study. In his article on the Baumgartner study he had this to say:

---

[*] The age range of the subjects – eight years to seventeen years – was also usually ignored. The stories invariably focused on poor hypertensive eight-year-olds about to expire from heart attacks while cavorting in the playground!

[†] A funny aside: Baumgartner noted that the girls' pressure reductions were also "a cause for alarm!" Anyone think that if it were found that an apple a day slightly lowered kids' blood pressure that it would be called "a cause for alarm!"

> [T]his appears to be one of the most blatant examples of investigator bias in a research study that I've seen. You set out to examine the relationship between secondhand smoke exposure and blood pressure. You find that exposed boys have higher blood pressure and exposed girls have lower blood pressure. You conclude that your findings for males are accurate and valid and that your findings for females are inaccurate and invalid. If this doesn't appear to be manipulation of data interpretation in order to reach pre-determined conclusions, than I don't know what does.[304]

The news headlines and stories, as always, simply ignored the important caveats and went straight for the blood and gore: innocent little children toddling toward early deaths from heart disease because of heartless and cruel smoking parents. The stories also ignored the fact that simply combining the boys' and girls' figures would have resulted in a conclusion showing no risk at all!

In any event, to suggest that blood pressure differences of such small magnitudes are so important in terms of the lifetime health of the children involved, particularly once one considers the potential for all sorts of confounding factors, is barely short of outright quackery.

Speaking of quackery, never underestimate the ducklike attributes of antismoking researchers. Shortly after finishing the Baumgartner analysis, I found another study by a Dr. Giacomo Simonetti of the University of Bern in Switzerland. Dr. Simonetti found higher blood pressures in children of smokers as well, although the increases were only a 1.0 mmHg systolic increase and a 0.5 mmHg diastolic increase. The incredible insignificance of these figures was made even more egregious by the fact that the researchers claimed to find them in children as young as *four* years of age![305] My favorite headline of the batch for this study appeared in softpedia.com. The article was titled "Smoking Exposure Early in Life Causes High Blood Pressure," thus leading one to believe that such exposure would actually produce clinically significant blood pressure problems later in life.[306] Again, the intended public "flavor" of the study can be seen in the very first line of the Abstract: "Hypertension is the leading risk factor for cardiovascular disease."

Cardiovascular disease from hypertension? Because of an arguable half millimeter blood pressure difference in a four-year-old? To claim that such a microscopic difference in something like blood pressure could be isolated down to a single weak factor out of dozens of possible confounding factors is absurd; and to suggest that such a difference has any real meaning in predicting the future lives and health of these children a half century or more into the future is beyond absurd. It's like claiming that four-year-olds treated to ice cream once a month are doomed to a lifetime of obesity and an early death from heart disease unless the little pre-fatsos are immediately put solely on a diet of celery and carrot sticks. The only purpose of publicizing this sort of research is to play with people's emotions; abusing our love of our children to achieve political ends.

What drives this sort of research? How much of it is truly being done because of researchers' concerns about children, and how much of it is being done because the researchers want the grant money and know that if they phrase their grant proposals properly, and produce the desired results, that they'll get even more money in the future? In the case of thirdhand smoke we can easily find the same sort of thing that we noted in the earlier Klein study – promising potential grantors results that will support specific public policies favoring bans *provided* that the researchers get the grant money.

If you visit the TRDRP (Tobacco Related Disease Research Program) pages on the Internet you can find details about their grant applications. Here's the quote from the one about thirdhand smoke that caught my eye:

> **[O]ur proposed work will be a critical step in a timely assessment of whether the THS exposure is genetically harmful to exposed non-smokers, and the ensuing data will serve as the experimental evidence for framing and enforcing policies prohibiting smoking in homes, hotels, and cars in California in order to protect vulnerable people.**[307]

Note the important jump from determining *"whether* the THS exposure is genetically harmful" to promising that "the ensuing data *will* serve as the experimental evidence for framing and enforcing policies prohibiting smoking in homes, hotels, and cars...." [Emphases added.] Note also the use of the phrase "vulnerable people." Like children perhaps?

Sad. And the grant-seeking trend of framing research design and interpretation in order to emphasize our care for "vulnerable" populations seems to be repeated in the next analysis, *Koolz For Kidz.*

Koolz For Kidz

The only thing worse than deliberately selling regular cigarettes to children is deliberately selling menthol cigarettes to poor, underprivileged black children. And the only thing worse than getting a grant for a study that preys on both our love for our children and our dislike for racism is probably not having a grant for a study at all.

And that's what brings us to look at the 2011 study, "Targeted Advertising, Promotion, and Price For Menthol Cigarettes in California High School Neighborhoods," by Lisa Henriksen, *et al.*[308]

It's been known for decades that the United States black population disproportionately smokes menthol cigarettes. Just why that's true has never been determined. Some folks like to claim it's genetic; some like to claim it's a product of marketing; and some just shrug their shoulders and admit they don't know why. But whether the cart came before or after the horse isn't that relevant to the world of present-day advertising, and it's well known that tobacco companies of today disproportionately advertise menthol cigarettes in neighborhoods and venues with high black populations.

Racial brand preferences are old news. Getting a grant today for a study showing that makers of Kools or Newports had ad campaigns aimed at blacks would be about as hard as getting a grant for a study examining whether there were more Kentucky Fried Chicken outlets in poor black urban neighborhoods than in rich white suburbs. However, once we "drag in the children," it's a whole new ballgame. On June 24, 2011, Reuters News Service reported the following:

> Stanford School of Medicine researchers compared pricing and advertisements for [Newports and Marlboros] at 407 stores within walking distance of 91 schools in California in a 2006 survey. ... school neighborhoods were increasingly likely to have lower prices and more advertising for Newport cigarettes as the proportion of African-American students rose.[309]

Reuters then went on to quote the lead researcher, Lisa Henriksen, as saying this showed "the predatory marketing in school neighborhoods with higher concentrations of youth and African-American students" by Lorillard Tobacco Company.

Given the likelihood that high schools in California tend to have at least somewhat larger black student populations in neighborhoods where there are larger adult black populations, the finding can hardly be called surprising. It's almost as big a no-brainer as some of the earlier study results we examined where more smoke was found in bars where people smoked than in bars where people didn't. But by focusing on the children, the researchers were able to get grant money for their work, were able to generate headlines, and were able to, as always, target evil Big Tobacco for still another dastardly scheme: addicting disadvantaged black children to menthol.

Oh. A final note. While I'm sure it was just a simple happenstance of pure coincidence, this research was made public just as the FDA was considering the possibility of banning menthol cigarettes.

Crystal Clear Distinctions

In December of 2010, the *Journal of Respiratory and Critical Care Medicine* published "Threshold of Biologic Responses of the Small Airway Epithelium to Low Levels of Tobacco Smoke" by Dr. Ronald Crystal *et al*.[310] The study claimed that even "very low-level exposure" to secondhand smoke could cause "changes in the function of [lung] genes [that] are the first evidence of 'biological disease' in the lungs." In speaking to the media about his research, Dr. Crystal expanded on this to state clearly that "this is further evidence supporting the banning of smoking in public places. … It doesn't matter if you are walking into a cocktail party where other people are smoking or if you smoke one cigarette a week. No matter what level of exposure you have, your lung cells know it and they are responding."[311] *Physorg.com* went on to state that Dr. Crystal described his work as giving "further evidence supporting the banning of smoking in public places, where non-smokers, and employees of businesses that allow smoking, are put at risk for future lung disease."[312]

The problem with Dr. Crystal's study is the same as that which has plagued many of the other studies we've examined so far. The raw research appears to have actually been done reasonably well. The researchers divided the population up into three groups: "smokers, who showed the highest level of the tobacco metabolites; nonsmokers, who showed none of these compounds; and a low-exposure group who fell in between." The problem arose when the researchers and the news articles began concentrating their statements on the "low exposure" group while neglecting to mention one rather important fact: *28 of the 36 "low exposure" subjects were actually smokers themselves!* Yes, you read that correctly; almost 80% of the group presented to the media as simply having "low exposure" to tobacco smoke were actually light or occasional smokers themselves. And when one actually examines the raw data, it becomes immediately apparent that the eight truly nonsmoking subjects in the low-exposure group were clumped tightly over on one side of the graphs with the data pool of the non-exposed nonsmokers, while the twenty-eight light/social smokers were mainly over on the other side, right next to the more regular smokers.

There was no scientific justification at all for the division of the subjects into those three categories (regular smokers, nonsmokers, low exposures), rather than either just two (smokers and nonsmokers), or four (regular smokers, light/social smokers, exposed nonsmokers, and non-exposed nonsmokers). Dividing it in the basic two categories would have shown the effect of smoking compared to not smoking; and dividing it in four would have shown, accurately and truly, that there was very little real difference between the unexposed and exposed nonsmokers.

What the division into three groups *did* accomplish was a propaganda trick making it sound as though a study had been done showing the danger of even "very low exposures" to secondhand smoke when no such thing had been found. The study certainly made no finding that could be reasonably interpreted as showing that passing exposures to secondhand smoke – the sort of thing most people would think was meant by the phrase "very low levels of exposure" – caused any harm at all to anyone. But it was presented to the public in a way that would make people worry that just briefly walking into a party where someone lit a cigarette, or perhaps just walking past a smoker on the street, might result in a deadly gene mutation leading to an agonizing death.

Think about why the division was made in this way. Can you imagine any reason for it *other* than to present and encourage a lie to the public for the purpose of procuring grants from antismoking funding sources or furthering an unjustified fear of low levels of exposure to secondary tobacco smoke?

The Sound of Smoke

Teenagers are a prime target for Antismokers, just as they were for the tobacco companies in years past, and as they still are for Big Alcohol today.* In the 1990s, antismoking organizations began realizing that their chants of doom and gloom about gory deaths from lung cancer, heart attacks, and emphysema among eighty-year-olds simply weren't ringing a bell with the youthful audience they wanted to reach.

Their solution? Create studies that would make teenage boys worry about their erections, teenage girls worry about their skin texture, and teens of both sexes worry about things like acne or whether or not their preferred sex partners would want to kiss them if they smoked. While I won't go into them all here – they're a bit peripheral to my main focus on secondhand smoke and smoking bans – antismoking grants actually did produce bogus studies claiming that smoking worked better than saltpeter in orphanage porridge when it came to turning boys into overcooked noodles; in turning girls into horror-show nightmares with catchers'-mitt faces for their proms; and in turning lonely kids of both sexes into wallflowers that no one would kiss without a gas mask.

But one of the more unusual studies I've seen involved just secondhand smoke and a particularly large and juicy target audience within the teenage world – those who enjoy listening to music. In early 2011, Dr. Anil Lalwani published "Secondhand Smoke and Sensori-neural Hearing Loss (SNHL) in Adolescents" and claimed that teens who spent time around smokers would no longer be able to appreciate all the wonderful tones of their favorite rock stars.[313]

Even a quick look at the study itself reveals a damning omission: there was no correction for social behavior differences among the smokers and nonsmokers studied. Obviously those teens who spent

---

* As I'm doing my final editing of this section today, I've just seen Budweiser's 2013 Super Bowl ad, starring a lovable little pony who grows up to be a huge and loving Budweiser Clydesdale. Can you imagine the high holy uproar that would ensue if RJR ran a Super Bowl ad with a cuddly little camel growing up to be a big lovable Joe Camel character? And yet smoking is legal at age 18 while drinking a Budweiser is off-limits until age 21.

most of their weekends sitting at home, quietly poring over stodgy textbooks safely away from wisps of secondhand smoke, would have less hearing loss than their counterparts who head-banged away in smoky mosh pits in front of thousand-decibel speakers. Any rational researchers initially noting the small observed differences in hearing acuity would attribute them to the abuse of heavy metal on the eardrums, but that effect was already well known and unlikely to garner grant money.

The researchers deserve some credit for not ignoring such a confounder completely, but their treatment of it, hidden right near the end – a trick we've seen before – while simply noting that "an indeterminate number of these individuals may have had CHL* instead of SNHL," was woefully inadequate. Note the phrase, "an indeterminate number," which, translated, means that literally *all* of the observed loss could be purely from too many nights of partying.

Such a reservation quickly disappeared as the authors basically brushed it off by noting that if there was a statistical association between smoke exposure and loud rock music, then it should be OK to largely ignore the music part and just look at the smoke – even if the physical exposure to the smoke had nothing to do with the loss! In the immediately following sentence they state, "This study demonstrates, to our knowledge for the first time, a relationship between tobacco smoke exposure and hearing loss among adolescents in the United States." At the risk of being repetitious, remember that they had just, immediately before that statement, indicated that the relationship might have *nothing at all* to do with causality.

It's as though researchers carried out a study funded by PETA (People for the Ethical Treatment of Animals) that determined that wearing leather jackets causes early deaths from fractured skulls – while quietly noting near the end that an "indeterminate number" of those deaths involved motorcycle accidents. Still, by attributing hearing differences to secondhand smoke exposure, the researchers were

---

* CHL, Conductive Hearing Loss, is the sort of damage that would occur from loud repetitive noise. SNHL, SensoriNeural Hearing Loss, is what they had hypothesized would be ETS related.

able to ensure good headlines and more research grants for future research silliness while also doing their bit to "Save The Children."[*]

As to how the study was used in the media, my favorite related story featured the headlined complaint that "Newspapers Aren't Warning Young People About Possibly Going Deaf From Smoking."[314]

Why does such a headline not surprise me?

Brain Fever: The Deadly Touch of a Smoker

The summer of 2010 brought us a new, bright, and happy study from researchers in Australia. The study – "published in international medical journals"[†] according to articles in the *Deccan Chronicle* and at the smoking-quit.info website – warned parents that if they allowed smokers to even *touch* their babies, that it could triple the younglings' chances of early and painful deaths from dreaded "brain fever."[315, 316]

The study itself actually gave a fairly reasonable explanation for the mechanism of contamination, noting that smokers might have a higher concentration of certain bacteria in their throats that are related to a rare condition known in Australia as "brain fever." It wasn't until I read it a second time though, that the numbers being cited as support suddenly caught my attention.

Basically what the researchers were saying, on the basis of this single study, was that if a baby in the backwoods of Australia was living with and being cuddled by smokers every day of its young life, there *might* be *one extra chance in three million* that such a disease – but only if you lived in the backwoods of Australia. Meanwhile, the all-important opening statement of the article ignores all of this to warn that parents, *all* parents, "will have to choose between their daily nicotine fix and the pleasure of kissing and cuddling their children," if they want to save those children from the horrors of brain fever.

---

[*] While probably irrelevant, it should be noted that this research was partially supported by the Zausmer Foundation, a major donor to the Settlement Music School in Philadelphia. It's rather unlikely the music school has any big stake in punk rock and heavy metal, but it's still an interesting coincidence.

[†] Unfortunately, as of June 2013 my best research efforts failed to turn up the "international medical journals" this scare story was supposedly based upon.

Is it right, is it moral, is it even barely excusable to frighten people about such extraordinarily unlikely risks in light of the likely repercussions involving deep human hurt, conflict, and injury to family love or social interactions? I believe that this study, in the way it was presented through the media and seemingly as it was presented *to* the media by the authors themselves, is little different than an outright crime.

And yet, instead of getting better, things simply get worse. In August of 2010, the University of California's Tobacco-Related Disease Research Program gave Professor Manuela Martins-Green a grant for a quarter of a million dollars to study the effects of thirdhand smoke exposure on wound healing.[317] Dr. Michael Siegel commented on this particular grant and similar ones in a critical blog entry on his website, saying,

> **Interestingly, a request for applications for research on thirdhand smoke put out by the Tobacco-Related Disease Research Program itself raises the issue of how objective the program is. The RFA "anticipates" the findings of the research in advance, arguing that the results will lead to policy enactment and that the research will demonstrate that thirdhand smoke contains "disease-causing toxicants." This is before the research has been conducted.[318]**

In this case, in the press release announcing the grants, the professor stated that in cases of regular exposure, she expected to find that "when injured, the skin will not heal normally and could even result in wounds that become chronic."[319]

Wounds that become chronic? From merely being touched by a smoker or their clothing? So, not only is the concern focused on simple brain fever or the mere disintegration of the skin of the tender infant in Granddad or Grandmom's arms, but now also on the threat that if the child has any sort of wound or abrasion on its body, that it might never heal after even a brief, but lethal, embrace! In a very real sense, "The Touch of a Smoker" has now been proposed to rival in the public mind "The Touch of a Leper."

At least in the case of Father Damien's leprous patients it was pretty widely recognized that touch was needed for the spread of the disease. But unlike the plague-bearers of the Middle Ages, mere avoidance of touch will not save today's youngsters from the contagion of the evil smokers. Even smokers who avoid touching their children and also go to pains to only smoke outside evidently pose a threat approaching modern biological warfare.

On October 28, 2011, headlines broke on NBC and other national outlets based on a press release – from the respected Center for Health Policy Research of UCLA – that sent chills down spines throughout the country: *"2.5M California Children Plagued By Secondhand Smoke."*[320]

It's not until you hit the fifth paragraph of the story that you'll discover that the qualification for over three quarters of those Plague Children is simply having a parent who smokes at all – even if only outside, with no nasty touching necessarily involved at all! Although the TV networks somehow seem to have missed covering the story, it appears that there must be traffic jams of body carts being pulled down the streets of Haight-Ashbury by gaunt men dressed in tattered rags, ringing bells and chanting "Bring Out Yer Dead!" for the millions of stricken children.

But wait! Over a million of the sad and tiny corpses have been resurrected to walk among us yet again, as antismoking researchers worked their mathematical magic. About a week after the story broke, a "correction" appeared.

> On Wednesday, researchers updated the findings, revising the total number ... of kids who have at least one family member that smokes outside dropped from 1.9 million to 742,000. "The errors were related to pooling three cycles of CHIS data. The center regrets the error."

Evidently, someone had discovered that the original bloated figures had a problem with "pooling three cycles of CHIS data."[321] It was just an innocent mistake of course, sort of like if you turned in your 1040 and counted your twins as six deductions instead of two. The Center's director of communications went on to assure the media and the public that "Such an error is rare."[322] Try that excuse with the IRS next time and see how far it gets you. After seeing the results of our Medical Examiner's dissections over the last hundred and fifty pages, I think you might agree that we don't all share the same dictionary when words such as "rare" are used by the antismoking research community.

Meanwhile, of course, most folks will simply remember the two-and-a-half-million Plague Children and their murderous parents while never knowing that the final number (for both one- and two-parent situations) was cut by over a million. Nor will they remember that the "plague" simply meant living in a home with a parent who smoked at all – even if only on the garage roof at the stroke of midnight on alternate Tuesdays. Most readers will never even realize that they've had their thinking "adjusted" to accept the concept that smoking in the garden is equivalent in some way to letting loose one of the Four Horsemen of the Apocalypse![323]

The Transmogrification of Tobacco Smoke

The Medical Examiner pauses and shakes his head as he considers the minds of those who could compare Europe's Black Death with the experience of children whose parents merely smoked an occasional cigarette on the back porch. The main corpses had been bad enough, but these tangled viscera are starting to unsettle even his own normally stable digestion. He is down to almost the last Giblet, and he looks at it with particular distaste. It has a mottled and ropy appearance, seeming to dangle tentacles that are still groping for sustenance as they decay in front of his eyes. It is clearly not an organ that simply survived on its own. It had been cancerous, and had been bleeding into the remains of the Giblets around it even as it lay murkily at the bottom of the pile. He hadn't thought that much could be worse than the diseased and leprous messes he'd just disposed of, but he thinks this monstrosity just might win the prize for worst of the evening's work.

People in general are familiar with tobacco smoke. They've seen it, they've smelled it, and many of them have spent good portions of their waking lives in some degree of contact with it. They know what it is.

So if I applied for a grant for a ban-promoting study that simply said I wanted to determine if there was more smoke in bars where people were smoking than in bars where smoking had been banned, I would probably end up with an empty bowl for my supper. It would be like asking for grant money to determine if people six feet tall were taller than people who were four feet tall; anyone over the age of five could probably provide the answer for free.

How then could I get at that money and at the same time strike yet another blow in the Crusade against smokers? Simple: I wouldn't say I was measuring smoke, I'd say I was measuring "toxic air pollution" instead! And that's exactly what a number of enterprising entrepreneurial researchers did in the first decade of the 21th Century.

Tobacco smoke is made up of many elements, some of which, – like nicotine – are almost unique to burning tobacco, and most of which – like nitrogen, carbon monoxide, or general particulates – exist at varying levels in air almost everywhere. Early studies of "tobacco smoke pollution" usually focused on nicotine or carbon monoxide in the air and, from a propaganda standpoint, produced disappointing results. Yes, the levels were higher in areas where tobacco was being smoked, but they were still far below any levels that had ever been shown to be even remotely harmful to human beings – hardly the sort of results that would support smoking bans.

However, one particularly well-defined and almost unique element of any sort of smoke is PM 2.5, particulate matter[*] which is made up of particles that are 2.5 microns or smaller in diameter. By way of comparison, the width of a human hair is roughly 100 microns. PM 2.5 has a particularly wonderful advantage for antismoking researchers in that it is produced in fairly large quantities by burning cigarettes. It's a good bit of what we "see" when we see smoke in the air and – quite important in terms a public relations standpoint – it is also one of the six "signal pollutants" measured by the EPA in determining general outdoor air quality and in warning people about unusual concentrations of air pollution.

Intrepid, grant-hungry researchers began traveling around to states, cities, and townships, wherever the smoking ban issue was hot and wherever someone would pay them to perform the research. For $70,000 or so (at least according to the Minneapolis Star-Tribune if ASH and Big Tobacco are accepted as reliable opposing sources in agreement on the figure) they'd happily march into town, hide their little sniffer monitors in their backpacks and sample the air inside of smoking and nonsmoking bars and restaurants to determine the comparative levels of PM 2.5.[324, 325]

And what did they find? They found that there was up to 99% less smoke in the air when no one is smoking! All they needed to do now was show that the smoke in the air of the smoky places was

---

[*] Sometimes you will see the term FPM 2.5, but that can be a bit misleading since the "F" can stand for "Fluorescing" – indicating a particular chemical type of particulate matter – or simply "Fine" – as in small. Using the simpler "PM 2.5" helps to avoid this confusion.

actually threatening the lives of the poor souls "forced" to work in them.*

For that of course they initially turned to OSHA – the Occupational Safety and Health Administration – the generally recognized arbiter of workplace health and safety issues in the US. Unfortunately for the Antismokers, OSHA has refused to recognize that the levels of smoke or of any of the elements of tobacco smoke in the workplace represent any kind of real risk to safety or health – even for people who breathe it regularly for eight hours a day, for 300ish days a year, for forty years. OSHA regulates the maximum safe exposure of workers to all sorts of things found in workplaces, sets safety limits for eight hours a day of continuous exposure, and even sets limits for peak one-hour exposures. OSHA has set such limits for many of the chemicals of usual concern in tobacco smoke, and generally a decently ventilated bar would have to literally stuff thousands or millions of smokers through its doors to even begin to approach those limits.

However, when OSHA was actually considering the general regulation of workplace tobacco smoke in the 1990s, some antismoking advocates felt it was taking too long in reaching a conclusion. The activist group ASH actually sued them in 2001 in an attempt to speed up their production of a standard. Rather than giving in, OSHA informed ASH that if it did produce a standard, it would likely be a standard that allowed for reasonable ventilation and smoking! ASH quickly dropped the lawsuit.

To hide their embarrassment over the reversal on their lawsuit effort, ASH's John Banzhaf issued a statement that tried to make it sound as though ASH actually had never wanted OSHA to pass a regulation. Banzhaf stated, "We might now be even more successful in persuading states and localities to finally ban smoking on their own once they no longer have the OSHA workplace smoking rulemaking to

---

* Interesting aside: for all the noise Repace *et al.* make about "no risk-free level of exposure," and the dangers of *any* level suddenly causing "acute cardiovascular events and thrombosis," they send their sniffer people to these deadly places after clearly stating in their official documentation that "No risk is expected to volunteers in collecting the data or to anyone in the restaurants during data collection via the air monitor." http://www.co.marquette.mi.us/departments/health_department/docs/Marquette_Air_Re port_Repace_2011.pdf.

hide behind." Since there's no real evidence that states and localities were ever hiding behind OSHA rulemaking, this seems to be a classic case of making lemonade from some rather sour lemons![326]

Thus we've been left with workplace safety standards that are set only for the components of smoke, all of which are well within safe OSHA limits even in the smokiest dives in the armpits of the dregs of the sleaziest skid row. So what did antismoking activists do? They simply decided they could ignore OSHA standards altogether and just appeal to a higher authority – the EPA.

This is where the transmogrification of tobacco smoke came about. It changed from merely being an element in the indoor air that might be harmful in sufficient concentrations, to being the equivalent of general environmental concerns about the deadliness of outdoor air pollution. Since the EPA has defined specific levels of PM 2.5 in outdoor air that are thought to be healthy, unhealthy, or even dangerous when breathed constantly over periods ranging from 24 hours to a full year, using it as a marker allowed antismoking advocates to appeal to the EPA's perceived authority – even higher in the general public eye than that of OSHA. And since PM 2.5 is generally almost non-existent in indoor air absent a general air pollution emergency, the copious amounts produced by even small amounts of burning vegetable matter can create impressive "pollution increase" claims.

If the grant-hungry researchers and idealistic antismoking activists simply talked vaguely about there being less smoke in the air when no one was smoking, the grant money wouldn't flow and the smoking bans would never materialize. This is where these researchers and activists pull the EPA into regions where angels would fear to tread. They take their one measurement of PM 2.5 and simply call it "air pollution" – as though it were actually the same chemical mix of automotive and industrial nastiness that chokes the air of our cities for days on end. Of course the PM 2.5 in outdoor air, mainly produced by automotive exhaust and general smog, has a very different chemical makeup than the PM 2.5 produced by the quiet burning of a few leaves wrapped in a thin sheath of paper. Equating the two is almost like saying that a teaspoon of sugar crystals is as deadly as a teaspoon of arsenic crystals because the crystals are the same size. But such a fine distinction can be overlooked if one simply wants to make a comparison designed to frighten non-scientists.

When the PM 2.5 levels in outdoor air reach above certain levels, it is generally a sign that other pollutants are also climbing. The EPA has set standards for six specific air pollutants that it believes fairly represent general air quality conditions that can adversely impact the health of either healthy or unusually vulnerable populations: ground-level ozone, fine particulate matter, carbon monoxide, sulfur dioxide, nitrogen dioxide, and lead. The EPA then assigns descriptive health ratings to these standards that range over six levels: Healthy, Moderate, Unhealthy for Sensitive Groups, Unhealthy for All, Very Unhealthy, and Hazardous. These ratings are generalized guidelines and are designed very specifically to be used only as averages over extended periods of time – exposures that individuals have no real escape from even when they are sleeping soundly in their beds dreaming of sugarplums and candy canes. And, as we saw in the earlier analysis of studies involving smoking in cars, the EPA is quite clear in warning researchers not to pretend that measurements of exposures of less than 24 hours to these pollutants have any meaning at all when compared to EPA's 24-hour guidelines.

Bars without smoking in them would usually have PM 2.5 levels in the 24-hour Healthy or Moderate ranges, unless the outside air itself had higher levels. Bars that allowed smoking would rarely be in the Healthy range for this particular measurement, and during peak periods might even find themselves climbing outside the Moderate range into one of the Unhealthy ranges unless they had a high-end ventilation/filtration system in place. The first step in the Unhealthy range, "Unhealthy for Sensitive Groups" states that at this level, "Persons with heart and lung disease, older adults* and children are at greater risk from the presence of particles in the air," while the second step, "Unhealthy for All," states that for 24-hour continuous exposures, "Everyone may begin to experience some adverse health effects, and members of the sensitive groups may experience more serious effects." Note two things very specifically:

---

* "Older" is not clearly defined, but as we saw earlier, 69-year-olds are now considered just to be "middle-aged."

(1) These "Health" labels refer *only* to 24 hours of continuous exposure and have absolutely *no* meaning or defined implications for exposures of an hour or two or even sixteen hours a day every single day. And,

(2) The EPA does not define what the adverse health effects might be (they could be as innocent as a runny nose or a scratchy throat), nor does it make any attempt to specify any probability attached to the vague word "may."

By ignoring the importance of time averaging, by ignoring the importance of defining terms like "adverse health effects," and by ignoring the essential difference between the simple burning of a few leaves and the complex chemical pollution caused by automotive and industrial processes, antismoking researchers can then go on to write up a publishable study about the "Percent Of Reduction In Air Pollution In Smoke-Free Bars In (Anytown) USA," collect their grant money or fees from the local MSA tax-funded antismoking group, and pack up their snake-oil show to perform in front of another set of gullible rubes in another gullible town with pockets deep enough to pay their fee. One small element, PM 2.5, commonly and copiously produced by any burning organic material (even blown-out birthday candles!), has been transformed into being equivalent to deadly air pollution in general despite its likely lack of chemical similarity to the actual atmospheric PM 2.5 that is the EPA's true concern.

>>> Cleanup 7 <<<

The Examiner wipes his brow in weariness. There is one final lump that he's been avoiding. It had been largely hidden among the assorted aorta and organs he'd been working on, but it is now all that is left on the slab and he can no longer ignore it. It looks heavy, and seems to be filled with a dense, grayish matter that promised to be both disgusting and, by its appearance, possibly even poisonous.

With grim determination, he reaches in to cup it in both hands and bring it over to the rarely used eighth slab in the morgue – the one usually reserved for disease victims who were suspected to have died from virulent pathogens. He gives a surprised grunt upon lifting it, however: although it appeared heavy and solid, it weighed almost nothing. It couldn't have weighed less if it had been nothing but a soap bubble! He reaches up to switch the morgue's ventilation and toxic fume analysis filters to high and pulls down his face mask, but his startlement causes him to be a little less careful than his norm. As he settles the globule onto Slab VIII *he fails to notice a discarded scalpel blade beneath it.*

As soon as the film of the bubble touches the sharp edge of the scalpel, it bursts, and its lack of weight is explained. The gray matter within is not a solid mass at all, but some form of deadly looking vapor! It expands quickly, and since he'd been leaning closely over the slab, a significant tendril of it shoots right up under his face mask. He jerks back reflexively, already knowing it is too late. Whatever it is, he has certainly inhaled enough to give himself a frightening dose. He instantly tenses as he listens for the alarm bells of the fume sensors and can't understand why he doesn't hear them. Was the vapor so fast acting that it has already destroyed his auditory nerve pathways? Is he about to die?

A second ticks by, then two. As he stands there, frozen in place, he is amazed to note that the thick gray vapor has almost instantly disappeared. Not even the morgue's advanced ventilation systems should have been efficient enough to achieve that. Just a week ago, the obnoxious cigar-chomping police chief had bulled his way in to look at a corpse and it had taken the system almost a full minute to totally eliminate the smoke. But all of the vapor is already gone, his ears hear no alarms, and his nose had registered nothing more than the ghost of a slight sweetness as he'd inhaled the concentrated gray vapor. His eyes move to the fume analysis dials, state-of-the-art devices that should have instantly registered anything their sensors detected. But, as usual, they simply showed the various levels of metabolic poisons he himself was breathing out, mixed with the usual gasses typically emitted in the early decomposition of human remains. For a moment it seemed there might have been a slight spike of an alkaloid, perhaps caffeine or nicotine, but whatever it was had barely registered at the nano-particle levels necessary for detection, and at those levels, even strychnine or cyanide would be harmless.

As his heart rate returns to normal, he takes a deep breath of relief and looks back at the spot on Slab VII *where the bubble had rested. He'd missed seeing it earlier, but under the bubble there had been a small label reading, "E-Cigarette Emission: control sample from a nearby non-crime scene." He has read about these e-cigs, and it has always seemed to him that, once the hysteria and histrionics were removed, nothing was left but insubstantial attacks on almost invisible and harmless traces of vapors. Still, it is his job to at least attempt an examination, so he reaches for his most delicate instrumentation to inspect the air-space where the evil-looking but innocent bubble had been just a moment earlier.*

Slab VIII

Of Vapors And Vapers

Social control is best managed through fear.

- **Michael Crichton,** *State of Fear*

Safe Cigarettes?

Back in the 1970s, there was a push to create safer cigarettes – ones that might still arguably be bad for one's health, but ones that would have far lower levels of the substances thought to be responsible for the higher rates of lung cancer and heart disease in smokers. Dr. Gio Batta Gori, Director of the Smoking and Health Division of the National Cancer Institute, was a strong proponent of the strategy at that time and is often identified as one of its leading figures.[327]

For a combination of reasons, partly legal* and partly due to pressures from an antismoking movement that rejected any long-term alternatives to an outright abolition of smoking altogether, the scientific search for safer cigarettes was largely shelved in the 1980s. There were a few exceptions – Philip Morris' Accord brand and RJR's Premier and Eclipse cigarettes were notable – but no major breakthroughs in terms of public marketing and acceptance were achieved.[328]

Suddenly, in the early 2000s, something new broke through to the public consciousness – cigarettes that didn't contain tobacco at all, but were actually designed to deliver nicotine and/or flavorings in a

---

* After all, if Big Tobacco legally admitted it could make cigarettes "safer," then it was opening itself to lawsuits by legally admitting it had been making "unsafe" cigarettes to begin with!

mist produced by heated propylene or vegetable glycol. These products came, not from the big American tobacco companies, but from inventors and entrepreneurs in China.[329] These new "smoking" devices were referred to as "electronic cigarettes" – or e-cigs for short – and while safety claims were usually carefully avoided in official promotional communications, they were generally pushed and perceived as much safer than traditionally smoked cigarettes. Few could argue that a "cigarette" that produced nothing except a pure and quickly dissipating mist with mild scents, flavorings, and sometimes small amounts of nicotine, was likely to be as harmful to health as the highly complex chemical output of burning vegetation.

Unfortunately, e-cigs ran into several road blocks. First of all, they were usually manufactured in China, a country that had received bad press in the US and Europe for problems with safety standards involving lead levels in toys for children[330] and unsafe melamine levels in chocolate candies.[331]

Secondly, since they were neither food nor drug nor, arguably, tobacco product, it was quite unclear who should be in charge of actually evaluating them and certifying them for safety. It wasn't even clear what "safety" meant with regard to the products; what if they were as "safe" as coffee with its dose of caffeine, or as "safe" as sodas with copious quantities of sugar? Opponents of e-cigs argued that even if they were a hundred times safer than smoked cigarettes, it would still be wrong to introduce them on the public market at a time when the professed goal of tobacco control policy was moving strongly toward eventually eliminating all types of tobacco use completely.

Thirdly, with a comparatively small and financially anemic funding base themselves, e-cigs came up against two well-established, powerful, and wealthy competitors who were not at all happy to see their entrance into the nicotine delivery system market: Big Tobacco and Big Pharma. Neither industry saw their interests well-served by these Chinese interlopers, and both threw their weight behind efforts to question, badmouth, and officially limit their new competitors.[*]

---

[*] A recent change in this scenario has been introduced as Lorillard Tobacco Company purchased a particular brand of e-cigs (Blu Cigs) and has rolled out a large-scale promotional campaign for them that includes TV advertising.

And finally, the e-cig ran into trouble because of its own success in emulating smoking and pleasing smokers. Although they emit almost no odor beyond slight traces of usually pleasing fragrances, part of their appeal to smokers is that they *appear* to produce a smoke-like vapor from a device that *looks* like a cigarette and which is *"smoked"* like a cigarette. Despite the absence of odor, that mere visual appearance caused a frenzy in an antismoking movement that was heavily emotionally invested in the war against smoking.

As I explained several years ago in *Dissecting Antismokers' Brains*, the antismoking movement is a hydra-headed beast composed of activists, researchers, and entrepreneurs fighting smoking for quite a varied set of reasons. For those Antismokers who are neurotically fearful of smoke, or those with strong emotional biases against smoking because of believed harms to themselves or loved ones caused by smoking, or those who simply enjoy using antismoking arguments as a way to exercise control over the behavior or morality of others, the e-cig was a blatant affront: something that could not be accepted or tolerated as a comfortable and approved public substitute for smoking.

The usual antismoking campaign arguments were trotted out, but they largely failed. The primary attack, of course, was dictated by the past success of Antismokers in playing the emotional "save the children" card. E-cigs were a big topic of conversation on Internet discussion boards and, since most of the e-cig companies were small operations working on a franchise-style model, there were a lot of amateur entrepreneurs working without real guidelines while trying to attract customers to their particular brands and websites. Whereas Big Tobacco bent over backwards to avoid any real appearances of Internet marketing, there was no similar degree of control over the actions of thousands of small sales operations by individuals. The idea that these products were being promoted to customers in a medium so clearly aimed at innocent children (since we all know that adults rarely use the Internet, right?) gave antismoking advocates an open avenue for attack.

Aside from their traditional "child modeling" complaints that the young and vulnerable might actually see people using e-cigs and engaging in an activity that looked like smoking, the anti-vaper* crowd had a rough time pushing the children argument though. The start-up prices for vaping were high since an e-cig was not something you could buy for a quarter or easily bum from a friend. While the liquids that produced the vapor in e-cigs were cheap, the e-cigarette device itself tended to have a price tag up in the fifty- to hundred-dollar range. It simply seemed highly unlikely that youngsters who'd never smoked would suddenly start laying out that kind of money in significant numbers and begin taking up this new habit on their own.

Enter: The Blu Cig

Enter the second decade of the twenty-first century however, and two new factors suddenly entered the fray: relatively cheap and disposable e-cigs promoted on a much larger and more easily accessible scale and, of course, "scientific studies" that would, blithely flying in the face of reality, attempt to prove that this new plague was infecting our schoolchildren.

After forty plus years of never seeing anything that resembled smoking portrayed in anything but a negative light during televised commercial breaks, football fans of the 2013 Super Bowl were treated to a commercial showing a good-looking actor happily puffing out clouds of what looked like smoke into the air while extolling the "Blu Cig."[†] Evidently, a traditional tobacco company, Lorillard, had seen the possibility of a truly safe cigarette in the future and had bought up the Blu Cig franchise for its own, possibly even hoping to transfer its primary business from the evil of smoking to the relative goodness of vaping!

You might have expected that the antismoking community would have been ecstatic at such news. But no, the new product *looked* like smoking, and that was quite enough to set off a wave of fury.

---

* E-cig users self-adopted the term "vaper" to describe themselves since, in reality, what they were inhaling and exhaling was far closer to being a simple and quickly dissipating vapor than a chemically complex combustion product.

† Not being a Bowl fan, I didn't get to see this myself, so I'm operating on secondhand accounts and YouTube replays here. I believe another brand, NJoy, may have grabbed an ad spot there as well.

Antismokers let loose a firestorm of complaints that children would now begin seeing people vaping and would be encouraged to take it up themselves, particularly since the advertised Blu Cigs were a variation on what had previously been offered: they were individually purchased at fairly low prices (roughly $10 per unit as of early 2013) and offered as many "puffs" as a traditional pack of cigarettes for that price.

Of course, following their past track record of success in molding public thought, antismoking advocates began looking to develop scientific evidence of one sort or another to back up their political contentions. One of the more prominently headlined studies along these lines was carried out and published in February, 2013, by Dautzenberg *et al.* in the *Open Journal of Respiratory Diseases*.

Dautzenberg surveyed young people in Paris and developed a seemingly impressive set of statistics regarding twelve- to fourteen-year-olds. The title of the study discloses its general orientation rather openly: "E-Cigarette: A New Tobacco Product for Schoolchildren in Paris."[332] The term "schoolchildren" usually calls forth the imagery of the six- to ten-year-olds whom we see playing in the schoolyards every day, and that imagery clearly seems to be what the authors intended to convey. The actual study sample mainly concerned a slightly older set of subjects, up to those hitting twenty years of age. The study abstract itself, however, which is all that most media reporters ever really read before writing their stories, makes no mention of the study having covered anyone over the age of seventeen.

Rather than go into great detail here, I'm going to just look at the most newsworthy statistic that was promoted on the basis of this study, and show how, just like all the other antismoking stratistics we saw earlier, it doesn't quite portray what it claims to. The stratistic in question is this one, "Among the twelve to fourteen-year-old school children, 64.4% of e-cigarette experimentation was by non-smokers." That claim, standing by itself, makes it seem as though e-cig use is spreading like a plague among nonsmoking children.

There are two things to consider about this particular figure, however. First of all, there's a trick dealing with age range. Most twelve- to fourteen-year-olds are generally nonsmokers. Therefore almost *any* adult-oriented activity we might do a survey about would show that "most" of the kids doing it would be nonsmokers as well. If we surveyed six-year-olds, the figures would go even higher: probably *98%* of six-year-olds who've "experimented with e-cigarettes" have been nonsmokers! Of course there are probably only a few dozen such six-year-olds in the world, but hey, don't let reality get in the way of a good stratistic, eh? That 98% number would be good and scary... just as the 64% one was.

Secondly, the stratistic is also weak in that, by itself, it says nothing about how experimenting with an e-cig goes on to affect future behavior with regard to actually smoking. The antismoking spin on the issue is that kids who try e-cigs will go on to a lifetime of sordid drug addiction as full-fledged tobacco smokers. In reality, the opposite argument could be made just as strongly; after experiencing the relatively pleasant and mild sensation of puffing water and glycerol vapor from an e-cig, it's quite possible that these experimenters would later try regular tobacco cigarettes and immediately find them, by comparison, deeply unpleasant. It's quite possible that young people will eventually be *less* likely to become regular smokers if they have gone through an early experience of trying e-cigs!

As a closing note on this study, I'll simply quote the entirety of their "Conclusion" as stated in the abstract:

> For teenager's[sic], e-cigarettes have become not a product to aid quit[sic] tobacco but a product for experimentation and initiation of cigarette use. Regulation is urgently needed to control the emergent use of this new tobacco product by children.

In addition to making the jump from a total lack of *any* statistical or scientific evidence tying e-cig experimentation to future initiation of smoking, the authors make the jump is made from teenagers to children, with children being in the prominent position of the traditional last word on the subject.[*] If the authors of the study had

---

[*] In fairness here, I should note that my concern about terminology may be an artifact of translation. The two grammatical errors noted in the block quote of the Conclusion, as well as difficult phrasings throughout the study, indicate that the language use may be inadvertent. Given the background history in this area, however, I hesitate to freely offer the benefit of the doubt in this case.

emphasized the complementary statistic – one which didn't get as much play in the media – that only about 6% of twelve- to fourteen-year-olds had ever actually tried e-cigs at all, the news stories wouldn't have had quite the same impact in fanning the flames against this newly perceived threat to the antismoking dynasty.

In that regard, it may also be worth noting two final points. First, the dating at the reference site seems to indicate that this study may have been somewhat rushed to press; less than one month separated its initial submission and its full publication. (Remember the ten-week wait Mr. Kuneman and I experienced in 2005 before getting even a simple rejection by the *BMJ* for our study disputing the smoking ban heart attack miracles?) Second, a trace on its funding seems to ultimately lead back to the French Ministry of Health, a Ministry that has experienced great difficulty in promoting acceptance of its antismoking efforts among the smoky French population. One has to wonder whether this study would ever have been published anywhere if it had concluded that e-cigs were of no concern and bore no relation at all to children and smoking.

Chemical Warfare

Aside from the universal propaganda tactic of dragging out the children, Antismokers ran into real difficulties in vilifying the e-cigs. Nonsmokers didn't notice any disturbing odors, fires were very unlikely unless someone grossly misused them (although there was one headlined story about an unfortunate soul who tried modifying the battery of an early e-cig model and ended up with it exploding in his face while he was experimenting with it), there were no butts to be stomped and left to litter the ground, and cancer – even among the most dedicated vapers[333] – was unlikely since there were virtually no significant levels of any known carcinogens in the vapors produced by the product*[334] and since it was highly unlikely that heart disease or

---

* Generally speaking, as noted in the reference indicated, you would have to take roughly 500 puffs from an e-cigarette to get the amount of TSNAs that you would get from a single puff of a regular tobacco cigarette.

other serious ailments would ever be notably affected by vaping – particularly in terms of any claims about "passive vaping."

The relative innocence of e-cigs, and their potentially huge role in the tobacco harm reduction strategy of inducing cigarette smokers to find safer alternatives for enjoying their nicotine, was ignored in the push to prevent their spread and quash their growing popularity. Toward that end, even before the introduction of the cheap disposables and their potential for casual use by teens, more studies were commissioned with the unstated, goal of finding justifications for banning the e-cig monster from sight and mind. As of early 2013, there have been at least two major studies of the chemistry of e-cigs that have made the headlines; one focusing on the potential for harm to the user, and one focusing on potential harm to the innocent bystander. The first was done by a generally unimpeachable American organization, and the second was carried out by researchers at a respected German scientific institute. Both, of course, cast the products in a negative light.

The FDA Study

To start with the most damning of the two, let's look first at the FDA analysis done in 2009.[335] The FDA examined two of the most popular brands of e-cigs at the time, NJoy and Smoking Everywhere, and eighteen varieties of the e-liquids* that were used in them. It is the FDA study that has been the basis for the widespread claim, mentioned in almost every news report I've ever heard discussing e-cigs, that e-cigs have antifreeze in them.

In reality, the study found that *one* of the eighteen varieties of e-liquid tested had very small amounts of diethylene glycol, a chemical that is not toxic in the quantities measured – about 1/10th of a milliliter (mL) in even a heavy daily use of 10mL by a vaper – but which is also used, just as water or salt is, in antifreeze. The actual toxic level would be about 100 times that amount, and since it would regularly be excreted far faster than absorbed, it would never build up to toxic

---

* E-liquid is the general term for the flavored mixtures of glycol, water, and often nicotine, that are vaporized to produce the visible emissions exhaled by vapers.

levels in vapers even if it was indeed normally present at such quantities in their products.

Unfortunately, the researchers do not seem to have performed what a normal person might consider to be the simple and responsible action of extending their research a little by purchasing a few more samples of the type of e-liquid that they had found the contaminant in. It would seem to have been a pretty obvious and important question to examine in the course of their study. Was such a contaminant a one-time occurrence? Or was it something likely to be found on a regular basis? Personally, I cannot imagine why any responsible researchers in such a situation with such a potential for significant public impact would not have simply ordered/purchased a few more samples from different sources to see if the unusual result was repeated. Since they were only testing for a very few elements, and since they had specially picked diethylene glycol to be one of those elements, it was obviously something that they expected they might find and which they must have been concerned about.

So why would they avoid just a little bit more research while they were all set up and running? Could it be that they *wanted* a negative finding about e-cigs? Could the diethylene glycol have just been a simple experimental error that could not be shown to be common? Could that be why the researchers deliberately avoided testing further samples? Could it be that each run of such a test is so unusually and so enormously expensive that their research grant couldn't pay for even a few extra tests on top of the hundreds* that they'd already run, despite the importance of the question?

I don't know the answer, though from what I've seen in this field over the years and have examined so far in these *Slabs*, I might have my suspicions. It's been almost three years since the FDA study was done. E-cig usage has grown to millions of consumers. And yet no one at the FDA or elsewhere seems to have replicated this result despite over a

---

* This study tested levels of over a dozen compounds in 18 separate brand samples of various e-liquid flavors, formulas, brands, and manufacturers. Running a test for just a single compound using a few extra samples of that one brand would certainly not have been onerous, and there is no indication anywhere in the report that the researchers might have tried to double-check the anomalous result in any way at all.

dozen further investigations by other researchers?* My concerns would would seem to be well-founded.

Now the FDA studied a number of other elements as well, but an analysis of their report seems to speak far more to the safety of the e-liquids than to their dangers with regard to these elements. For example, one measurement was of a tobacco smoke element known as anabasine. The LD 50 (i.e. lethal dose for half the population) of anabasine for a 110-pound person is about fifty milligrams. The FDA study declared that it was able to detect at least 14 parts per billion (ppb) of anabasine in about half the samples tested. Even a heavy vaper going through a full 10 mL bottle of e-liquid per day would be getting at most only 0.14 picograms. To reach fifty milligrams a vaper would have to vape roughly three billion such bottles in a day. As a threat, compare that to the few dozen ordinary aspirin tablets needed for the same effect.[336]

A standard garden hose puts out perhaps ten gallons of water a minute at the highest pressure. If we substituted e-liquid for water in someone's water system and then tied them down and taped a garden hose into their mouth, how long would it take for us to pump a 50% lethal dose into their system if we turned the faucet on full blast? If I've done my figures correctly, it would take roughly three million minutes. That's 50,000 hours, or about 2,000 days... over five years. At that point, about half the subjects tested – if they had all been prevented from going to the bathroom while that water gushed into their bellies at a rate of close to 15,000 gallons a day for five years – would have died from the amount of anabasine found in the vapor of some of the e-cigs. For the rest of the e-cig users it might take considerably longer.

So here, within the single most "damning" study presented to the public as part of the drive to condemn e-cigs, we see just how weak the evidence really is. It's exactly the same sort of game we've seen repeated over and over and over again with regard to claims about secondhand and thirdhand smoke, but now it's being aimed at people who are trying to make their lives healthier by quitting smoking and switching to e-cigs for enjoying the nicotine and sensation of smoking

---

* As referenced in the overview by Cahn and Siegel, which will be discussed in a few pages.

instead. And it's being aimed with the same manipulative intent and the same disregard for true human welfare – and all because people who are vaping simply *look* like people who are smoking.

The Schripp Study

The second heavily covered chemical analysis study on e-cigs was one done in mid-2012 by Dr. Tobias Schripp of the Fraunhofer Wilhelm-Klauditz-Institut in Germany. Dr. Schripp published an analysis of the elements that someone in a room with an e-cig vaper might inhale.[337] The conclusions reached in his study are presaged in the very first paragraph of its abstract where it states that "'passive vaping' must be expected from the consumption of e-cigarettes." Although his general results showed virtually nothing that could reasonably be construed as active, passive, or even other-dimensional significant exposure to any chemicals of concern, his conclusion was indeed that "passive vaping" existed and needed further serious examination.

His research was broken into four parts:

(1) "Smoking" several e-cigs in a small chamber (eight cubic meters, or about six feet wide, long, and high) with three air changes per hour – roughly the air exchange rate of a very poorly ventilated bar, or perhaps 1/10th that of a well-ventilated bar or restaurant.

(2) Having a real cigarette smoked in that chamber.

(3) Having the vapors from a single inhaled puff of the e-cig or tobacco cigarette exhaled directly into a sampling bottle with a volume of ten liters (i.e., about 1/3rd of a cubic foot).*

---

* This would provide an exposure roughly equivalent to having one's head inside of a single sealed "helmet" the size of a milk crate and sharing that space with the head of a friend who exhaled two full puffs from an e-cig. A rather unusual exposure unless you and your friend have a "thing" for milk crates and e-cigs. (Exposure calculated on the basis of an average thirty-liter 13" x 13" x 11" crate with volume reduced by the presence of two good-sized human heads displacing about five liters apiece.)

(4) Having undiluted e-cig vapor puffed via machine directly into the ten liter bottle.

Dr. Schripp attempted to measure about twenty separate elements in the air of the chambers after the vaping of three different types of e-cigs and the smoking of one tobacco cigarette. The general results indicated that the researchers were generally unable to obtain any meaningful increase at all for the chemical compounds they were looking for. The chemicals tested for included such standard second-hand smoke "scare" chemicals as formaldehyde, benzene, acetaldehyde, isoprene, and toluene. They did find some increase in acetone and acetic acid in that chamber during the e-cig phase of the testing, but, as the authors themselves admit in the body of the study, "The rising concentrations of acetic acid and acetone during e-cigarette operation may also be attributed to the metabolism of the consumer." In any event, although acetone is usually noted by Antismokers as being "used in nail polish remover," it is also commonly found in fruits and vegetables... a fact usually overlook for some reason. And acetic acid is just the vinegary part of vinegar and many people enjoy sprinkling not just a few micrograms, but many billions of micrograms of this deadly tobacco smoke component on their supposedly healthy salads.

In other words, any "pollution" given off by the e-cigs was generally so low and inconsequential that it was pretty much totally indistinguishable from the pollution one would be exposed to from simply breathing in a room where another person was also breathing. It could well be argued that "passive breathing" in a room with two or three friends, even without the real and additional threat of biological pathogens involved, might be far more dangerous than passive vaping while sharing a room with a single vaper!

The researchers also included one element that was expected to be quite high for e-cigs in particular: 1,2-propanediol. Even when measuring for that very e-cig-specific element, they were unable to measure it at any level above one microgram per cubic meter – the lowest limit of their measuring equipment – in the "room chamber" for the e-cigs, although they *were* able to measure it at 112 micrograms per cubic meter for the tobacco cigarette.

When the puffs were blown directly into the 1/3rd of a milk crate-sized bottle by a machine with no chance of any absorption by a person actually vaping, the researchers had better luck. They were finally able to detect and measure significant amounts of 1,2-propanediol at perhaps 1/10th the concentration that one might find in a full-sized test chamber after a tobacco cigarette was smoked. Two other VOCs (volatile organic chemicals), 1,2,3-propanetriol and 3-methylbutyl-3-methylbutanoate, were also measurable in amounts that would translate to about one-half of a microgram per cubic meter and about five nanograms per cubic meter in the larger chamber. Both such measurements would have been below the one-microgram concentration limit of detectability of the researchers' equipment in that larger chamber. One other VOC, 2-butanone, was measured at a level of two micrograms per cubic meter in the test chamber after the e-cigs were used. That's a level roughly one million times safer than OSHA requirements, but is still noted by the researchers as one of the VOCs of concern for "passive vaping." Evidently when workers are told by OSHA that they are safe, they should not only be concerned, but be a million times concerned!

Meanwhile, roughly five micrograms of nicotine per cubic meter was measured in the test bottle – equivalent to about eight nanograms per cubic meter if it could actually have been measured at all in the larger chamber. At the standard half cubic meter or so per hour relaxed normal human inhalation rate, a researcher would have to sit around reading this book to an e-cig user in such a chamber for roughly 200,000 hours (about a hundred standard worker-years) to inhale the nicotine that a smoker would get from a single standard Marlboro cigarette.

The researchers also measured aerosol particle numbers in the milk crate chamber and found them to be about ten times less and 50% smaller than those for the tobacco cigarette. Of course the chemical composition of those particles would be quite different. The researchers believed that the aerosol from the e-cigarettes was almost exclusively the comparatively quite safe supersaturated 1,2-propanediol, while the particles from the tobacco cigarette contained hundreds of complex compounds thought to be harmful.

Despite the fact that the researchers were almost totally unable to measure anything at all within even a small and poorly ventilated room with e-cigs being used, their conclusion, based largely upon the milk crate part of the experiment, was that "Overall, the e-cigarette is a new source of VOCs and ultra-fine/fine particles in the indoor environment. Therefore, the question of 'passive vaping' can be answered in the affirmative."

The milk crate experiment did show that *something* existed, even if it was generally too small to be measured as distinct from ordinary exhaled human breath with its waste products. Thus indeed, "passive vaping" exists ... even if it's much safer than "passive breathing," with its bonus load of pathogens. In terms of comparing it to "passive smoking" ... well, as noted just a few paragraphs earlier, be prepared to sit around for a century or so to get the full-flavored enjoyment of that single Marlboro.

Unless you're in a well-ventilated bar of course. Then it might take you closer to a thousand or even ten thousand years, and by that time the bartender might have thrown you out unless you ordered a second wine spritzer while pointing your little sniffer tube around at other customers.

The Cahn/Siegel Literature Review

As with secondhand smoke studies, there are several different types of published research works dealing with e-cigs. We have just examined two of what would be called "primary studies," i.e., studies where researchers actually perform experiments or measurements according to a scientific design and report their results and conclusions. Beyond that level, there are two "overview" types of studies: meta-analyses – widely used with regard to secondhand smoke studies – and literature overviews.

Meta-analyses take a large number of smaller studies covering populations that are sometimes of very tiny or very different sizes and types where the subjects are exposed to differing amounts of a variable over different periods of time, and then analyzed with different corrective factors for different additional, and possibly confounding, variables aside from the main one being studied. If all of that sounds a little squishy, there's a good reason for it. Meta-analyses are indeed squishy, and while they can sometimes be useful in seeing a larger picture, they're rarely meaningful or reliable enough by themselves to serve as a real basis for public policy decisions. So far there have not been enough studies of the right sort on e-cigs for anyone, to my knowledge, to have attempted a real meta-analysis on their chemical output.

However, in 2011, two researchers, Zachary Cahn and Michael Siegel[*], examined over a dozen primary chemical analysis studies and produced the second type of overview study: a literature review.[338] They found no other instances of diethylene glycol present in e-cig liquids or vapors. Nor did they find any levels of any substance at all that would normally be considered troubling from a health standpoint. The only substances of any possible concern appeared to be the tobacco-specific nitrosamines (TSNAs) that had also been pinpointed in the FDA work (e.g., the butanone and anabasine mentioned earlier). Siegel and Cahn noted that the levels measured in the direct vapor inhaled by the e-cig aficionados were generally on the order of a hundred to a thousand times less than the levels inhaled by a regular smoker or absorbed by an oral tobacco user. At those levels, obviously, any concerns about exposures due to passive vaping are about as realistic as worries about getting skin cancer from starlight.[339†]

However, quite aside from any scientific analysis, there is always a purely political question when any question regarding anything related to anything that looks like, smells like, or possibly even makes one think about the concept of smoking. Private enterprises and governmental bodies are free to make all sorts of arbitrary rules about behavior, as long as those rules don't contravene constitutionally protected rights. Thus, we've seen New Jersey ban the use of e-cigs in workplaces in 2010,[340] we've seen moves toward banning them throughout the entire European Union,[341] and we've seen private

---

[*] Yes, it's the same Dr. Siegel. Although he's an Antismoker at heart in terms of workplace bans, he's also honest. In this case his honesty has led him to support e-cigs as a form of tobacco harm reduction and has thus unjustly earned him new accusations of being supported by Big Tobac... er.... I mean Big E-Cig.
[†] Worries about the dangers of starlight may not be too far off in the future. Just as I was giving this manuscript its final reading while waiting for format conversion of the cover art I discovered that July 3rd was being celebrated as National Stay Out Of The Sun Day with people being encouraged to "enjoy indoor fun." Seriously. Now you see why I'm giving up writing satire in this area. There's simply no point to it anymore. I surrender.

restaurants and airlines ban them. In one 2011 incident on an Allegiant Airlines flight a passenger tried to convince the crew that his e-cig was not actually a real cigarette and thus should not be covered under their smoking ban regulations. To buttress his argument with the concerned flight attendant and make the distinction between something real and something that simply looked like something real, he used his cell phone's Internet connection to call up a picture of razor blades. The crew looked at the picture, looked at his e-cig, and quickly diverted the plane to an unscheduled emergency landing at Norfolk Airport. Evidently, things that look like cigarettes and pictures of things like razor blades are now given treatment similar to that given to outright terrorism.[342]

Some later online discussions of this event brought up the fact that some airlines allow e-cigs unless another passenger objects on the basis that it looks like smoking. It seems that children sometimes fly on planes and one of them might see someone vaping and thenceforth be doomed to trod the dreary path toward a lifetime of suffering and addiction.

We were forced to wonder though, if a Muslim or a Mormon passenger saw someone drinking a cup of water and felt that the water looked like vodka, could that passenger then demand the removal of the offending cup?

>>> **Cleanup 8** <<<

The Medical Examiner steps back from Slab VIII. *He is happy. The last chunk has been removed from the bucket and he is free to go home.*

But as he picks the bloody pail up by the handle, his eye catches something strange and small and glistening in the bottom, peeking out through a final bit of muck. He doesn't recall ever having seen anything like it before. He reaches in with his forceps and delicately lifts it out. It is an **egg**! *And even more notable, it appears to be an Antismoking Egg! An unfertilized but beckoning grant, just waiting for an intrepid, sperm-like researcher to pierce its distended skin and fertilize the nutritious dollars bulging within, bringing it to full and evil life. The appearance of the egg, so bloated and ripe, prompts the Examiner to put it under the microscope lens before heading out, despite the waiting Dexter and his exquisite knives.*

The strangest and most repulsive thing about the egg is that its innards are deformed in a visually oscillating and unsettling Mobius-type convolution. The nutritious lucre filling it to bursting actually has a shape that reminds the Examiner of old manuscripts illumined with the world-enveloping, tail-eating snake that promised to go on for ever and ever while devouring itself. And in this case, the Examiner muses, the image is certainly apt, for each grant gotten, each study born, will surely yield up more grants and more fatally flawed studies ad infinitum *for as long as the Antismoking Crusade lives.*

Stifling his gag reflex, he slices open the egg's swollen membrane and forces himself to examine its innards more closely...

L'Oeve Ouroboros

In June of 2010, a new golden ring was hung out for the grabbing by antismoking grant grabbers. The Tobacco-Related Disease Research Program of the University of California announced it wanted to give away $3.75 million to enterprising researchers willing to cook up studies examining the deadly twin threats of thirdhand smoke and cigarette butt environmental pollution![343, 344]

Almost four million dollars to examine two of the silliest concepts in the entire arsenal of antismoking research – thirdhand smoke – which is about a thousand times more innocent than the effectively nonexistent threat of secondhand smoke – and littered cigarette butts which wouldn't kill off the majority of exquisitely ultrasensitive water fleas even if every smoker in the world dumped their butts in the water for *twenty-five million years!*[345, 346]* They were basically telling researchers: "Come up with some good study designs to scare the bejeezus out of people regardless of the facts, and you can walk right home with almost four million Washingtons snuggling nice 'n comfy 'n warm in your pocket!" (Who says thar ain't no Santa Claus?)

In all seriousness, this was indeed what was being asked for and offered. Antismoking organizations saw the twin threats of thirdhand smoke and environmental pollution as potentially powerful new weapons in their arsenal that could be developed quickly and convincingly to play upon public fears. And what better way to manufacture the needed evidence than to dangle millions of dollars in front of eager grant-seekers?

The Examiner shakes his head sadly as he places the egg in the disposal bucket with the rest of the slop. There is no point in trying to destroy it, since he knows it will simply be replicated. As he turns out the lights and leaves for a warm home with the simple joys of Dexter and his smokingly sexy sister, he breathes a small prayer that perhaps sometime in his life, the beast responsible for laying these diseased eggs will be slain. He can only hope.

---

* The origin of the standard sound bite about butt pollution involves putting a tenth of a butt in a liter of water and observing the death rate of water fleas. The sound bite breaks down when you determine how many butts it would take to pollute the world's 1.4 sextillion liters of water. At current smoking rates it actually would take 25,000,000 years. If the research examined health effects upon edible fish instead of water fleas, the time frame of concern moves up into the billions of years!

CONCLUSION

Well, you've now seen dissections of a good sampling of the types of studies that Antismokers claim offer support to their drive for smoking bans. True, there are thousands more out there, all greedily sucking from the putrid ponds of corporate, tax, and grant money, but they're all basically the same. If they promise psychic visions that will prove the need for widespread government-mandated smoking bans, there are almost always hidden levers and fog machines buried deep away inside them. If nothing else, I hope I have at least left you with the realization that such studies most definitely do not form a strong foundation for laws that are ultimately so divisive and harmful to the weave of our social fabric.

But there's one more question to seriously take under consideration. Have I perhaps engaged in the same sort of cherry-picking that I've accused Antismokers of doing? Could it be that the studies I selected for analysis and criticism just happen to be the weak and deformed runts of the litter? Is it possible that my claims about their importance and their headline status are simply exaggerated?

How can I prove this is not so? One Internet debater repeatedly stated that I was not worth listening to unless I dissected *all* the antismoking studies supporting bans. Obviously, as noted at the start of *The Giblets,* such a task is impossible without an army of researchers and millions of dollars, but there is another way I have found to defend my honor and my assertions.

Anyone visiting the Internet and Googling my full name will find well over a thousand references popping up; largely various news boards where I openly identify myself near the beginning of any thread where I'm seeking to expose the lies behind the press releases and headlined studies. My activities on those news boards have given me the opportunity to make repeated challenges[*] – not just three or four

---

[*] These challenges will be examined in more detail in *Slings And Arrows.*

times, but hundreds of times – to those writing in opposition. One of my most basic challenges is fairly simple: Pick the three absolutely best, most powerful, and most respected studies you can find that show real harm to people's health from the level of smoke exposure that would normally be encountered in any decently ventilated venue today, and be willing to defend those studies against criticism.

This challenge has rarely been answered with anything other than expletives, *ad hominems*, and references to bathroom habits.* Antismokers generally don't react well when they're asked to back up and defend their claims and sound bites with real research.

I think it is most appropriate to close this section with a simple, but very sad, statement by a former editor of the prestigious *New England Journal of Medicine,* Marcia Angell.[347]

> *It is simply no longer possible to believe much of the clinical research that is published, or to rely on the judgment of trusted physicians or authoritative medical guidelines. I take no pleasure in this conclusion, which I reached slowly and reluctantly over my two decades as an editor of* The New England Journal of Medicine.
>
> - Dr. Marcia Angell

---

* This book began with a discussion of Godwin's Law. With a hat tip to Mr. Godwin, I was inspired at one point to create "McFadden's Law." McFadden's Law closely parallels the structure of Godwin's, but states, *"As a Usenet discussion on smoking grows longer, the probability of a comparison involving urination or defecation approaches One."* Anyone who's spent much time reading Internet smoking discussions will appreciate the truth of this.

Tired of Antismokers Taking A Bite Out Of Your Life?

Then DO Something About It!

Slings And Arrows...

To be, or not to be: that is the question:
Whether 'tis nobler in the mind to suffer
The slings and arrows of outrageous fortune,
Or to take arms against a sea of troubles,
And by opposing end them?

- Hamlet, Act 3, Scene 1

Peace is indisputably preferable to war, but sometimes conflict is pushed upon those who would otherwise seek to avoid it. Fortunately, aside from isolated incidences to this date, most of the conflict between smokers and Antismokers has avoided violence, but it still clearly exists.

Children will say "Sticks and stones may break my bones but words will never hurt me," but adults know better: sometimes words can be the most effective slings and arrows of all the weapons in the political arena if they're used wisely and effectively. This section of *TobakkoNacht – The Antismoking Endgame* presents a number of examples of how these nonviolent weapons of social change can be used "to take arms against" an enemy who, on the face of things, seems to be far more powerful than oneself.

Forget "nobler in the mind to suffer." Fight.

One of the greatest problems those of us in the Free Choice movement have had to struggle with over the years has been an almost total lack of funding to get our information and message out to the wider public. In the 1990s, the big tobacco companies experimented briefly with supporting grassroots efforts and with setting up larger groups such as the three million-member National Smokers Alliance,[348] but popular support for such groups seemed to disappear with the Master Settlement Agreement and with Philip Morris pulling their funding support in 1999, as PM "announced that it would withdraw funding after the NSA made an ethics complaint about John McCain."[349] Most grassroots Free Choice groups since then have generally stayed away from Big Tobacco money, perhaps feeling the money hurt more than helped, as Antismokers were always quite quick to point their fingers and accuse local spokespeople of simply being mouthpieces, shills, or puppets of the companies. Meanwhile, most ordinary everyday smokers see no need to support activists fighting bans since they simply assume, wrongly, that they *are* well-funded through the tobacco companies!

The fact that the antismoking movement gets most of its power from mountains of confiscated Big Tobacco dollars directed to it every year from the Master Settlement Agreement goes totally unnoticed by the media and the general public. Antismokers' statements and the spokespeople making them rarely have their content and motivations questioned despite their almost total dependence on such funding. The antismoking lobby has hundreds of millions of dollars a year to waste, and they have spent a lot of it on buying thousands of hours of TV commercial airtime to pound their sound bites relentlessly into our brains from cradle to grave.[350]

Those wonderfully produced commercials, artistically pleasing billboards, and all the newspaper and TV ads that appear before ban votes are not the product of grassroots volunteers working for free; they are the result of work by people who are paid very well to influence our thinking and our votes. The political power of the workers, lawyers, and high-salaried CEOs employed at Crusading organizations flows directly from such payments – it is not as altruistic a system as it appears at first glance. Indeed, when some funding cuts for New Jersey's antismoking efforts were being considered about ten years ago, Paul Wallner of the Medical Society of New Jersey said *"Everything stops. There is no money,"* despite the fact that NJ Breathes, their main advocacy group, was still slated to get $14 million for its activities in 2002.[351] When fourteen million dollars is referred to as "no money," there is clearly something very, very wrong.

And just one month after the World Trade Center disaster, an antismoking group's report to Congress recommended that they "triple the federal excise tax on tobacco to fund anti-smoking efforts."[352] Somehow, I believe most Americans were thinking about our government using money for more pressing priorities at that time. Yet in that single year, 2001, the AMA documented 883 million dollars being given to tobacco control through the MSA.[353] The total has decreased somewhat in recent years, but is still, as of 2012, on the order of 500 million dollars annually, and is well-supplemented by more funds from the NicoGummyPatchyProductPushers who see smoking bans as a natural boon for sales.[354]

This sort of funding is not available to those of us fighting on the other side of the issue and so such fighters have largely had to depend on either very cheap or completely free forms of communication to get their message out. Three of the best such avenues of communication open to Free Choicers are the letters columns of newspapers, the blogs and news-comment boards of the Internet, and direct communications to lawmakers and news reporters themselves.

How do those avenues compare? Well, Internet postings generally don't reach a lot of people (I would guess an individual post probably averages less than a hundred or so viewings over its lifetime, with many never being viewed by anyone but the writer.) but they still can reach dozens or even hundreds in the right places. Letters to the editor, although being read by many more, are usually limited to a hundred or so words being offered in response to full-length, headlined articles espousing the opposite point of view and are often never even printed. And communications to lawmakers and reporters, while occasionally effective, generally seem to simply be ignored as they are drowned out by the greater numbers produced by the professionals at the large and well-funded pro-ban organizations.

Still, for the price, such avenues are about all we've got, so we've learned to make the most of them. The "enemy" may have entire tank battalions attacking from the darkness, but we've got slings and arrows that we can launch into that night. Individually they may generally not do much damage, but occasionally one can land precisely right and have an effect, landing like a flaming arrow smack in the middle of a munitions dump and blowing their plans all to hell 'n back for a while!

Even aside from those individual starbursts, every effort represents a grain of sand thrown into the treads of an advancing tank column. With enough grains gumming up their gears, eventually that column will be brought to a grinding halt.

This section will offer examples of such weapons and show how useful they can be. Not everyone has a million dollars in their pocket or a tank in their backyard, but a pen or a keyboard can do a helluva lot of damage to an enemy that depends on ignorance and lies for its power. *Slings And Arrows...* showcases several varieties of missiles.

(1) Those *Launched Into The Abyss* are letters to newspapers whose editors seem to resolutely avoid giving the microphone to any who seriously question their paper's editorial opinion on this subject. While it does not stand alone in this, my experience has shown that *The New York Times* is one of the most egregious offenders and the letters in this section were all sent to, and ignored by, that paper.

(2) Those *Launched Among The Stars* are letters to papers that take their responsibility of representing citizens' voices on their letters pages more seriously, even if those voices might land a significant counterblow to their main editorials. I offer several dozen examples, addressing different issues in different ways, many of which actually did see publication in various papers around the world. I hope they will provide both ideas and aid for Free Choice proponents as they respond to the sound bite arguments of the Antismokers.

(3) Those *Launched In The Trenches* was originally meant to be a collection of arguments on Internet boards that would offer examples of counters to antismoking claims and sound bites. After completing *The Abyss* and *The Stars* however, I realized such examples had already been pretty well laid out. So,

instead, I decided to examine some of what one finds on those blogs and news boards in terms of hatred against smokers and censorship of Free Choice postings, while also offering some thoughts on effective styles of communication in such forums and an examination of one particularly effective type of such posting that nicely shows Anti-smokers' inability to answer even the simplest and most straightforward challenges to their sound bites and claims.

(4) Those *Launched O'er The Ramparts* are examples of extended and official communications to the sort of government bodies that often believe their official status as "the good guys" will render them immune from criticism and questioning. The first example marked my first effort at formal testimony to a government body in the Free Choice battle, and was presented on May 31st, 2000, to Philadelphia's City Council. The second was a presentation to the British House of Lords as they considered granting Wales unlimited power to ban public indoor smoking. The third, and last, was a sharp response to the Department of Health in Findlay, Ohio. All three exposed the truly execrable bases beneath smoking bans and, while staying carefully polite, did so without pulling any punches.

Welcome to the School of Free Choice Weaponry! Whether you're firing your verbal projectiles into an abyss, among the stars, in the trenches, or on the ramparts of government castles, just remember: we may not be able to afford hi-tech televised communications, but the pen (and the voice) can be mightier than the sword!

Launched Into The Abyss

As noted in the introduction to this section, my first selection of writings is drawn from letters I've written over the past ten years to the *New York Times*. The *Times* has a long, though sometimes disputed, reputation of being a "Paper Of Record," supposedly an even-handed dispenser of objective news that keeps its opinions where they belong – on the editorial page.

Editorials are the privilege given to those with the money, skills, and ambition to start or buy newspapers and magazines. While the bulk of their publications are given over to the news and articles that attract their readers, editors reserve a special space for themselves in which they can pontificate to the masses with the historical freedom of any pope or king. Newspaper editorials are widely read because most people are looking to see what the smart people think about the issues of the day. They look to the thoughts of the editors to provide, not necessarily the "right" opinion, but at least an educated opinion that they can openly agree with and repeat without looking foolish or, on occasion, disagree with while appearing both brave and intelligent in being able to mount an argument against such an important and official source.

Ideally, editorial views are supposed to stay strictly within those editorial pages while news department sub-editors act independently and with the single goal of reporting the news accurately. In reality, that rarely seems to happen. The sub-editors know that their long-term job security depends upon pleasing their liberal or conservative bosses, so their news stories often reflect those leanings to some degree. Some newspapers are better than others in not letting that slant spill blatantly over into their news reporting – and some are worse.

Unfortunately, when speaking about smoking, smoking studies, smoking bans, and tobacco control in general, the respect for a solid division between slant and news seems to have broken down badly at the *Times*. In the section examining thirdhand smoke, you saw how that tilt not only sometimes produces yellow journalism aimed at the basest sorts of fears and prejudices, but also how it seems that the editors and publishers are unwilling to offer a correction to that slant even when called clearly on it in public. They clearly know that such calls are unlikely to reach beyond an extremely small fraction of their original readership, so they're fairly safe to ignore.

The letter I wrote to the *Times* on thirdhand smoke was not an isolated instance. Over the past ten years, I have written dozens of letters to the *New York Times*. My average record in getting letters published by newspapers other than the *Times* and the *Washington Post* (another offender in this area) is about one publication for every four attempts. The *Times* gets far more letters and far better written letters than most newspapers, so I would not expect to see one out of four of my efforts published in their pages. I would feel quite honored if even one out of eight were published. But if I wrote sixteen and saw not a word in print, I would begin to feel slighted. And by the time I hit thirty-two I would feel that the *Times* had abdicated its responsibility to offer reasonable space to a responsible voice that was being raised against its editorial position.

My fifty or so letters, written, revised, and carefully edited, were shot into the air and fell into an abyss... a black hole where they were seen by no one other than those responsible for rejecting them.

Still, that is a newspaper's prerogative. Maybe my letters were not as good as I imagined them to be, or maybe they were simply unimportant or uninformative, or perhaps just poorly written or boring. Maybe I should take up creative doodling in my spare time rather than sweat over the placement of commas and semicolons and the selection of various phrases.

I'll let you be the judge. Here is a small selection of some of the letters I have written to the *New York Times* on smoking issues in the twenty-first century. This book may not reach an audience of the size available through the letters pages of the *Times*, but perhaps it will reach an audience that may make better use of the ideas and infor-mation offered. Read these letters and see if you agree with the *Times*' editorial rejections of them and their companions. If you feel those rejections were wrong, maybe you'll think twice about where to seek unbiased news and a fair balance of opinion in the future.

The SCHIP Of The Damned

One of Barack Obama's first official acts as president, and one that he hailed proudly as one of the most important that would be carried out under his leadership, was signing the SCHIP bill for children's health care.

The SCHIP bill seemed to be something few could fault. It would provide health care to millions of uninsured children, and best of all, it wouldn't cost anyone a penny... except smokers. Regular smokers got hit with a 150% federal tax increase to fund SCHIP, and one sub-group of smokers – a group that was arguably the poorest well-defined minority in the entire country, those smokers forced to roll their own from scraps of paper and shreds of tobacco – got blasted with a 2,150% tax increase from about a dollar a pound to over twenty-four dollars a pound.

This sort of blatant unfairness – hitting an extraordinarily poor minority group with a tax that would inordinately simply transfer their money to those often far better off (SCHIP has been criticized for subsidizing families making up to $75,000 a year) – would have sparked revolutions among other groups in other times in other countries, but American smokers accepted it with barely a whimper, largely because the media rarely offered any support for criticizing the source of the funding.

And, as I noted in the *Author's Preface*, when the President relegated smokers to the status of vermin by denying he had raised any taxes on any "people" since taking office, the *Today Show* interviewer didn't even raise an eyebrow.

Of course once a group accepts extortion without resistance, it's a clear target for a repeat – and now in 2013, that's exactly what Obama is planning with his SCHIP 2 and its planned 94 cent hike on pack taxes and a further doubling of the taxes on loose tobacco.

No Room for a Minority?

Dear Editor,

Your article, "House Votes to Expand Children's Health Care," devoted only two sentences out of thirty-six to concern about how SCHIP will be funded. Neither sentence noted that part of the funding will come from a two thousand percent increase in taxes on the poorest of the poor in America, those who cannot even afford to buy a pack of cigarettes at current prices but must roll their own from loose tobacco instead.

A two thousand percent tax increase on an unorganized, unrepresented minority group that is, on average, probably far poorer than such groups as blacks, hispanics, seniors, or single moms – and not space for a single mention of it in an 850-word story by the Times.

You should be ashamed.

Puffing On Polonium?

While Polonium 210 was the *demon du jour* when the *Times* ran the story on thirdhand smoke covered earlier in this book, the evil element had actually been featured in the *Times* three years previously, on December 1, 2006, and it also featured an emphasis on that hapless Russian spy. In this case, however, "Puffing On Polonium" was written by a prominent Antismoker who seemed to be seeking absolution from the antismoking community for his somewhat "inconvenient" book, *The Nazi War On Cancer*.[355]

Dr. Robert Proctor's book, published in 2001, presented well-researched and indisputable details on one of the largest antismoking movements in modern history: the one implemented by Hitler's Third Reich as part of his drive to perfect the ultimate "Healthy Aryan Race."

No one could fault Proctor's book for its research, but it was a profound embarrassment to the United States antismoking community at a time when it was trying to pooh-pooh the idea that its push for smoking bans was somehow too authoritarian and would spread to private clubs, the open air, or even into people's private homes. In the early 2000s, American Antismokers were getting desperate. They'd gotten undreamed of amounts of money from the Master Settlement Agreement's "invisible tax" on smokers, but their push for smoking bans seemed stuck in California. The last thing they needed was a reminder of their Nazi roots.

So, when the Russian spy was dosed in November of 2006, Dr. Robert Proctor must have been quite pleased to be able to write a highly profiled Op-Ed piece for the *New York Times* that would help reestablish his credentials as a card-carrying Antismoker worthy of future grant money and the respect of his annoyed colleagues.[356]

The Radioactive Russian

Dear Editor,

Dr. Robert Proctor's opportunistic slap at smokers following the KGB assassination of a Russian bureaucrat is a classic example of the scientific distortion routinely employed to promote smoking bans.

He correctly notes that a cigarette emits .04 picocuries of radiation, but then goes on to speak of the Litvinenko assassination while adding up all the cigarettes smoked in the entire world to develop a scary picture. Examining individuals produces a very different story.

According to MSNBC, Litvinenko was dosed with five millicuries, or 125 billion cigarettes' worth of smoke. A worker in a decently ventilated modern casino would inhale at most about 1/10th of a cigarette per shift, requiring 1.25 trillion days to reach that 5 millicurie dose. At fifty workweeks per year, that translates to working for five billion years to get the Russian's dose.

But wait: polonium's half-life is 138 days, so we'd have to suspend the laws of physics itself or the dose would simply *never* be reached!

So when Dr. Proctor glibly claims, "No one knows how many people may be dying from the polonium part of tobacco," he's correct only in the sense that it's too outlandish a thought for anyone but an avowed Antismoker to even consider.

Through Rose Colored Glasses

In March of 2004, Antismokers were declaring New York's smoking ban a blinding success. Hyperboles flew with wild abandon and the ban was hailed as being more popular than the NY Yankees and healthier than peanut butter and jelly on rye. To listen to the media running story after story featuring happy owners, healthy and smiling workers, and shopkeepers with overflowing tills, it seemed that Bloomberg's Ban put the storming of New York by the Beatles forty years earlier to shame.

I wasn't in New York to witness how the ban was going myself, but I was very active on an email list with over a hundred bar owners in the city, and it was clear that the real picture was far from the fairy tale being pandered off to the masses. Unlike the antismoking groups putting out press releases and setting up media opportunities with little children in brightly colored antismoking T-shirts, the bar owners had only one concern: that their bottom line was being blasted to hell and back by Bloomberg's Folly. The problem lay in finding a way to communicate that fact to the general population.

Once more, I turned to the *Times* in the hope that maybe my previous rejections had just reflected the vagaries of editorial discretion or the quality of my entries. And once more my letter never saw the light of day.

A Smoking Ban Solution!

Dear Editor,

I believe I have found a solution for the smoking ban debate that will make all sides happy.

If the Antismokers are telling us the truth, there should be no problem at all with simply eliminating the official ban. According to them, business has boomed under the ban, establishments are making more money than ever, and the owners and patrons of almost every bar in the state are in love with "clean air."

If the Antismokers are telling us the truth, very few bar owners would opt to lose money by allowing smoking if the ban was lifted, and the bars that remain smoke-free would see profits soar as customers fled such smoky dens.

If the Antismokers are telling us the truth ... but, of course, we know they are not. They lie about this, just as they lie about the "deadly threat" of even the merest wisps of secondhand smoke.

The lies and exaggerations of these Antismoking Crusaders need to be fully exposed by a responsible media that rises above scare-mongering. Challenge the Antismokers by proposing the elimination of the ban. If they are NOT lying, they should be just as happy without it as they are with it.

Antismokers: Made, Born, Or Both?

While researching material for *Dissecting Antismokers' Brains*, I came across an article about some particularly interesting new research. It indicated that a certain gene tended to be defective more often in nonsmokers than in smokers.[357] Evidently, this gene allows smokers to better metabolize nicotine so that they enjoy smoking. When this gene is defective people who try smoking are less likely to find it enjoyable, so they go on to become non-smokers, ex-smokers, or perhaps even Antismokers! Of course this sort of thinking wasn't very highly approved of, but eventually, over a decade later, one clever researcher, sponsored by the NicoGummyPatchy folks at the Robert Woods Johnson Foundation, came out with work indicating that tax increases were failing to reduce smoking "for individuals with specific genotypes," and went on to suggest that "alternative methods may be needed to further reduce use."[358]

After *Brains* was published, another study came out.[359] This one, written by a Senior Systems Engineer named Jay Schrand, followed up on earlier research by Nasir Naqvi of the University of Iowa Carver College of Medicine.[360] Both Schrand's and Naqvi's work indicated that smokers who suffered a particular type of brain injury (similar to an insular stroke) tended to no longer enjoy smoking and often went on to quit.

Combining these pieces of research with the increasingly rabid desire of parents to produce perfect children who will go on to lead perfect lives free of imperfect vices produced the following letter.

The Final Solution To The Smoking Problem?

Dear Editor,

Benedict Carey's Jan. 25th article, "Scientists Tie Part of Brain to Urge to Smoke," notes that a specific type of brain injury to the insula can help a smoker quit smoking, but it oddly ignores the closely related question of whether people with such brain injuries are more likely to be nonsmokers in the first place.

Previous research has indicated that nonsmokers are more likely to have a defect in a particular gene that helps the body metabolize nicotine.

Combining the two approaches – inflicting brain injuries and inducing genetic defects in our offspring – might go a long way toward producing a generation that doesn't smoke at all. And after that, we can deal with teenage promiscuity and a fondness for drinking Irish whiskey.

The wonders of modern science.

When Used As Directed...

It's unlikely that there is an adult American alive today who has not heard that "Smoking causes more than 400,000 deaths a year!" But very few have any idea just where that number comes from.

Basically, it's simply an educated, but still fairly wild, estimate spit out by the SAMMEC computer program discussed earlier in *Stratistics*. It's not based on real death certificates or mounds of body bags, and the number can go flying up or down with just the simple change of a few assumptions, variables, or formulae. Antismokers have long been proud of being able to display the highest "body count" of any societal affliction, as it assures them of ever-increased funding, and SAMMEC gives them that body count in spades.

However, a few years ago their pretty little applecart was in danger of being upset. The Centers for Disease Control came out with a report indicating that the computer-estimated body count from obesity might actually have grown higher than that for smoking – Deaths Due To Eating!

Well, hell hath no fury like a woman scorned and neither doth it have the bite of an antismoking lobby that feels its bankroll threatened. Leading Antismokers wrote letters and applied appropriate pressures and quickly got the official obesity body count "corrected" to reflect their desires. Thus the following letter....

Deaths Due To Eating

Dear Editor,

Your 11/24 article, "Data on Deaths From Obesity...," noted that Antismokers are "furious" that obesity might gain top billing as a health threat and eat away at the hundreds of millions of dollars allocated to fight tobacco each year. It claimed the obesity death figure is exaggerated.

The contrary is actually far more likely, at least when one compares the "Deaths Due To Eating" to "Deaths Due To Smoking" overall. Obesity-driven heart disease is only one aspect of the puzzle. Cancer, diabetes, accidents, post-operative recovery, breathing problems, and traumatic bodily injuries are also often related to individual dietary habits.

Over one million Americans die annually from heart disease, and it's generally accepted that close to half of these deaths are at least partially caused by the fat, cholesterol, and salt in our diets, regardless of obesity. In addition, the Harvard Reports on Cancer Prevention have estimated that 30% of all cancer deaths are diet related. That adds an additional 215,000 eating-related cancer deaths to the 500,000 from heart disease. Adding together the cancer, heart disease, previously mentioned problems and additional deaths due to such things as allergic reactions, choking, and food poisonings, could well bring the total figure for "Deaths Due To Eating" near the one million mark.

Now THAT figure would get antismoking advocates furious indeed. It might even give some of them heart attacks... deaths from "secondhand eating!"

Everything I Say Is A Lie

One of the classic contradictions found everywhere that smoking bans are pushed is the self-contradictory insistence of Antismokers that, (1) The ban will be good for business, while at the same time claiming that, (2) It is necessary to protect businesses in areas with a ban from the competition of businesses in areas without a ban, thus requiring statewide bans to provide a "level playing field."

Obviously if #1 is true, then there's no need for the concerns expressed in #2, although there have actually been cases where a single, nonconforming bar has been accused of ruining the business of all the compliant bars in town! I knew it was futile, but I fired off another letter to the *Times* when this concept once again reared its ugly head during the Atlantic City ban debate in 2006.

Antismoking Pinocchios

Dear Editor,

Laura Mansnerus' 11/29 article, "As Atlantic City Eyes Smoking Ban...," notes that ban supporters say, "Restaurants and bars in other cities have not just survived the restrictions but have thrived," despite the fact that bar and restaurant owners in Atlantic City claim that with the casinos exempt from the ban, their businesses cannot compete.

There is clearly something wrong here. If bans make bars and restaurants "thrive," then how can a casino exemption be harmful to bars and restaurants near casinos? If ban supporters are telling the truth, the nonsmoking bars should be thriving in double overtime as they fill their seats with smoke-avoiding refugees from the casinos!

Could the bar owners be lying? What motive would they have? Could it be that they prefer to make less money? Doubtful.

Could the ban supporters be lying? What motive would they have? Could it possibly be that lying about the economic damage suffered by bars and restaurants supports their goal of smoking bans? That's far more likely.

The antismoking Pinocchios have very long noses.

It's Not A Tax. It's Just A Fee.

The next two letters are presented together as they both address the greed of the State for the tax money of its citizens. Both attack the excuses given for oppressive taxation levels and the further excuses given for limiting interstate purchasing of tobacco. As any American knows, it has long been standard practice for citizens of one state to order goods from other states through such channels as the iconic Sears Roebuck catalogue and never have a worry about paying state taxes on their purchases. The Constitution itself would seem to protect such practices in its sections forbidding restrictions on interstate commerce.

But the antismoking movement actually seems to have become more powerful than the increasingly flimsy paper of the Constitution in this regard, again using the cry of "It's for the CHILDREN!" to overwhelming effect.

These two letters, written in 2003 and 2009, represent just a small part of my unsuccessful effort to get that message out to the readers of the *New York Times*.

Abuse Of Our Children

Dear Editor,

It was with interest that I read the American Cancer Society's claim that the new law banning online cigarette sales was not a tax scam, but was meant to prevent children from starting an evil habit by buying cigarettes online.

It is hard for me to imagine many children using their parents' credit cards to order multiple cartons of cigarettes at once (the usual minimum for online orders is five cartons) and then trusting that their parents will never notice and also won't be at home when the smokes are secretly delivered. However, it is NOT hard for me to imagine seventeen-year-old smokers driving out of state to buy cigarettes cheaply and then having deadly car accidents on the drive home.

Tax scams in the name of our children that are simultaneously at the expense of their lives should result in jail time, not re-election.

Editorial Hypocrisy....

Dear Editor,

Your Editorial on roll-your-own pipe tobacco being sold for cigarettes reeks of hypocrisy.

You speak of 30 million dollars a month being "lost," while never mentioning that smokers still pay well over 100x that amount in taxes. You speak of children attracted to the "gimmickry" of rolling shreds of tobacco after all your years of editorializing about Big Tobacco's brand advertising being the cause of children smoking. You play the children card again by emphasizing "flavored tobacco – now banned in packaged cigarettes," while never mentioning that almost none of those banned cigarettes were coming from Big Tobacco or noting that the one major child-friendly flavoring those companies actually do use, menthol, was given a built-in exemption all of its own.

You ask, "What's the record for shutting a loophole?" How about asking, "What's the record for the largest single tax increase upon a minority group?" Does RYO tobacco's 2,150% tax increase upon the poorest segment of American smokers sound like the right answer?

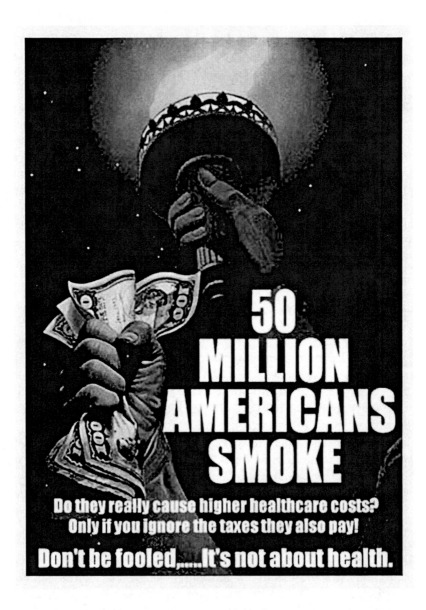

Launched Toward The Stars

Fortunately, there are thousands of newspapers not named the *New York Times* around the country and across the seas, and many of them have a more balanced attitude toward reporting the news from the various fronts in the War On Smokers. I've shot a lot of arrows into the air that have actually sailed up and created some starbursts, and a lot of others that have flown true and hit a target. That's far more satisfying than having them lost in the dark underbelly of New York's Old Gray Lady.

A lot of those letters were a bit repetitive – there are only so many ways one can describe a 2,150% tax increase – but I've picked out a selection that I felt spoke to particularly important issues, imparted useful information, or that I just plain liked. About half the letters in this section saw publication, while half did not. Each letter or group of letters will be preceded by a short introduction describing either the background of the issue or describing the particular stories that the letters are responding to. I hope you enjoy reading them as much as I enjoyed writing them, and I hope their ideas, arguments, and information will serve as inspiration, weaponry, or both.

Inhospitable Hospitals

While most US hospitals have had strong restrictions on smoking for the last twenty years, hospitals in the UK were far more relaxed about puffing patients, visitors, and staff, until the turn of the twenty-first century. The incoming bans have not been greeted with enthusiasm. When hospitals institute such bans, they not only ignore the discomfort they will be causing some of their patients, staff, and visitors, but they also ignore the problems that such policies create.

These artificially created problems could be easily solved by the provision of clean, comfortable, separately ventilated smoking lounges within the hospitals. Would such lounges cause some expense? Certainly... but very little when one thinks of the problems solved and the goodwill created. It could even be argued that patients' health would benefit as they no longer have to sneak outdoors in the cold and rain to grab a smoke or freeze to death when a door snaps shut behind them during such a break. (And no, that is not a fantasy; it has already happened more than once.)

Three particular problems resulting from the UK hospital bans – harassment, bugs, and litter – inspired letters from me. The first dealt with Antismokers' liking for preaching their gospel whether the listener wants to hear it or not. The second dealt with a staff problem where hospital workers were leaving a back door propped open so they could pop out for a smoke while little visitors quietly popped in for some fun. And the third was a response to one of those annoyingly shrill news stories that want to gripe about a problem by putting all the blame on smokers while ignoring the fact that at least part of the problem exists because of Antismokers.

The first two never made it to print, but the third surprised me and did.

For Your Own Good...

Dear Editor,

Cherry Thomas' March 6th article, "Stop Smoking Advisers on Hospital Duty" notes that "Stop smoking advisers will be in the main hospital corridor all morning on No Smoking Day ... talking to patients, visitors and staff..."

There is a fine line between "talking to" and "harassing," and I'll bet these "advisers" will be stepping quite a bit over it. Hospital staff, unless they are deaf, dumb, and blind, have certainly heard all there is to hear about smoking so many times that they'll need to sign up for a stomach pumping after being so "advised." Patients in the hospital for virtually anything more than an inflamed bunion have probably also heard the stop-smoking lecture more often than they've had their bedpans changed. And visitors who are worried and frantic about sick loved ones could almost be excused if they put insistent advisers into hospital beds for a slightly longer stint than a "No Smoking Day."

There is no good reason for banning outdoor smoking on the hospital grounds and the hospital administration knows that. So to pump up support for the move, they have activities like this to quiet dissent and instill the "proper" attitude amongst all concerned. I can almost hear the intercom buzzing now: "Calling Dr. Orwell! Dr. Orwell! Visitor needs a dose of NewThink in Ward # 9!"

It's bad enough we've let the crazy folks take over half the hospitals in the United States. There's no reason why the UK should follow suit.

A Buggy Headline...

Dear Editor,

Helen Branswell's Feb 4th article, "Another Reason Not to Smoke in Hospitals: Cockroaches," had a very misleading headline. It should actually have been, "Another Reason to Provide Smoking Lounges in Hospitals: Cockroaches."

The bugs did not come from smoking in a hospital, they came because of the hospital's unreasonably strict smoking ban. Smoking bans always have "unintended consequences" such as lost income, smokers gathering on nearby properties or outside doorways, fires from hidden smoking, and now, cockroaches sneaking into hospitals. Those consequences may not be "intended" but they are real, and the solution is simple: provide comfortable, separately ventilated, friendly indoor spaces where smoking patients, staff, and their friends can gather together and relax. The provision of even a few such spaces would put a simple halt to back doors being propped open while people smoke and provide an unintended open highway for invading insects.

Of course, those pushing for bans will never accept such a solution. It doesn't fit in with their real fundamental goal of "denormalizing" smokers. Fires and cockroaches are seen by such extremists as simply being regrettable, but fully acceptable, costs of smoking bans.

Litter Legacy of What?

Dear Editor,

The headline of your Feb. 12th story, "Litter Legacy of Smokers at the Royal Bolton Hospital," was misleading. The proper headline would have been, "Litter Legacy of the Smoking Ban at Royal Bolton Hospital."

As part of the story, your reporter interviewed a groundskeeper who had been working at the hospital for sixteen years. It would have been very easy, and a very obvious reportorial responsibility, to ask him the simple question, "Did the litter problem change once the hospital banned all indoor smoking?"

The answer would most definitely have been a very resounding "Yes!" ... but I guess it wouldn't have been very politically correct to point out that the hospital had created its own problem by refusing to provide even a few decent and separately ventilated smoking lounges for its patients, staff, and visitors.

Easier to just blame the smokers who've been thrown out into the cold, right? And certainly more in tune with the nasty little mechanics of social engineering.

An Appetite Never Sated

After banning smoking in airplanes, schools, hospitals, stores, offices, restaurants, and even casinos and bars, the antismoking movement began looking greedily toward apartments, row homes and the great outdoors, but they were a little afraid of taking that next big jump. They needed something intermediate to coax public opinion along, and they found two nice wedges in pushing for smoking bans in cars with child passengers and around outdoor playgrounds.

Their original car-oriented efforts were based on the concept of smoking drivers causing accidents, but, as we saw earlier, the statistics didn't support that contention very well. So they were forced to reach into the bag and pull out one of their oldest tricks: "Save The Children!" Stories and statements and press releases and TV ads began picturing innocent children, strapped into rolling gas chambers with no defense, and a baby-step was taken toward reaching into the private family within its home.

Canada moved first in this pioneering effort, but was soon followed by a number of US states. While car smoking bans are still far from being the norm, they show signs of growing... after all, who would argue against protecting innocent young pink lungs locked inside rolling metallic death boxes?

And hey, why stop at indoors? We *are* talking about protecting the children here, right? Anyone who sets *any* limits on such protection is clearly a cad. Onward to the playgrounds and parks! After all, those evil smoking parents are all throwing their butts in the playground sandboxes, right?

Big Brother In Our Cars...

Dear Editor,

In your Feb. 2nd article, "Car Ciggy Ban Gathers Speed," the three quotes from Ottawa City Councilor Diane Deans are revealing. She says "there's no safe level of secondhand smoke," while failing to note that there's also no safe level of exposure to the car exhaust on the highways you'd be driving your children on. She claims that "smoking in an enclosed motor vehicle is far more potent than in a home," while neglecting to mention that this is generally only true if you're smoking with the windows all rolled up – something few smokers would do with their offspring in a car. She also overlooks the fact that very few children live for 24 hours a day in a car with smokers.

Finally, and most importantly, she admits that the ban is "as much an educational tool as anything else." So for the sake of an "educational tool," she plans to enact a law that clearly infringes strongly upon personal freedoms and parental autonomy? Think how much more easily such values will be thrown out the window in the future when issues more important than "educational tools" are brought up and the government wants to pry into your home.

Should parents smoke in cars with children while keeping the windows rolled up? Obviously not – to say the least, it's inconsiderate. But should people in Ottawa or anywhere else support a Big Brother approach with a law for this? Use some common sense instead of jumping out into the daisy field to dance with the crazy folks – smoking in a car with the windows open is NOT equivalent to child beating!

A Law That May Kill...

Dear Editor,

Banning smoking in cars with kids may sound like a good idea at first, but once you actually read the studies the law is supposedly based on, the shine wears off. Yes, kids exposed to smoke all day long in the home tend to get more earaches and colds; but no, there are no studies showing that short exposures to normal levels of smoke are harmful in any way at all unless a child already has smoke-specific asthma.

Despite the lurid pictures from antismoking lobbyists, virtually no smokers drive around with their car windows rolled up tight while babies strapped in bassinets are smothered in clouds of smoke. And driving kids around in a car to begin with will give them far more exposure to deadly chemicals from general vehicle exhaust than anything they'd get from a few minutes of a driver's smoking.

Car smoking bans pose an additional risk. There are an increasing number of psychologically borderline individuals out there who feel they're part of a holy Antismoking Crusade. It's not at all hard to imagine such people engaging in unsafe, high-speed driving to get license numbers while simultaneously trying to dial a cell phone to report an illegal smoker. Some, feeling emboldened by the law, might even swerve around and drive up next to smokers so they can roll down their windows and yell, "You're breaking the law!" thereby setting the stage for road rage and accidents. While it's pretty doubtful a car-ban law would actually save any lives, it could very easily set the stage for accidents that will kill.

Smoking bans in general are bad laws based on lies, and this law, despite its superficial appeal and innocence, may be among the worst in its immediate effects on loss of life.

Playgrounds and Common Sense

Dear Editor,

Your article about keeping smoking parents far away from playgrounds quoted ban supporters as saying, "Ten metres is probably sufficient to get the message across (and) make the city healthier and cleaner, while ... protecting the welfare of its children."

Is the ban about health or about "getting the message across"? What message? And why is sending smokers 33 feet from the edge of a playground, maybe over 50 feet from where their children are playing and possibly slicing themselves on a sharp piece of glass or falling from a swing suddenly pushed too high or being threatened/hit by a bully, described as "protecting the welfare" of those children?

Thunder Bay citizens need to understand what we've gradually realized in the States: the antismoking extremists are lying to you and feel perfectly justified in doing so "for your own good." No child's "welfare" will be harmed by a rare wisp of smoke from a parent at a playground. The city will not be "healthier" by moving parents away from their children. And shuffling smokers off into wide circles around playgrounds will not make the city cleaner.

A few comfortable benches a meter or two outside the playground's edge near a proper ashcan actually *would* make the parks cleaner, while safeguarding the children's welfare. Almost all parental smoking would occur on those benches and any small amount that might actually encroach the playground itself would, in practical terms, be a smaller threat to those children than a mosquito flying into the airspace.

Casting Down The Gauntlet

Letters to the Editor have to walk a fine line. If they're dull, boring, or simply preachy, they're unlikely to get printed. On the other hand, letters that are too challenging may scare the editors off.

I've submitted over a dozen public challenges to antismoking groups and politicians in letters where I ask them to stand behind their promises and claims with more than simple lip service.* You would think that newspapers would love to print such challenges – after all, the blood 'n guts that would ensue while these figures tried to dodge and weave would certainly provide enough fun to sell a few more copies! Regrettably, such has not proven to be the case.

My challenges took various forms, sometimes focusing on promises about monetary consequences of bans by asking officials to back up their integrity with personal legal commitments to cover "unexpected" ban business losses, and sometimes simply challenging the need for a continued legal ban if indeed businesses were doing so well under the bans.

Two examples follow, one from Philadelphia, and one from Chicago. The Philadelphia effort made it to print, and to this date it remains the only one of my "$ Challenge" letters to do so. Neither of these challenges, nor any of the many similar ones I've sent or posted around the Internet have ever accepted or even acknowledged by their targets.

Nonetheless, as far as slings and arrows go, I have always been proud of these particular missiles for being especially pointy.

---

* I'll be exploring these challenges in more detail later on, in *Launched In The Trenches.*

A Smoking Challenge

Dear Editor,

Your June 16th editorial, "Clearing the Air," quotes the CEO of the Chamber of Commerce as saying that the "facts" show that a smoking ban "will not adversely affect business."

If Philadelphia's Mayor Street believes that and wants to sign the smoking ban, then how about showing a little good faith and guaranteeing that claim? If he and other ban supporters truly believe that bars will not suffer losses, then why not back that up with something more solid than mere words?

All the city and antismoking lobby groups have to do to in order to prove their honesty is commit to covering any resulting business losses in the year after the ban takes effect. Most of the opposition would dissolve and the ban would be a shoo-in!

After all, if they're telling the truth, there's nothing to lose, right?

Smoking Ban Needed?

Dear Editor,

Your March 19th editorial stated "It's hard to argue with Breathe Easy Washington." I disagree. Antismokers claim business improves for bars and restaurants after mandated bans, and you evidently believe them. But I ask you to think for a moment: if those claims are true, then New York, where the ban has existed for a year, should no longer NEED a ban. Bar owners are not stupid people who open businesses because they love losing money. Despite wacky antismoking claims, they are not secret agents of "Big Tobacco."

People want to make money in their businesses. That's it. Period. If Antismokers are truthful, and these people are enjoying wild success because of New York's ban, then clearly they would continue to ban smoking in their establishments even without the law.

So will we see Antismokers accept a challenge to repeal the ban and show the world how those happy businesses keep it in place on their own?

Of course we won't. Why not? Simple: because they are lying, and they know they are lying. Don't make the mistake for a single moment of believing them.

The Dose Makes The Poison

It's perfectly true that secondary tobacco smoke is a mix of over 4,000 deadly toxins, chemicals, and carcinogens. It's also perfectly true that your nice healthy dinner salad contains thousands of deadly toxins, chemicals, and carcinogens. In neither case would any well-informed and reasonably sane person worry overmuch about the incredibly small quantities involved unless they were deliberately misled.

Despite the fact that a Google search on "Surgeon General + no safe level" turns up over 100,000 hits attributing that phrase to the 2006 Surgeon General's Report, the phrase actually does not even exist in the main body of the Report at all! The closest it offers is a statement on page 11, stating "The scientific evidence indicates that there is no risk-free level of exposure to secondhand smoke," without any citation indicating just where such evidence might be, how strong it might be, and without any qualification for the word "indicates."[361] Also note: "risk-free" is *not* the same thing as safe. Nothing in life is truly 100% risk-free, yet we use the word safe for many things.

Basically, it is an opinion judgment, an educated opinion certainly, but not the same thing at all as the finding of a definitive scientific study. In the proper world of real epidemiology, absolutist statements are never stated in ways that might cause unreasonable fears and societal damage, and I believe the Surgeon General clearly left the bounds of that world with his statement, and dragged the public media in his wake.

This "no safe level" concept is again explored in the following three letters that explore the submicroscopic "risks" that normal exposures to others' smoke might entail. The first two address the general question, while the third follows up on the flurry of news stories reassuring parents that the amount of formaldehyde in baby shampoos is far too small to worry about.

Formaldehyde? In baby shampoos? Smoking ban advocates usually describe it as "a deadly chemical used to preserve corpses forced upon innocent nonsmokers." On Internet chat boards antismoking advocates will stridently ask, "Should I be allowed to just walk into restaurants and spray everyone with formaldehyde and asbestos?" Yet the *Washington Post* appeared almost eager to pass along Big Pharma's reassurances that there was nothing for parents to worry about due to the very low concentrations found in the shampoos. The smokophobic *New York Times* actually quoted an expert's opinion that, "When you look at the extremely low levels the report found, it turns out that we are exposed to these chemicals every day in food, air and even in shower water, all without apparent ill effect."[362] Extremely low levels? Wait till you see the figures...

Then take your pick.

No Safe Level?

Dear Editor,

The smoking ban law is based upon the scientifically ridiculous concept of there being "no safe level" of exposure to tobacco smoke.

Under such a measure, smoking on a traffic island in the middle of a street could be ruled to be "unsafe" for the tenants of buildings on either side of that street since molecules of that deadly smoke would disperse into the air and find their way into those buildings. A similar law based on such a concept could bar restaurants from serving carafes of wine with meals since carcinogenic and highly volatile alcohol fumes will bubble out of those carafes and into the lungs of fellow diners and their children.[363] Another such law might ban serviced patio dining, since such dining forces staff to venture out under the deadly ultraviolet rays of sunshine. After all, sunscreen and awnings only reduce exposure, not eliminate it, so they are simply unacceptable as solutions.

There has never yet been a scientific study showing long-term harm from the low levels of exposure to secondhand smoke that would exist in any decently ventilated modern establishment. Smoking bans are based on a desire for behavior modification and a wish for a reduction in smoking rather than upon any true basis of concern for workers' health.

So, is the Surgeon General's statement about "no safe level of exposure" more of a scientific statement or more of a political one?

Take your pick.

Don't Drink The Water!

Dear Editor,

Your Oct. 23rd poll asked if Columbus, Ohio citizens felt their tap water was "safe to drink."

We've all seen the antismoking ads on TV telling us that the arsenic from a friend's cigarette is going to kill us. The government "safe" standard for arsenic in tap water is ten million picograms per quart. A nonsmoker in a decently ventilated smoking environment inhales about 30 picograms of arsenic per hour. You would have to sit next to that "dangerous" smoking friend for over 300,000 hours in order to get the same dose of arsenic as you would get from a "safe" quart of water.[*]

Either the government is lying to us about the safety of our water or the Antismokers are lying to us about the dangers of secondhand smoke.

Take your pick.

---

[*] Remember, although I'm only doing it occasionally in these reproductions, whenever I write letters to newspapers where I state scientific facts, calculations, or derivations, the basic references are always set out fully and clearly in an addendum. Given prevailing attitudes and misguided "folk knowledge" in this area, I certainly don't expect editors to simply take my word without proper documentation!

Smoky Shampoos...

Dear Editor,

Your article on formaldehyde in baby shampoo claimed the concentrations of 610 parts per million (ppm) were "tiny" amounts of "no concern."

Really? Let's consider a small (400 cubic meters) and rather poorly ventilated (six air changes per hour) restaurant with ten smokers, each lighting up twice per hour. Would you be worried about taking a child there after all the scary news and antismoking advertisements about formaldehyde in cigarette smoke? You'd probably whisk your baby out of there faster than a waiter could pick up a tip.

According to government figures those ten smokers will produce a total of 17 mg of formaldehyde per hour, diluted in 2400 cubic meters of air, giving a concentration of .007 parts per million.

That "deadly threat" you'd be fleeing from has an 87,000 times lower concentration of formaldehyde in its air than you'll find in the baby shampoos described as having "trace amounts" of "no concern." Of course, smoke has other chemicals, but their "threat concentrations" are generally even less than formaldehyde.

There's clearly a problem here. Either the threat of wisps of smoke has been greatly exaggerated, or the 87,000-times-more-deadly baby shampoos should have wiped out almost every child in America.

Take your pick.

In The Land Of Churchill

In addition to the earlier-noted ships, hospitals, and public transit, the UK has now banned smoking in its pubs. Ban advocates promised that everyone would be happy and that there'd be no economic harm, but in the years since the ban the news of pub closures has done almost nothing but go up, and up, and up.

It's hard to get reliable figures and impossible to say just how many of the closures are due to the smoking ban, but the causal relationship is pretty clear. British pubs were closing at a rate of about three per week before the ban.[364] In the first few months after the ban, that rate increased to twenty-seven per week,[365] then a year later the figure hopped up to thirty-nine per week,[366] and by July of 2009, the British Beer & Pub Association pegged the closings at an incredible *fifty-two per week!*[367]

Antismoking advocates offer all sorts of explanations for this. Sometimes they blame "the recession," while ignoring the fact that the jumps to twenty-seven and thirty-nine per week occurred *before* the recession. Sometimes they blame "cheap supermarket booze," while failing to note that it's always been cheaper to buy booze to drink at home. And finally, one of the funniest and most clueless of their excuses, as noted in a 2008 *Sunday Mirror* article, was simply that "many people nowadays prefer to drink at home."[368] Somehow, they fail to realize that maybe the reason "many people nowadays prefer to drink at home" is simply because they or their friends don't get thrown out into the cold to smoke!

Three letters follow. The first was written in late 2005 and warned Britain of antismoking duplicity and exhorted the Brits to resist their oncoming ban. The second was meant to serve as a warning to one of Britain's long-associated Commonwealth members, Barbados, as it contemplated a ban of its own. And the third spoke directly to the closing of the British pubs and to the economic harm those closings have caused.

Fight Them On The Beaches...

Dear Editor,

November 25th's article "Medical Chief's Anger at Partial Smoke Ban," made repeated references to figures indicating that 90% of the public wanted a universal smoking ban. On closer inspection, it was revealed that these figures came from surveys conducted at medical centers under the eyes of health personnel; clearly not an unbiased random sampling source. Taking such a figure seriously would be like accepting a Catholic School survey of the "self-touching" habits of altar boys that was conducted by the nuns as being the gospel truth.

Looking a bit farther afield, another figure emerges. Deborah Arnott, head of UK's Action on Smoking and Health, recently admitted in an interview in the *Publican* that 59% of the public is against such a universal ban. Given the source of this damning admission (ASH) it's likely that the real opposition is even stronger.

A powerful and tax-rich antismoking lobby has created the perception that demand for government-mandated, universal bans is at a fever pitch and that their universal implementation is on an unstoppable path: "Resistance Is Futile! Bans Are Inevitable!" It's a cheap trick, but it worked for Attila the Hun in Asia, for the Borg in Star Trek, and for the Antismokers here in the United States.

Don't let it work in Britain.

The New Math

Dear Editor,

I live in Pennsylvania, where a government-imposed smoking ban is still fairly new, but I have already seen the harm it has done to bars and restaurants that have not gotten exemptions. I could only grit my teeth in frustration as I saw the leader of Barbados' antismoking forces, Pastor Victor Roach, quoted as saying that reports from other countries showed "no loss of business" as a result of smoking bans.

The Pastor has made the mistake of reading only those reports that support his view. In England, the average pub-closure rate during the two years before their ban was three per week. Three. In the one year since their ban began, that rate has shot up to thirty-nine per week. Thirty-nine.

From three closures to thirty-nine closures per week is an increase of 1,100% in business failings and yet the Pastor says there have been "no reports of loss of business." In actual fact, almost every economic study not funded by antismoking interests has shown economic harms from smoking bans, whether local or national.

No loss of business? If this is an example of "new math," I'm glad I never learned it. I hope Barbados never has to learn it either.

Lead The Way, Britannia!

Dear Editor,

I'm responding to the Mirror's story, "Credit Crunch: One in Six Pubs to Close By 2012."

Our taverns here in the United States have been closing as well, long before any "credit crunch" excuse. Smoking bans kill small pubs no matter how many lies by no matter how many "authorities" you hear to the contrary.

You noted that in the first five years of the UK ban, you will have lost 7,000 pubs. Even if we assume these are mainly small country pubs – the heart and soul of Britain's social life – they still must average at least five employees per pub. That's a total of 35,000 people who will have lost their livelihoods thanks to the Health Ministry's "Cost-Free" smoking ban.

If we accept the United States EPA's cancer figures as real, the ban might, eventually, forty or fifty years from now, save about twenty of those 35,000 from lung cancer. In the meantime, I guess the other 34,980 out-of-work Brits can just go starve for all the antismoking fanatics care. And take their kids with them.

The people who created this ban should be run out of office, fired from their positions, stripped of their titles, face full liability for mistruths spoken in bringing it about, and live in disgrace for the rest of their lives.

Will they? Or will you simply sit back and let one of the most basic units of your social community be destroyed by fanatics? These bans have destroyed our social lives in the States in piecemeal fashion, enabling the Antismokers to hide the damage. Don't let the same thing happen to you.

We need help. Lead the way Britannia… we're depending on you!

Of Freedom And Its Loss

If there is a single word that best defines America in the minds of its citizens, that word would likely be Freedom. When we think of our Constitution, the first things that come to mind are usually "Freedom of Speech" or "Freedom of Religion." We go to war "to fight for Freedom"; the Statue of Liberty holds up "The Torch Of Freedom" to welcome immigrants fleeing from oppressors; and we jealously protect our "right to bear arms" despite thousands of shootings every year.

Unreasonable, government-imposed smoking bans abuse and restrict that freedom. Yes, it may seem like it's only in a small way, but as Justice William Douglas reminded us in the opening quote of this book, *"We must be aware of change in the air, however slight, lest we become unwitting victims of the darkness."**

Smoking is a small freedom, but if we don't hold on to the small ones, we'll eventually lose the larger ones. The constant din of the antismoking movement has made people feel afraid and insecure when they see even the faintest touches of smoke, but as Benjamin Franklin warned Americans over 300 years ago, those who are willing to trade their freedom for passing feelings of security deserve neither.

The first letter in this group speaks to the basic concept of freedom and offers a warning. The second letter replies to the absurd Orwellian notion proffered by some Antismokers that smoking bans actually increase, rather than decrease, freedom; that they're not really bans at all!† The third and fourth letters pose some thoughtful questions, while the fifth looks at some rather ungrateful attitudes expressed by prominent antismoking advocates toward those responsible for preserving our American freedoms.

---

* That particular quote served as my closing quote in *Dissecting Antismokers' Brains* almost ten years ago. In terms of our freedoms, things have gotten a lot darker, in very many ways, in the intervening decade.

† An example from American Cancer Society official, Shawn Cox: "This is not a smoking ban. This is a comprehensive indoor clean air effort." ("Smokeless Advocates Concerned" by Cindy Schroeder. *Cincinnati Enquirer* September 10, 2008, p. B1) and another from L.A. Councilman Parks saying that their ban was "not an outright ban" because it only applied to places where "there is an expectation of people being present." (http://reason.com/blog/2008/08/13/ bernie-parks-encourages-smokin)

Safe In Our Homes...

Dear Editor,

The most amazing and frightening thing about your July 25th article, "San Diego Suburb Bans Smoking," was the note that the bill specifically still "allowed" people "to smoke in their homes." Somehow I find the idea that the government has to add a specific exemption to allow one to do something perfectly legal while within the privacy of one's own home to be deeply disturbing.

Of course, anyone following the smoking ban battles shouldn't be too surprised. Once the antismoking lobby controlled restaurants, they went for bars. Once they got the bars, they went for the casinos and private clubs, then outdoor patios, then the beaches and parks; so it's really only to be expected that they'll now look toward our cars, our condos, apartments and homes, and then, eventually, even the public sidewalks.

It has been clear for quite a while that this was coming. As the founder of Action on Smoking and Health's John Banzhaf put it back in 2006, "Here we are literally reaching into the last frontier — right into the home... No longer can you argue, 'My home is my castle. I've got the right to smoke.' "[369]

Should we be thankful that we still have some rights in our own homes? Or should we be start being very, very afraid of the power we've given these extremists?

Of Freedom and Noses...

Dear Editor,

Your Dec. 14th commentary, "Sweet Land of Liberty?" makes a solid argument against wanton government interference in our private lives, but also spoke of not having the right to punch someone else in the nose. Antismoking advocates often equate smoking with such punching.

The comparison is actually very poor. Few would regularly go to a bar where patrons were generally allowed to punch each other in the nose as part of the evening's festivities. But for centuries, many have willingly, and even happily, gone to places where other people were allowed to smoke.

"Freedom from being exposed to smoke" is far closer to "Freedom from being exposed to annoying chatter that might arguably raise your blood pressure." If a bar chatterbox at Murky Mike's Mescaloozer insists on bending your ear, you're free to move to another section of the bar or even pop down to a quiet booth at Blinkin' Bob's Boozatorium. Government has no role beyond seeing that OSHA safety standards are met, and OSHA has consistently refused to rule that normal exposure to secondhand smoke – even when it's for eight hours a day, five days a week, for forty straight years of work – constitutes a significant workplace hazard.

Freedom is a vital and precious thing: our children will never forgive us for allowing well-funded lobbies to chip it away to nothing.

Stinkier Than Smoke!

Dear Editor,

Tuesday's editorial on the state smoking ban asked, "Why shouldn't a municipality adopt a tougher ordinance if its residents want one?"

Indeed! And why shouldn't a municipality be able to adopt a milder ordinance if its residents want one?

And, for that matter, why shouldn't a smoking bar or restaurant owner, with only smoking staff and serving only smoking customers and their willing friends, be allowed to have smoking?

To forbid such a choice is so un-American it stinks a lot worse than any smoke ever could.

Of Parks and Prohibitions...

Dear Editor,

I was a bit puzzled when I read your column on a smoking ban in the parks. I noted the words "not allowed," "prohibits," and "enforcement" used freely as part of this policy which you make pointedly clear is "not a city law." I then pondered the ramifications.

Under the United States Constitution, Congress "shall make no law" regarding an establishment of religion. Nevertheless, it seems that, following the Duluth model, Congress will now be welcome to "prohibit" Judaism, "not allow" Rastafarianism, and perhaps even bring some "enforcement" into play against those furshlugginer Catholics tippling wine at Sunday Services.

Rule by intimidation was practiced by a certain tyrant-to-be in Europe a few years ago before he got the full legal powers he sought. Fortunately, there were enough freedom lovers to stop him – eventually – but it cost a lot of lives.

It's sad to see that so many of today's Americans are willing to fall into lockstep against a newly hated minority at the expense of the edges of their freedoms.

A Modest Proposal....[*]

Dear Editor,

January 5ths' article, "Winooski Hearing Set On Smoking Ban," quotes Councilman Clark justifying his desire to boot smoking veterans out into the cold by saying, "This seems like a good way to honor our veterans, to prolong their lives."

The Councilman would probably agree with the Lung Association's spokesman, Joel Africk, who urged people not to send smokes to soldiers in Iraq even if those soldiers asked for them, saying, "Tobacco use presents an immediate and real danger for our soldiers who are on the lines today... our troops should be sent care packages that don't kill."

I would like to suggest that Messrs. Clark and Africk spend some time with these fighting men personally and discuss such issues of honor and danger with them as they stand in the snow outside their veterans' halls or huddle in Afghanistan's trenches. I'm sure they'd find the discussions most enlightening and I think that all Americans would benefit from the end result.

---

[*] I'll admit to being quite surprised when this particularly pointed letter actually made it to print in Vermont's *Burlington Free Press*. Kudos to Vermont and the *BFP*!

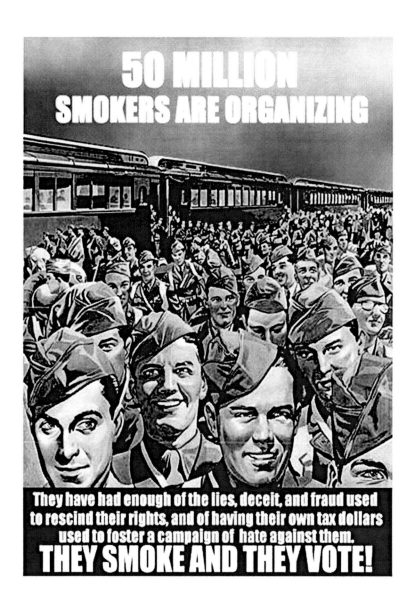

Taxation Without Representation
{a.k.a. "The Master Settlement Agreement"}

We're all familiar with the statement, "The power to tax is the power to destroy." Nowhere has that statement had more relevance than in tobacco control. Twenty years ago, the average price of a pack of cigarettes was less than $1.75 with just 20 cents of that going to taxes. [370] Today, *just* the taxes (including the MSA) on a pack of cigarettes in New York city approach $8.00 – forty times the average tax of 1992.

For decades, legislators were afraid to do more than apply little nudges to cigarette taxes for several reasons.

(1) They were afraid of Big Tobacco.

(2) They were afraid of the voting bloc of smokers.

(3) They were afraid of creating criminal black markets.

(4) They were afraid of increasing the Consumer Price Index (CPI), since cigarettes were part of the average family's expenditures.

(5) They were afraid the "raise taxes to reduce smoking" argument would be immediately seen as nonsense since Europeans smoked far more than Americans despite paying, at that time, outrageously higher taxes.

But the fear of Big Tobacco declined as it scrambled to save its butt from lawsuits. The fear of smokers as a voting bloc disappeared when politicians saw how placidly they accepted the unlegislated Master Settlement Agreement's doubling of taxes on cigarettes in 1998. The fear of the black market was softened by antismoking reassurances that Americans were law abiding and a significant black market wouldn't develop. (Yeah, right. Just count the dead bodies and the filled jail cells due to cigarette smuggling and turf wars nowadays.) The CPI problem was fixed by removing tobacco from the list of "normal" purchases. And finally, the fear that smoking reduction claims would be seen as nonsense evaporated since their other claims were nonsense anyway.

Tax Evading Grannies?

Dear Editor,

Regarding the attack on websites offering interstate shipment of cigarettes without taxation, the problem is the Constitution and its annoying restriction on interstate tariffs making such taxation difficult.

The real solution is to simply get rid of that silly old document! Then we can also zoom in on terrorist grandmoms who evade taxes through TV's Home Shopping Club and Sears-Roebuck Catalog ordering. We can grab Sears' tax records for the last seven years and tell the grandmoms to either pay up or we'll let them rot in jail.

Finally, if we're really going to crack down on such things, perhaps the Governor of Delaware should be thrown in jail for putting those big signs on the highways from surrounding states that announce, "Welcome to Delaware! State of Tax-Free Shopping!"

And don't even get me started about interstate wine, book, or computer shipments![*]

---

[*] This letter was written in the early 2000s. Today, in 2013, we are now seeing sales taxes estimated and added to bills from online sellers like Amazon.com. I wonder what Sears Roebuck is doing nowadays? And I wonder if, as part of dealing with constant political falls off of fiscal cliffs, we *will* start going for those tax-evading grannies?

Taxes For The Children...

Dear Editor,

Kentucky's new cigarette tax will represent over a 3,000% tax increase on a targeted minority group in less than ten years.

How about doing the same for alcohol and gasoline users? A price doubling of gasoline would certainly be "for the children." Just think of the young lives spared from maulings under Detroit's Death Machines as children pedal their bicycles on paper routes. As for alcohol, consider the hospital beds freed up from rotting livers, the wives spared drunken beatings, and the infants saved from Daddy's "Beer Breath" after a night's bar-boozing.

There's a lot you can do with tax increases aside from simply raising money off defenseless minorities: you can socially engineer an entire population. Of course, you'll also be building the black market and the disrespect for government that's so beloved by anarchists, but hey, it's for the children, right?

Magical Taxes...

Dear Editor,

Your Nov. 9, 2007 "Decrease in Smoking Rates Levels Off" makes the interesting claim that "The relatively unchanged price of cigarettes since 2002 is considered important" in this leveling. The reason this claim is so interesting is that just five days later, the Director of Tobacco Free Kids, bragging about his success in raising cigarette taxes, publicly stated that "Since 2002, 44 states, Puerto Rico and D.C. have increased their cigarette-tax rates more than 75 times."

We've known for a long time that cigarette smoke itself is magical – able to travel against hurricane winds while defeating air cleaning systems that handle trucks, jet planes, and bubonic plague units in hospitals – but this is the first evidence we've seen that cigarette prices are also magical! Even if they are raised more than 75 times, they somehow manage to stay about the same! If only we could transfer such financial magic to the rest of the economy!

I'm Confused...

Dear Editor,

James Nash's article on a "cigarette surcharge" left me feeling a bit confused.

In 1997 the *New England Journal of Medicine* analyzed the cost of smoking and compared it to what were then very low cigarette taxes. Even with such low taxes, they concluded that smokers were already paying for their own health care plus a fair amount extra for the health care of nonsmokers.

Then in 1998 the Government wanted money from Big Tobacco to pay for the sick smokers who'd already paid for themselves, but BT claimed poverty and made a deal with State Attorney Generals called The Master Settlement Agreement. The MSA added a new 50-cent nationwide cigarette "tax" so smokers themselves could then pay for their health care a second time. Smokers didn't get a say in this, of course; they were just the sheep the wolves were having for dinner.

So smokers paid for their health care not once but twice, and watched that money spent on wild and wonderful things like golf courses, roads, and lots and lots of ads saying that smokers are smelly and dirty and are killing little children.

But now, in 2008, the antismoking groups who've been getting fat off smokers' money all these years want smokers to pay the same bill a THIRD time to pay for promoting ideas like firing smoking employees for smoking at home.[*]

Meanwhile, Antismokers still blame smokers for driving up health care costs while ignoring the fact that smokers specifically pay extra surcharges for their insurance plans. And if smokers lie about their smoking status? Well, if they work for Whirlpool, they may find themselves failing a "drug" test and end up on the unemployment line – while they're *still* paying for everyone else's healthcare when they buy their smokes.

So that's why I'm confused. Maybe I just need to smoke more?

---

[*] Flash forward to 2010: regular smokers are hit with a 150% federal tax increase and roll-your-own smokers are hit with a 2,150% increase, but this time it was for the health insurance of children that nonsmokers were too selfish to pay for themselves. And now, in 2013, they're discussing a SCHIP 2—raising the RYO tax to 4,300% over 2009 levels while also raising the tax on pipe tobacco from $2/lb to $50/lb! Now I'm *beyond* confused!

Where There's Smoke...

The image of innocent children being burned alive in horrific fires caused by soulless nicotine drug addicts is powerful, and it's been played to the hilt by Antismokers seeking to inspire guilt among smokers while increasing outrage against "those filthy murdering scum" by nonsmokers.

The fact that most such deaths are among the smokers themselves (usually with the aid of too much alcohol) is ignored, as is the fact that many of the fires headlined as being caused by "smoking materials" may have resulted not from smoking, but from unattended matches* left where kids might play with them. There's also the sad fact that at least some smoking-caused fires are actually due to bans themselves, as they encourage hidden smoking without proper ashtrays and the hasty disposal of incriminating butts in the face of approaching authority. These facts are simply disregarded while the focus shines purely upon the evil addicts.

The first letter concerns a fire sadly similar to one that I wrote about in *Brains*, a fire in an adult-care setting where smoking was generally forbidden. When such fires occur, the possibility of deaths greatly escalates as elderly residents, fearful of eviction, waste precious moments trying to fight and cover up a small blaze on their own rather than pull an incriminating fire alarm and face being thrown out on the street.

Just another "unintended consequence" brought to you by your friendly neighborhood Antismokers.

---

* Matches which may have only been used for candles or to create "healthy smoke" from medicinal marijuana.

Of Fires And Foolishness

Dear Editor,

So the *Journal* believes that "The impetus for the (smoking ban) was a fatal fire in March at a Mocksville adult-care home that was caused by a resident smoking in her room."

Is that true, or is that just spin put out by the antismoking lobby seeking to ban smoking everywhere by everyone? Let's test it out.

If the above was true, then the law would forbid smoking in private rooms and insist it be done safely in supervised public areas. It would certainly not ban public smoking, since that would lead to an increase in hidden smoking in private rooms, thus causing future fires.

So is that how the law was written? Of course not: they went for a blanket ban that will actually put the lives of our senior citizens at increased risk. Hastily disposed butts, disposed of without ashtrays by frightened and confused seniors frantically trying to avoid eviction onto the streets for breaking iron-clad smoking ban rules, are FAR more likely to cause fires than a rule allowing smoking in some designated, comfortable, and separately ventilated lounge areas.

The Downside Of Smoking Bans...

Dear Editor,

Daniel Dasey's "Patients flout smoking bans" ignores a negative aspect of bans that Antismokers never mention – the fact that extreme bans may actually increase, rather than decrease, the danger of fires and injuries.

How can that be? Simple: before bans, reasonable smoking rules allowed comfortable lounges where smokers had access to ashtrays and the time and attention needed to extinguish their butts properly and safely. Smokers generally heeded no smoking signs because they knew such signs honestly indicated hazardous conditions.

Extremist bans encourage covert rebellion where illegal cigarettes are more likely to be hastily disposed of with a quick flip or stamp – and then remain smoldering under a couch, in a stairwell, or at the bottom of a paper-filled trash can.

Antismokers never like to acknowledge this "unintended side effect." When I once complained to an AeroMexico representative about their new ban, he justified it by referencing a smoking-related fire on a Canadian airplane. When I pointed out that the fire in question was actually caused by a sneaked bathroom smoke *after* a ban and that the airline might actually be putting its passengers at increased risk, the rep simply refused to continue the discussion.

Smoking bans are bad laws based largely upon lies and distortions. When they are resisted or flouted, it's as much the fault of the lawmakers as that of the smokers: citizens of free countries defy unreasonable laws. That is, in the long run, what keeps those countries free.

Foreseen Consequences...

Dear Editor,

Dr. John Dent wrote about the "unforeseen consequences" of fires and other problems caused by the UK smoking ban. It is unfortunate that the NHS neglected proper research prior to the ban, because there was nothing at all "unforeseen" about such consequences. I had warned of such fires five years ago in *Dissecting Antismokers' Brains* and that warning has been repeated many times around the world on the Internet.

The UK ban was implemented with callous disregard for the lives of those it would impact, from the pub owners and veterans seeking a warm evening over a pint down to the elderly and patients in hospices and psychiatric facilities. No objection or voice of moderation was listened to. The goal of a total ban was seen as all-important by its supporters, and all aspects of "collateral damage" were willfully and reprehensibly ignored.

Dent's concern for his psychiatric patients is doubly justified since many of them self-medicate with nicotine to calm themselves, deal with swings of depression, and equalize their moods. The most surprising thing may simply be that the NHS has not yet been hit with crushing lawsuits over this.

Thousands of businesses closing; tens of thousands of people put on the dole; smokers filling sidewalks and lounging on neighboring properties; fires in hospitals, care-homes, dormitories, and airplanes; muggings and rapes of workers on smoke breaks in alleyways; a tax-driven black market that happily sells to children – all of these were fully foreseen consequences that are now passed off as unexpected by the same folks who touted the ban as being "cost-free."

They lied about the cost and they lied about the health risks – and they should be held fully accountable for the resulting damage. Smoking bans everywhere need re-evaluation and amendment for the good of all. The Great Smoking Prohibition Experiment has been tried and it has failed, just as America's Great Alcohol Prohibition Experiment of ninety years ago tried and failed. The only difference is that this time, the politicians were fully warned of the costs to come and chose to blunder onward regardless.

Slaves For Hire?

One of the great forces against widespread smoking bans in bars, clubs, and even restaurants was the fear that they would simply be ignored. If everyone ignored the law, the State would be helpless. It could not afford the police, the time, or the court costs to process thousands of defiant smokers. Sadly, they found a way around this. Even though non-military, government-imposed involuntary servitude in the United States is blatantly illegal and unconstitutional, smoking ban laws were shaped to intimidate business owners into acting as unpaid, uninsured, untrained, and unarmed citizen-vigilante law enforcers by threatening them with fines or closures if they "permitted" smoking in their establishments.

Realization of this mechanism led me to understand better what happened in Ireland when the Irish caved in to their ban with barely a ripple. I had been quite puzzled by that as I knew the Irish loved their pubs, and their pub life was clearly being disrupted and destroyed by the smoking ban. So why did these hardy folks who had defied the mightiest naval power on earth for centuries decide to so meekly surrender to the smoking Gestapo?

The answer lay in that clever focus of penalties on the hapless pubkeeper. Regular pub-goers see their pubkeepers as friends they've known and loved for years. Few smokers wanted to put their friends in jeopardy — and thus simply toddled outside or stayed home with their smokes, their telly, and a six-pack.

Of course, both in Ireland and in the United States, things have not *always* gone smoothly. The first of the following letters looks at what happened in Ireland when the ban hit in an environment where the "protect a friend" element wasn't applicable,[371] while the second responds to stories of enforcement problems around the United States.[372] One surprising statistic came from a 2004 study done right in the antismoking heart of the United States that found 50% of California bars to be in some degree of violation even after five full years of their supposedly popular and successful ban. California is not alone; undercover TV "stings" have regularly shown bans being flouted all over the country – resulting in high-profile, but always temporary, crackdowns.[373]

We'll start with the Irish as they toddle off to work on the tram, and then return to the United States.

Dangerous Donuts?

Dear Editor,

Doyle's column noted that for the transit smoking crackdown, "two plainclothes inspectors board buses, backed up two uniformed inspectors and a driver. In areas where there may be a risk to the safety of inspectors a Garda presence accompanies the inspection team."

Now I don't know how it is in Ireland, but here in America, most busses and trolleys also supposedly prohibit eating or drinking. If it's that way in Ireland as well, will you be sending out six-man hit squads to go after the morning coffee slurpers and donut chompers? Will a six-man force be sufficient to take down someone armed with both a hot Starbucks and a launchable chocolate creme croissant?

I've been saddened to read about the destruction of Ireland's small pubs and the resulting social damage caused by the New Puritans that have taken over my ancestral country. Yes, I know you blame the Americans for your woes, but don't blame all of us: some of us have fought valiantly, and are still fighting valiantly. For the moment, though, we both need to look to the Dutch and Germans for guidance in regaining our freedoms. They at least have largely united against ridiculous and intrusive laws.

Police Without Portfolios...

Dear Editor,

So the health supervisor's officers not only refuse to approach people to ask them to stop smoking but will not even dare to identify such people unless they are "accompanied by a police officer"?

Meanwhile, they expect bartenders and waitresses to plunge right into confronting smokers to enforce the law on their own. His suggestion that staff should call the police if they expect to have problems is laughable. Do the police want to be called EVERY time a staff person sees someone smoking in a bar? Does Madison want to hire that many extra police officers? Would workers end up getting fined themselves for making nuisance calls?

If the "smoking police" themselves fear to confront smokers, how can they possibly insist that bar workers put their own lives on the line?

Statistical Skullduggery

We're all familiar with the phrase, "Lies, Damned Lies, And Statistics." We're familiar with it because it carries such a pithy essence of raw truth that it has been repeated millions of times throughout the English-speaking world.

People tell lies. Some people tell really nasty and outrageous lies. In both cases, anyone with a sharp eye or ear can usually uncover those lies with a little investigation.

Statistics are different. They can be true, or they can be lies, or they can even be a confusing mixture of both (as we saw in the earlier section on *Stratistics*), but few people feel competent or brave enough to take on the task of entering the statistical mine field to battle "experts" who purportedly know what they're doing. We saw an example of how ordinary people are treated in my earlier writing on the employment study where the researcher brushed off my statistical concerns by saying their choice of analysis was simply "the most appropriate" choice, while implying that any who questioned such choices must be in the pay of Big Tobacco. And a year later, when I posed a question to her about what seemed to be a very obvious flaw in the newly published second stage of that research, she basically just recommended that I read a statistics textbook.[374]

My general rule is to only use statistics in arguments when I can keep them at a level understandable by a standard high school graduate. My college and graduate work with statistics left me with a deeply ingrained mistrust of anything that goes much beyond basic p-values and confidence intervals – and taught me to keep a sharp eye out for confounders. It's just too easy to lie when you can justify it with a spate of formulas replete with weird Greek letters, while brushing off questioners with recommendations that they go back to school or read a textbook.

The following examples try to make some solid points about the Pennsylvania and Ohio smoking bans by using simple and straightforward statistics-based arguments.

Of Bans and Fanatics...

Dear Editor,

Adam Brandolph leads off his Nov. 8th article about exemptions to the Pennsylvania smoking ban with the editorial note "So much for the smoking ban," indicating that those exemptions make the ban completely worthless.

There are roughly 2,000 exemptions,[*] mainly small bars employing at most an average of ten people, so we're looking at exemptions that allow about 20,000 Pennsylvanians to work in smoking establishments.

There are about four million working adults in Pennsylvania. A little math shows that this smoking ban, which is supposedly "filled with loopholes," is protecting 3,980,000 of 4,000,000 workers. That's 99.5% of the working population. Only one half of a single percent of Pennsylvania's working adults are allowed to work in Free Choice establishments. Since about 25% of Pennsylvania adults smoke at least occasionally and might like to be able to smoke on the job, that means the exemptions should actually be expanded by roughly 50 times, or 5,000%, in order to give the workers a fair shake at having Free Choice.

And yet that measly one half of one percent exemption makes antismoking fanatics so enraged that they simply dismiss the entire ban as being worthless, filled with loopholes, and in urgent need of strong revision.

If there are any redefinitions of guidelines to be made, I'd say the most urgent one is a redefinition of the word "fanaticism" so that it more clearly points to the antismoking lobby as a prime example.

---

[*] After writing this letter, I discovered that there are actually close to 3,000 exemptions, but my overall point remains valid even if only 99 and 44/100% of Pennsylvania's workers are included. If it was good enough for Ivory Soap, it's good enough for the Keystone State!

Statistical Nonsense

Dear Editor,

Garry Pincock's statement that Pennsylvania's smoking ban exemptions will cost "thousands of lives a year" is absolute nonsense and should not be allowed to stand unchallenged.

Even the wildest antismoking extremists only try to claim about 50,000 ETS deaths per year nationwide, with 80% of those being from home exposures. That leaves us about 10,000 ETS-related workplace deaths nationwide. Pennsylvania's population is 12 million — roughly 3% of the nation's total. Therefore Pennsylvania's total for workplace deaths from secondhand smoke would be about 3% of 10,000 or just 300 deaths. The Pennsylvania bill would ban smoking in the workplaces of literally 99.5% of all Pennsylvanians, leaving only .5% — one and a half deaths — with even those one and a half existing only in the wild estimates of heavily disputed computer models.

The fact that the CEO of the American Cancer Society can make an official public statement magnifying less than two theorized deaths into "thousands of lives a year" shows just how much these nonsensical statistics are worth.

Smoking bans are based on distortions, exaggerations, and outright lies. Americans should have the right to patronize or work in smoking or nonsmoking establishments by their own free choice, not coerced by a government mandate driven by extremists who can't tell the difference between 2 and 2,000!

Rip Off The Mask!

Dear Editor,

In your editorial on amending the smoking ban, you wrote that Ohio's Issue 4 "masqueraded as a smoking ban but was so riddled with exemptions and restrictions that it didn't amount to much of a prohibition."

Issue 4 would have eliminated secondhand smoke exposure for over 95% of workers and yet you claim it "didn't amount to much of a prohibition"? If your city council passed a law prohibiting 95% of passenger cars from entering your city, would you say it "didn't amount to much of a prohibition"? Or if there was a law prohibiting wine in 95% of your full service restaurants or closing 95% of your bars, would you say it "didn't amount to much of a prohibition"?

Or is your thinking on this issue possibly just a bit biased by the thousands of tax-funded TV ads and press releases from antismoking organizations in the past decade? Remember, these organizations could have been using those dollars to cure cancer and make our lives better instead of dividing our communities.

The amendment to the smoking ban that Ohio is currently considering will bring the state closer to the norm established by Pennsylvania, allowing small businesses and private clubs to live instead of die, and allowing workers and patrons some degree of Free Choice in deciding their preference for smoking or non-smoking.

Ohio's voters clearly voted for a ban that exempted family-owned businesses and private clubs. They were robbed of their decision for Free Choice by powerful special interests masquerading as the saviors of workers and children. Rip off their masks and give people back their freedom.

Figures Lie And Liars Figure...

Lies with numbers mingle with lies with words, and New York's Mayor Bloomberg has a special talent for saying the most outrageously untrue things about his smoking ban right in the face of TV cameras and microphones and getting away with it because he's "The Mayor" and it's not polite to point out that he's either a liar or just simply nuts. One particular *Toledo Blade* article caught my eye in 2004 because in a single interview, he managed to come out with not just one or two humdingers, but with *four* of them!

That article, along with one by Chris Stirewalt of the *Washington Examiner*, created the incentive for the first two letters in this selection, while the sheer Orwellian redefinition of legal language around the terms "Employer" and "Employee" brought about the third. While the issue was a bit too complex to examine in the context of a Letter to the Editor, one thing I found when checking up on the legal terminology in that third letter was particularly unsettling. As noted earlier in the section on outdoor tobacco smoke, a Googling of terms defining "Employee" as one who "performs services for an Employer with or without compensation" turns up about 8,000 with about 7,999 referring only to smoking ban laws. Evidently we here have a case where the fundamental meaning of words has not only been consciously changed for the benefit of the Antismokers, but it has been fully accepted and enshrined within our laws, all without anyone even noticing such a fundamental alteration!

The Blade, The Bloom, And The Truth...

Dear Editor,

In the *Blade*'s Sept. 2nd "Reporters' Notebook," four statements are made by Mayor Bloomberg. All four should be seen for what they are: little more than lies.

Bloomberg says bar owners enjoy not having to have "separate smoking and nonsmoking sections." The truth, however, is that most bars never HAD nonsmoking sections and very few of their customers or workers cared.

Bloomberg says bar owners are worried about being sued by nonsmokers. The truth, however, is that the only ones proclaiming such worries are the Antismoking groups trying to convince bar owners to ban smoking. No bar owner I have *ever* spoken to was worried about such suits.

Bloomberg says Big Tobacco supports the bans "because they realize their future is in other products." The truth, however, is that Big Tobacco is now simply supporting anything and everything that might help it squeeze out of lawsuits by playing the "good corporate citizen."

Bloomberg says that as he drives by, smokers outside bars extend their arms and hands to wave their cigarettes at him "with a smile." No human being in their right mind could possibly believe that smokers ENJOY being exiled to the streets. The truth, however, is that the "extensions" he's seeing from the smokers' raised hands are most certainly something other than cigarettes.

The Antismoking Crusaders lie. They do it consistently, they do it with intent, they do it "for your own good," but they lie and the *Blade* should see to it that anyone reading about the issue of smoking bans should be fully and completely aware of that. If you do not, you are doing your readers and your community a grave disservice.

Deep And Serious Trouble

Dear Editor,

Chris Stirewalt's "Free Markets Go Up in Smoke" does a good job pointing out West Virginia's hypocrisy when it refuses to ban smoking in the gambling meccas where such bans would hurt its own pocketbook. At the same time that government preaches to businesses that smoking bans will actually be good for them, that same government knows full well that it's a lie and exempts itself from coverage.

That sort of doublethink is not unique to West Virginia. Rhode Island gave its video gaming parlors a complete exemption (the state gets 250 million dollars a year from gaming taxes), and New York exempted a good number of its off-track betting parlors. Most other "smoke-free" states practice variations of the same double standard.

But I would disagree with Stirewalt's statement that the real problem is constitutional. I think the real problem lies with simple honesty. Ban supporters know perfectly well that thousands of small business owners will suffer because of the bans, but they'll stand up in the media spotlight and blatantly deny it. The Emperor has no clothes, but almost no one in a position of power is willing to admit it.

New York's Mayor Bloomberg himself has been publicly challenged to prove his honesty in proclaiming that his ban is a roaring success and that businesses are thriving. All he has to do to prove he's telling the truth is to lift the ban. If he's lying, then we'll see lots of places going back to allowing smoking. If he's truthful, then we won't.

Has he accepted the challenge? Of course not. Will he ever? Of course not. He knows he's lying, everyone knows he's lying, and he'll continue to lie, and people will simply accept it as par for the course.

A democratic republic that allows its policies to be built on the basis of lies, and a citizenry that accepts those lies as being the norm, is a republic and a citizenry in very deep and serious trouble.

Am I Your Employer?

Dear Editor,

The *Toledo Blade*'s November 20th editorial, "Smoke Bombs," noted that "Private clubs already are exempt from the smoking ban, as long as they have no employees."

Can you name three? I'd guess you can't, although your editorial tone clearly implies that many such places are thriving as voters clearly intended. I'd also guess you're fully aware that "employees" has been redefined for this travesty of a law as anyone "who performs services for an Employer, with or without compensation." And "Employer" is anyone who utilizes the services of such a person, even if they don't pay them a penny.

In reality, just about anyone, anywhere, who does anything indoors for anyone else, is covered under Ohio's present ban... even if it's just your wife getting you a cold beer from the fridge.

I believe you know perfectly well how misleading the ballot language was and are continuing that deception by pretending there are indeed many "private clubs" that are exempt. I challenge you again: name three.

And while you're at it, name just three of the thousands of "family-owned and operated places of business" that Ohio voters voted to be exempt that were actually given exemptions.

Can you? I doubt it; but if you did, would that make you my employee under Ohio's smoking ban law?

Who Would Love A Litterbug?

One of the hallmarks of the Antismoking Crusade, and one of the reasons for its great success, is the awareness of Crusaders that they're fighting a war on many different levels with many different fronts. Rather than just concentrate on smoking's damage to one's health, they've expanded into promoting worries about secondhand and thirdhand smoke, working to allay concerns about economic damage from bans, counting momentary images of ashtray on tables in a movie, encouraging employers to preferentially hire – or even *only* hire – nonsmokers, and encouraging lawsuits by tenants to force condominium association boards and landlords to ban smoking. And remember, they will always try to bring children into the picture no matter what argument they're making.

Littering provides an easy avenue for raising ire against smokers since they produce a unique and easily identifiable form of litter – cigarette butts. Throw in pictures of cute little kids collecting bucketfuls of butts or, for the gross-out factor, toddlers picking up discarded butts and putting them in their mouths, and you've got a surefire propaganda winner.

Surefire, that is, unless someone points out the game they're playing.

Butt Not What It Seems...

Dear Editor,

Peter Slevin's article repeated a sound bite that should instead have been questioned.

Antismoking extremists love to push for beach smoking bans on the grounds of "butt litter," while magnifying both its quantity and its supposed dangers. The article correctly noted that 30,000 butts were found by a cleanup group. But when you realize it took 1,300 volunteers working for three hours to find them, things look a little different. Each volunteer was able to scour up a bit less than eight cigarette butts for every hour of searching. That's not exactly the picture of massive litter that the 30,000 number itself conjures.

As for the danger to children, a quick records check will reveal that of all the untold millions of children who've played on beaches for untold millions of hours, none have been killed by cigarette butts, while many have died from broken glass and other forms of toxic litter – not to mention, drowning.

An additional note of importance is that sanitation officials now believe most butts found on beaches are *not* coming from beach-goers. Instead, they are coming from urban smokers who've been banished to the sidewalks and whose butts get washed to the beaches in storm drain runoffs.

While littering can never be excused anywhere, it seems that smoking bans are the main root of this particular problem, not beach smoking. And, of course, the obvious solution to that – amending the bans to allow smokers some comfortable indoor areas where they can relax with their friends and smoke if they wish to – is never, ever, mentioned in articles like Peter Slevin's.

In reality, concerns over litter or children aren't the real motives for those behind the beach ban push, no more than workers' health is the real concern behind bar bans. The goal is simple social engineering – the "Denormalization" of smokers by whatever means are necessary and using whatever emotive hooks they can come up with. The "Environment" and the "Children" are just their tools, nothing more and nothing less.

KrazyWorld!

Step right in! Sit right down! Open your eyes and ears to a new and psychedelic experience! Welcome to KrazyWorld, where even the craziest statements and arguments can be made and then taken with utter seriousness by newspapers, TV commentators, and, ultimately, the innocent public. The first letter in the lineup looks at yet another Orwellian mechanism at work, the making of unhistory as applied to smoking and global warming. The second looks at a comparison similar to equating birthday candles to forest fires.

The third and final letter in this collection of the absurd takes us to the world of Shakespeare, a playwright who, if he lived today, would almost certainly have something quite quotable to say about smoking bans. Up on stages filled with hot lighting and extensive ventilation systems designed to swoosh all that hot glare up and away from the performers and actresses and out through roof vents; up on stages where figures may be dancing around with flaming, oil-soaked, and very smoky torches; up on stages where a lavish production may even include a central set with a working fireplace or actual bonfire; up on those stages, the Antismokers have swooped in and declared that a lit cigarette will threaten theater-goers dozens or hundreds of feet downwind.

Even sadder than the fact that they've been successful in getting such bans is the fact that they've succeeded in so brainwashing the population that occurrences such as the one motivating the final letter in this batch actually happen!

Wacko Waxman & Winston Smith...

Dear Editor,

Last month it was reported that Al Gore gave a UN speech blaming smokers for causing global warming. And now Lisa Friedman ("Smoke Blows..." November 24th, 2006) quotes Congressman Waxman as saying of smokers, "They feel they have the right over everyone else to use up the air."

When will it be recognized that these people have simply gone beyond the bounds of sanity? They're dancing in the fields surrounding the crazy farms after busting out of their cells. They're taking children away from smoking parents in custody battles,[375] throwing smokers out of apartment buildings, and firing smokers from lifelong jobs unless they quit and submit to random nicotine drug tests. How much further does it have to go before they are stopped?

What's even scarier is when media "handlers" protect the nuts from their own excesses in the spotlight. This same story was reprinted verbatim in the *Pasadena Star-News* with one small change. The damning last sentence, Waxman's quote, was removed and made into "unhistory" by some dutiful Winston Smith.

I guess the idea of "useless breathers" using up our air was seen as going just a bit too far over the edge to reflect favorably upon Mr. Waxman.

Eight and a Half Billion?

Dear Editor,

The push for banning Lambert Airport's smoking lounge shows just how silly the Smoking Prohibitionists have become. The idea that workers or passengers may be getting a dangerous dose of secondhand tobacco smoke from a small, closed, and separately ventilated cubicle is nothing short of absurd.

A "dangerous dose" of toxins? In an airport? An airport where, every single day, hundreds of planes freely spew jet fuel fumes into the terminals' air intakes. Yes, those intakes have filters, but the Prohibitionists have told us repeatedly that those filters are supposedly not even adequate to handle the smoke from just a few grams of quietly burning leaves wrapped in tissue paper.

Looking at just two of the emissions that jets and cigarettes have in common shows how ridiculous this is. According to the Surgeon General, a cigarette puts out a total of 3 mg of nitrogen oxide (NO) and 40 mg of carbon monoxide (CO). Meanwhile, a 1995 EPA study on airplane emissions cites a single 747 takeoff/ landing at about 115 pounds of NO and 32 pounds of CO; that's 52 million mg of NO and 14 million mg of CO, if you do the math.

A little more math shows that for a typical 500 flights per day, all that nice, clean, smoke free air being pumped into those terminals has the CO equivalent of over 160 million cigarettes and the NO of eight and a half BILLION cigarettes.[*] All of which is being shwooshed right into the lungs of travelers who are supposedly being "poisoned" by a few cigarettes being puffed in a secluded, separately ventilated, and sealed-off terminal area.

This would be funny if it weren't so sad.

---

[*] As usual, this was submitted with full supportive documentation. I'll admit, however, that I was both pleased and surprised to see it immediately accepted for publication in the pages of *USA Today*!

Smoke, Magic, and Neuroses...

Dear Editor,

Peter Goodman's story about actors being forbidden to smoke on stage during a production where the playwright clearly intended for smoking to be part of the play shows the extent to which actual mental illness has truly begun to dominate the Antismoking Crusade.

Goodman claims that someone in the audience "was sickened" by smoke from the stage. Has anyone reading this ever been in a play, under the blazingly hot stage lights? What happens to hot air? It RISES! Was the smoke somehow Magickal, defying the laws of physics to reach out against the incoming wind to target audience members?

Whatever incredibly miniscule amount of smoke might possibly have fought its way out to an audience member, it's highly doubtful that there exist any people on this planet who could actually, physically, "be sickened" by such exposure. On the other hand, due to the ever more nonsensical propaganda about secondhand smoke, there are probably millions of people who are "sickened" purely by a neurotic fear arising from the sight of a lit cigarette.

Have you ever been to a Shakespearean play where there are actual burning torches on a stage? How is it that droves of audience members over the centuries were not overcome by the deadly fumes? How about plays that use actual wood burning fireplaces? Where are the corpses in the aisles?

The Antismoking fetish has exceeded any semblance of rationality. It's driven by hundreds of millions of dollars and it is being used to brainwash an entire nation of normal people into becoming neurotics about everything in their environments.

The Other Side Of Love

From craziness it's just a short hop to fear, and then an even shorter hop to hate. In the epilogue to the opening story of this book, we saw how people's fears have already extended to the utterly unreasonable actions of skipping around sidewalks or facing the million-fold greater "threat" of crossing a street in order to avoid a momentary wisp of smoke. In *TobakkoNacht!* itself, we saw the sort of path such hate could lead us down, no matter how unlikely the ultimate scenario might seem. And in a section yet to come, *Into The Trenches*, you'll see enough concentrated hate to make your hair stand on end. The following two letters address the subject of engineered hate, and neither one is particularly pleasant.

The first letter was in response to an article by Michael Kinsley and did not get printed.[376] However, I would like to give some credit to the normally antismoking *Thunder Bay Source* for printing the second one. This was one instance where the other side did get at least a bit of a hearing.

As is usually my practice, I submitted the first letter with a set of full documentation s to the claims made in it. They're not fantasies or suppositions or imaginings, but are all actual and verified new reports that have made local papers but rarely garner national attention. All were properly referenced to serious media sources for the editor of the paper it was submitted to. That should have raised enough eyebrows to bring it to print, but it did not.

Hate Is Not Cute...

Dear Editor,

Michael Kinsley's column treated smoker discrimination as a "cute" sort of thing, rather than being deadly serious. Smokers have lost jobs and housing, been denied medical treatment, and have even been beaten, shot, or tortured simply because they were smoking.

Beaten? Shot? Tortured? Am I joking? Sadly, no. Some recent incidents in the news have included a teen blasted with a homemade flamethrower for smoking near a pregnant friend, a daughter punished by having her face burnt with a hot iron for her nasty habit, a pregnant woman shot after refusing to put out her cigarette in a parking lot, a twelve-year-old girl murdered by an outraged antismoking neighbor, and a thirteen-year-old smoker beaten to death by a fifteen-year-old for supposedly giving a cigarette to his brother.

All true, and all probably just the tip of an iceberg whose size is hinted at every day on the Internet where hate is freely spewed toward Kinsley's cutely termed "pariahs."

Smokers Face Discrimination

Dear Editor,

Canada seems to see nothing amiss in sending their oldest and most vulnerable citizens out into the cold and snow to enjoy what for many of them has been a lifelong, if arguably unhealthy, comfort. One writer showed the degree of intolerance underlying your antismoking campaign by insisting that smoking areas should be "out of sight." Of course: no one wants a reminder of what's being done to the victims.

Your government spends millions of dollars for ads about "breathing easier" after a ban and pays secret police cops to enforce it. Your people have no idea that they are simply part of an expensive social-engineering campaign designed to ensure unresisting compliance and cooperation.

Let me ask this: If your ban is truly so well-supported, why do you need to spend so many millions advertising and enforcing it?

Potpourri

The previous letters have largely addressed matters that made repeated headlines. But there are also topics that don't come up often enough to fit into pretty, bow-tied packages, and this section will end with a half dozen examples of such.

The first letter concerns a recently developed problem as schools are pressured by Antismokers to drop their tobacco stock investments while forbidding staff from accepting tobacco money for research.

The second looks to a future where other activist groups will follow the road to hell paved by the good intentions of Antismokers.

The third entry is the longest, more of an Op-Ed than a letter, and examines several contrasts to exaggerated fears of curls of smoke in the air.

The fourth shows how easily Antismokers change horses from moment to moment as they use our love and concern for children to push whatever their cause of the day is.

The fifth speaks to the lies about worker protection used to pass New York's smoking ban. Most bar/restaurant workers have always been against these bans. They know a good bit of their tips derive from the leisurely satisfaction of a smoke over a drink. Workers' health was merely used as the excuse to quiet anti-ban protests, and the lie was plainly seen when small and completely owner-operated bars came up against the power of the law.

The sixth and final letter is one of my shortest but is also one of my favorites, inspired by the sad closing of Platinum Showgirls. I had never been to Platinum, but I did have the good fortune to meet its wonderful and motherly owner and a dozen or so of her very beautiful performers at a 2006 New Jersey anti-ban demonstration that attracted over 1,500 angry citizens. Feelings against that ban were so strong that the event attracted people from as far away as Massachusetts. Stephen Helfer, founder of "Cambridge Citizens Against A Ban," took a five-hour train ride from Boston simply to hold up his favorite "Tobacco Control Is Out Of Control!" sign for the TV cameras.

Unfortunately, the ban passed and Platinum Showgirls is now closed down, with its staff and owner thrown "safely" out of work – for their own good, of course, but also "to protect the children."

Of Bias And Bucks...

Dear Editor,

You state that that the University of Alberta's decision to refuse all tobacco company funding "will enhance the school's reputation as a center of independent research."

Actually, the opposite is true. Rejecting all studies that would be funded by those supporting one particular view while welcoming funding from those supporting the opposite view can hardly be called "an enhancement of independent research."

Unless the University wishes to display a shameful lack of ethics, it needs to adopt the same policy for grants from sources publicly devoted to advocating smoking bans, higher taxes on smokers, or any other such goals that would interfere with the true neutrality of its research. It should also insist that no government organization funded by tobacco taxes be allowed to provide grants, since that money comes directly from tobacco sales and since bodies funded by such taxes would clearly be biased against studies supporting tax reduction.

For example, imagine a study examining whether eliminating tobacco taxes might reduce underage smoking by wiping out black market sales that have no ID checks. What financial resource could a researcher at UA now have that would support such an investigation?

Or how about a study whose preliminary data showed an increase in heart attacks after a smoking ban? Does anyone seriously think that a Big Pharma company pocketing multi-mega-millions from NicoGummyPatchyProducts would fund such research?

Research studies are always inherently biased to some degree in their design, subject, data choice, and their ultimate interpretation – all with an eye to pleasing the potential funder. The argument could even be made that "no-strings" grants from Big Tobacco might actually be more impartial than those from advocacy groups that clearly state their grants are meant only to advance the goal of reducing smoking.

This push for universities to ban grants from "politically unacceptable" sources is both scientifically unethical and, in light of history, actually dangerous. Researchers should be free to seek support for their work wherever that search for support might take them; that is the only true way to advance scientific knowledge.

The Meat Of The Truth

Dear Editor,

I believe Truth Labels on meats should go beyond just focusing on animal conditions and slaughter. Using the same standards for "meat-associated diseases" that antismoking lobbyists have used so well, pictures of fat clogged arteries and cancerous organs could be mandated on up to 75% of meat packaging. In addition to using a "wrinkled smoker's face" on a pack of smokes, a grossly corpulent bikini-clad lardo could sport rolls of sagging flesh on every fattened up turkey package! It would work even better as such product pictures could be much larger and more detailed.

And just as we've enlisted children into the Antismoking Crusade with scary warnings about their parents dying young and frightening them with images of thirdhand smoke leaping out of their teddy bears to kill them, we can use them to change their parents' eating habits. Children seeing the packages in the supermarket will clearly and loudly demand the gross products be yanked back out of the cart!

Taking another page from antitobacco educators, schoolchildren could be tasked to create pro-vegetarian posters for extra credit or organized to lobby for higher taxes on meats! T-shirts could be distributed to kids with happy positive sayings like "My Mom Loves Me! She Doesn't Feed Me Dead Animals," "Kissing A Corpse-Eater Is Gross!", or just "Meat Eaters Stink!" Finally, college scholarships could be awarded to older students whose essays best demonstrated the virtues of mammalian goodthink!

A short hop further and we can ban BBQ joints and force restaurants to cook meats off-premises. After all, mere ventilation of a separate room in the same building provides no protection against burning tobacco or carcinogenic meat cooking fumes, right?

We can go a long way, baby! We just gotta use the right approaches!

Curing The Cause...

Dear Editor,

Your December 11th article, "Smoking Should Be Banned Out-side...", justifies the proposal by claiming that the mere sight of adults smoking outside is the major factor in causing children to smoke.

Odd. I thought seeing cigarettes on display was the major fac-tor? Or was it TV ads for cigarettes? Ads in *Newsweek* maybe? Wait! Billboards, that's it! And sports sponsorships! Hmm ... or was it Joe Camel? Or possibly the pretty colored packaging? Or maybe that people smoke on TV or in the movies? Might it have been that the evil tobacco companies put extra nicotine into tobacco plants to addict the children? Or was it that they reduced it so kids wouldn't get sick on their first few smokes? Or both? Or neither? Or that taxes on cigarettes are too low? Or that they're too high and have created a black market that doesn't bother checking ID cards? Maybe it's stores selling kids cigarettes? Or bars inviting children to come in and use their vending machines after school? Could it be because smoking is allowed in restaurants where children might go, or in the strip-clubs and veterans' posts where they hang out after school? Or possibly it's not enough funding for antismoking groups? Perhaps it's candy-flavored ci-gars? Hookahs? Bidis? Cloves? E-cigs? Wait... I know... kids smoke because they go on the Internet with their credit cards to buy cartons of cheap cigarettes from Indians!

Hmmm... do you get the feeling that the antismoking fanatics will say just about anything they darn well feel like saying in order to get whatever it is they want on any particular day? Maybe that's why we don't hear so much about peer pressure anymore: there's just no grant money in pushing that particular hustle!

When One Is One Too Many...

Dear Editor,

Steve Yablonski's *Oswego County Today* article about the smoking ban's effects on the bar business was very well done. He's one of the few journalists out there who don't feel obligated to repeat the politically correct nonsense about how "most" bars are doing just fine.

It was sad, though, to see the innocent but mistaken belief by one of the bar owners he interviewed. Rose Anthony, owner of Rosie's, believes that since the law was passed on the basis of "protecting the workers," she will be able to run her business herself, just on her own, with no employees, and allow her guests to smoke.

Sorry, Rose, but you're wrong. Just ask Mrs. Glanville of Costello's Bar up in Cayuga County. Her life's work and investments in her owner-operated and self-staffed bar were destroyed by the smoking ban.

The main justification for the smoking ban was fraudulent, a fact clearly seen last week in New York's Senate when they denied bars the chance to guarantee air CLEANER than the air in nonsmoking bars through state-of-the-art ventilation and filtration systems. Even with such a guarantee, the antismoking lobby fought and succeeded in blocking such a change in the law. So much for concern about workers' health, eh?

Mrs. Glanville, Rose Anthony, and many others like them are having their lives and livelihoods destroyed on the basis of a fraud designed to promote a social engineering goal. This is a disgrace to America and what America stands for.

Fumes & Fun, Sunshine & Popcorn

Dear Editor,

Antismokers like to say, "There is no safe level of exposure to secondhand smoke." They ignore the fact that the same is just as true for alcohol, sunshine, or all the other carcinogens we encounter every day in small quantities with nary a thought or care.

Sunshine? Is it truly that bad? Well, the UN claims 60,000 people die annually from its cancerous effects. Even a quick peek out a window may kill you – if we use the Surgeon General's standards for tobacco smoke.

Antismokers laugh and point out "There's no avoiding sunlight, but you can avoid smoke by having bans." They forget smoking bans are passed mainly on grounds of "protecting the workers," since there's no research indicating any harm from the passing smoke exposures customers experience. There are many workers forced to serve self-centered Sunners lunching on restaurant patios. Why should those poor workers be "the only ones forced to work in a carcinogenic environment"? Should daytime patio dining be outlawed? After all, outdoor patios are no more intrinsic and necessary to eating at restaurants than indoor smoking is to drinking in pubs.

Speaking of drinking... Antismokers also like to say, "You're not forcing others to drink!" but you certainly are forcing a carcinogen on them. The sum total of discrete Class A Carcinogens in the smoke of a single cigarette masses only .0005 grams. A nice, fresh martini puts out roughly one full gram of alcohol vapor per hour, and ethyl alcohol is also a Class A Carcinogen. That one drink is putting as much known, discreet, human carcinogen into the air as 2,000 cigarettes, even though you neither see nor smell it.

Try it yourself. Pour a jigger (48gms) of grain alcohol into a martini glass. It'll be gone in two days. Where did it go? Well, unless your kittycat is a closet tippler, all that nice juicy carcinogen bubbled straight into the air to be inhaled by you and your family – almost 100,000 cigarettes' worth at the rate of about one gram for every hour.

Applying the same zero-tolerance rules to alcohol as extremists demand for smoke would force us to ban alcohol from restaurants "where people are forced to work." Alcoholics could be told, "Just step outside for a moment" to grab a few quick gulps of their fine merlot after a meal.

It's not just sunshine and martinis, though; consider the deadly popcorn fumes! Research now indicates that popcorn factory workers who are constantly exposed to that delicious buttery aroma can lose up to 80% of their lung capacity to bronchiolitis obliterans, a condition that literally obliterates the bronchioles – the lungs' tiniest airways.

If "firsthand exposure" among popcorn workers is that deadly, what about the "secondhand exposure" you and your children get at the movies or while munching microwaved popcorn on the couch?* Picture TV ads showing babies in the womb being force-fed butter-flavored popcorn while their tiny computer-animated lungs slowly wither and die. Picture parents being denied custody of their children or losing their jobs because they are "popcorn eaters."

The only thing unique about the "deadliness" of secondhand tobacco smoke is that a powerful lobby has focused our attention on it and magnified our fears of it. Having a patio lunch may actually be more dangerous than being inside with the smokers. Dining in a poorly ventilated, non-smoking, alcohol-friendly restaurant may be deadlier than dining in a well-ventilated smoking establishment. Going to the movies may be more lethal than both combined!

The hysteria surrounding secondhand smoke is deliberately created to pressure smokers to quit – no more, no less. It's a hysteria fed by media outlets making extra bucks from scary headlines. And it's a hysteria that has ruined the lives and livelihoods of many innocent people.

---

* When this was first written in the early 2000s, such "secondhand popcorn exposure" was pure fantasy, but in 2007, Reuters reported the first case of bronchiolitis obliterans in a heavy popcorn consumer! ("FDA to Probe Popcorn Link in Man's Lung Disease" by Julie Steenhuysen, Reuters, September 9, 2007)

I'd also like to note that, although some companies have discontinued their use of diacetyl (the suspected villain) in popcorn, it continues to be present at levels approaching two billion picograms in a single bottle of a nice chardonnay wine! (Don't worry... that's only two-thousandths of a single gram, and it's unlikely much of it evaporates into the air! Most sane folks would be a lot more worried about the *alcohol* in the bottle!)

Johnny's Little Lap Dance...

Dear Editor,

Your November 21st edition carried a story about Platinum Showgirls being fined $1,000 for violating the New Jersey state smoking ban. I found this interesting because I'd originally thought this ban was intended to "protect the children," since children were so often prominently displayed in the ban campaign.

If little Johnny is stopping into Platinum Showgirls to spend his lunch money on lap dances while on the way home from school, I think New Jersey has much larger problems than people smoking.

Figure 14

Stephen Helfer at a New Jersey Free Choice Demonstration, 2006

Launched In the Trenches

The antismoking industry has enormous financial resources that it can use to get its message out. In one section of *Dissecting Antismokers' Brains* I used the magic of TiVo to examine the programming of just one popular TV channel, MTV, over the course of a random 24-hour period. Despite complaints from antismoking groups that MTV was "filled with smoking," I found only about two minutes out of the 1440 in that 24 hour period that actually portrayed people smoking – not quite the image that "filled with smoking" brings to mind.[377]

I also found something else that was interesting: there were ten full minutes of anti-smoking commercials! If you stop to think for a moment how much a minute of advertising time on MTV costs, and then think about the thousands of other TV channels out there across the country that are also targeted by antismoking advocates, you can begin to get a grasp of how "Tobacco Control" can spend hundreds of millions of dollars a year of our tax money. Plus, there's a healthy chunk of antismoking advertising that never even makes it to the financial ledgers. TV stations are required to provide a certain amount of public service advertising for free, and many of them have found a clever way around that requirement. They simply insert pre-scripted antismoking messages into the plot lines of their top-rated shows and get credits that allow them to return station breaks to actual paying advertisers.

This sort of subliminal advertising through scripted character behaviors is a significantly nasty step beyond the simple product placement that many companies (Heh, except for the tobacco industry nowadays!) legally and frequently engage in. Simple product placement and designated advertisements during commercial breaks are things that viewers are aware of and can psychically defend themselves against. The deliberate subliminal manipulation of their attitudes and beliefs by the scripted actions and words of characters in their favorite shows is a different story; it's a powerful and, in my opinion, very perilous use of the mass media for social engineering. It's a form of "thought placement," or "correct attitude adjustment," and to the best of my knowledge, the only major peacetime use of it thus far has been in the service of the Great American Antismoking and Antidrug Crusades. But even within such a limited scope, its mere existence should be enough to horrify any decent citizen.*

Those of us fighting this machine don't have the money or power to buy such advertising or to influence the content of TV programming. We don't have even a thousandth of that kind of money or power. All US Free Choice groups combined probably spend no more than a few millionths of what's spent by their opposition, and pretty much all of that spending comes straight out of the pockets of the private activists in those groups. So what can we Smoky Freedom Fighters do?

As you saw in the preceding sections, one answer is to go to the letters pages of newspapers. If one is smart enough – which I obviously was not – to avoid wasting energy on tightly controlled organs like the *New York Times* and *Washington Post*, many local papers are open to solid letters representing contrary views. It's a very limited outreach tool, but when you're fighting with nothing but slings and arrows against mechanized tank brigades, you've got to work with what you've got. A single flaming arrow with the luck of landing in the midst of a munitions dump can do greatly disproportionate damage; but to get that one landing, with that element of luck, many have to be fired.

While writing letters to newspapers may sometimes feel like an exercise in futility when they don't see print, the Internet offers an alternative where Free Choice activists can be fairly sure that their efforts will at least reach a small audience. That audience, however, can be quite important when it contains people who may not have

---

* In wartime, Hollywood has often offered visions of patriotism, but how would you feel if a "pro-war" administration facing off against a pacifist political opponent began offering tax credits to TV networks to have scripted characters glorifying the war effort in order to sway future voters? Think how innocently it could be guised as simple patriotism—and how dangerous such political manipulation could be at the wrong times and in the wrong hands once the precedent for it had been well-established and accepted.

previously had any awareness at all that rational opposition against the Antismoking Crusaders even existed outside of Big Tobacco. Blogs and comment areas on news boards are free for the posting, and some of their readers may be inspired by your writings to stand up and speak at a Council hearing or be spurred to look around for more information to counter the sound bites of MTV commercials, paid press releases, and subliminal bits of TV dialogue.

Internet postings may not achieve the visibility of the heavens, but each and every one of them gets out there in the public eye to spread knowledge, raise consciousness, offer moral support, and sometimes actually change opinions. This is the trench warfare of the smoking ban fight. It can be nasty and dirty and painful at times, but it can also be fun! Antismokers are not used to having opponents armed with knowledge who are willing to stand up to them. They think that simply flinging words like "dirty," "smelly," "addicts," or "scum" will serve to make the smokers and their friends scurry to hide like vermin behind the dumpsters.

In recent years, they've gotten a surprise. Internet boards and blogs are now far more dominated by those fighting for Free Choice than by those opposed. And while those pushing smoking bans are usually doing nothing more than repeating expensively developed sound bites or casting slurs and accusations,* the people on our side of the aisle often know and communicate far more of the actual science than would typically be conveyed in dozens of MTV commercials.

In the trenches of the Internet, Free Choice rules. The few anti-smoking professionals who have jumped into this battlefield have usually had themselves and their arguments quickly exposed or out-classed. That's not to say there aren't any professional Antismokers earning paychecks for posting opinions on web boards, since there's little in the way of real oversight of how their money gets spent or used. Unlike the situation faced by today's tobacco companies where pretty much every sheet of toilet paper used in the executive wash-room has to be accounted for, antismoking planning and activities take place behind a dark curtain that's rarely penetrated.

---

* A favorite, naturally, is intimating that anyone opposed to bans is somehow in Big Tobacco's pocket by labeling them as lobbyists, puppets, or shills.

Note I said "rarely." In fact, the Antismokers do occasionally slip up, and when training new "grassroots" groups they'll sometimes be a bit too open about their tactics. There are a number of professionally and expensively developed "Strategy & Training Guides" floating around on antismoking sites and some of them have snippets of embarrassing advice. For example, in *Smoke-Free Outdoor Public Spaces: A Community Advocacy Toolkit,* the Physicians for a Smoke-Free Canada urged organizations to train their members to "submit at least two letters to the editor each month during the campaign, *under the names of different authors.*" [Emphasis added.] Not exactly the image of integrity you'd expect from public health groups, eh?[378]

Antismokers aren't happy with the Internet's level playing field and exposure of their activities, but so far there's been little they can do about it. Most webmasters refuse to take sides in disputes and actually love debates that get a bit bloody while still exchanging good arguments and information. That sort of action drives traffic to a website and encourages return visits to see how the battle is going, and that traffic, as any good website manager knows, is what drives Internet profits and power.

Of course, such Internet fairness is not always the rule. As we'll soon see in the section on censorship, Orwell's dreaded unhistory makers are alive and well and living in cyberspace. And the next time you read a news article about protecting "The Children" from internet excesses, think about how those tools may be developed to protect adults from disturbing political thoughts.

Originally, I was planning to fill the *Trenches* with examples of the general debate and mudslinging of smoking-related Internet news boards and blogs. I wanted it to be a microcosm of the fights between the foot soldiers on both sides of the issue on the digital battlefield. I thought there'd be good information for sharing and good illustrations of how the arguments of the Antismokers fall apart when space is given for equal and thoughtful response.

Eventually, I decided against that approach. It might be entertaining, but to really do it justice would extend into far too many pages, and, for the most part, the factual information covered would simply be a repetition of what's already been covered in the previous sections of this book. Instead, I'm going to examine four distinct but fairly limited areas of concern in public Internet communications. I hope that such an examination will help Free Choice activists as they use this wonderfully egalitarian tool to communicate while it's still openly available.

(1) Style
(2) Challenges
(3) Hate
(4) Censorship

Afterward, I'll conclude with a few more general thoughts.

Style

Debating an issue on the Internet can be one of the most productive forms of discussion in the world. Both sides of an issue are on equal footing. One side can't simply "shout over" the other side, and generally there's no overbearing moderator to swing the debate over to a favored view.

Attempts to dominate the discussion by being long-winded are worthless as readers end up simply clicking a key a few times to skip over all the verbiage; while attempts to dominate by bullying through nasty attacks or by TYPING IN ALL CAPS – the Internet equivalent of shouting – also get skipped over. In both instances, the bully's arguments are generally downgraded in the minds of readers rather than being accepted.

Attempts to win an argument by ignoring your opposition's main points fail miserably, since your opponent can quickly and efficiently point right back up the page to the points that were ignored without having to waste valuable time repeating them. Attempts to use "Straw Men" by altering an opponent's argument to one more easily attacked also fail, since the alteration can be speedily and clearly pointed out and the scurrilousness of the trick rebounded upon the trickster as an illustration of the weakness of their position and their desire to avoid honest argument.

Likewise with "Appeals To Authority" or "Ad Hominems" or most other fallacious arguments. These tricks make for a good "sound show" in a TV debate but they fall flat when they can be calmly pointed out as fallacies and the audience clearly directed to examine their use in previous paragraphs on the same page.

The Internet is the one medium where facts can be instantly and efficiently cross-checked by the audience while the debate is put on hold, and where that audience can easily and effectively judge which side of an argument was presented with the most merit and substance – as opposed to the simple volume and flashiness of style that typify televised verbal debate. On the Internet, brevity and accuracy rule. Reading posts on one's computer monitor is different than reading pages in a book. Readers tend to compensate for the lack of professional editorial judgment on the net by rapidly skimming through posts to look for the meat of the argument. Anything longer than a few paragraphs is generally viewed as likely to be either boring or just full of cut-and-paste junk – so it largely gets skipped over.

When people read a book or an Op-Ed piece in a newspaper, they're generally prepared to be preached at to some extent. On the net, though, they simply don't have the patience for it. When skimming the Internet looking for new information, new ideas, scintillating bursts of insight, and maybe some fun blood 'n guts being spilled between two sides of an argument. If they really wanted to read in-depth arguments on one side of an issue, they'd either get a book on it or go to a dedicated website that specifically devotes pages upon pages to the presentation and analysis of that side. They want to applaud their side and boo the other, or, if they're fairly neutral on the subject, at least get a quick sense of whether one side is throwing better punches than the other.

There are three unfortunate aspects to net discussions, though: hate, obscenity, and censorship. I'll be discussing hate and censorship in a while, but as for obscenity, whether mild or nasty, it's something you'll just have to deal with when you're out prowling in the slums of cyberspace. Some "neighborhoods" (the better run and organized blogs and professional news boards for example) have built-in protections to shield the eyes of the innocent from offensive words, but their effectiveness is spotty and they can also be annoyingly overzealous. When you're speaking about old school chums, you don't want it to appear as stories about your cl***mates, and if you want to want to say "I thought its assumption was hit perfectly," then you really don't want it to appear as "I though* *** ***umption wa**** perfectly."

Beyond that, there's the whole realm of nasty but not outright obscene wordings. I've been called "McDouchebag" and "McFaggot" and other amusing derivations by Antismokers on the net a whole sh*tload of times by various f*ckers!

But, as we've seen, there are good aspects to Internet debate as well. People can't easily run away from questions and challenges other than by changing anonymous identities (which reduces their ability to appear legitimate). And, when it comes to truly public figures – such as the heads of antismoking organizations or politicians who are pushing bans – such anonymity isn't an option. A wonderful example of such a situation can be seen in the back-and-forth battle I had with Mr. James Repace (the tornado chasing "Secondhand Smoke Consultant") on the Greenbelt Patch website where Mr. Repace made the tactical error of "dropping in" to comment on a trial where he'd testified as an expert witness. The exchange is too long to include here, but it can be seen in all its glory at the referenced website.[379]

A major point of interest in that exchange of postings extending over several weeks was Mr. Repace's complete inability to provide even the most basic scientific evidence to back up his initial claim that levels of smoke from a neighboring apartment could cause endothelial dysfunction. Eventually, after thirteen specific requests (mixed in and through comments from many others who'd joined in the party) Mr. Repace simply disappeared after a botched censorship appeal to the editors of the Greenbelt Patch painting me as a "tobacco industry spammer."[380] Aside from another short-lived attempt to paint me as a "tobacco industry mole,"[381] he hasn't, to my knowledge, been active on the Internet since, at least as of January, 2013.*

Such convenient offerings from the men behind the curtain aren't very common, though, so I try to create my own opportunities to pin antismoking advocates publicly to the wall by publicly calling them

---

* Repace's charges stayed on SmokeFree DC's website for months, while my defense remained hidden. Eventually, the site's legal minds decided discretion was the better part of valor and removed his message. Here is a segment, with the rest at the referenced iCyte: "*MOLE WARNING: [The Patch] has been attacked by semi-pro Tobacco Industry Spammers. ... McFadden, ... Mulvina, and Magnetic, among several others, invariably pollute the message boards ... with their fanatical flat-earth pseudo-science. ... some moderated sites have banned these industry moles.*"

to task on their statements to the press. That effort has resulted in my creating a series of "Challenges" made to antismoking advocates and politicians in public forums. They're similar to some of the challenges I've offered as letters to the editor, but on the Internet, there's room to lay them out in more detail. There's also the nice option of returning to them later and pointing out the lack of answers they've received from supposedly responsible public officials[382, 383] or journalists[384, 385] who have been fully notified of their existence. They're rarely answered, but the very lack of an answer serves as an answer in and of itself.

Challenges

In other mediums, it is difficult – and often sounds petulant or didactic – to bring up unanswered points from earlier debates with an opponent. On the Internet, though, calling people up on their exact words is more acceptable, partly because readers can so instantly and easily cross-check for themselves with a quick click up the page or hop to a referenced link, and partly because the nature of Internet argument tends to have more real substance that of the easily distracted, flashy, and side-tracked verbal duels on TV.

Of course, there are some individuals out there who will disappear for a year and then suddenly come back, repeating their views as something fresh and new and worthwhile; but there's nothing more satisfying than pointing out, with links, how they carried on the same long and annoying diatribe a year earlier and then ran away once their lies were exposed. The speed with which they again disappear can be wonderfully fulfilling … as well as being quite educational for any passersby who are simply skimming to see which side of an argument has more substance.

One Internet-friendly approach to arguments that I've become quite attached to is creating and presenting simple and forthright "Challenges" to Antismokers where I ask them to back up their sound bite / playbook-scripted claims with facts. Those sound bites and claims have often been created by advertising agencies and professional organizations to create good, short punches to the gut of the general public, but they also have a fatal weakness: because of their repetition it's easy to develop good responses. And I've found that it's a rare Antismoker indeed who can even begin to defend the sound bites he or she presents.

I've got five basic Challenges that I have put out frequently on the net, with the first two being direct responses to two of those sound bites. None of the five have ever been met with successful responses. Over 95% of the time, discussion board opponents either simply ignore them in hopes they'll get buried by other postings, run away in frustration, or respond with *ad hominems* that make it clear they have no real answers aside from making personal attacks.

(1) **The $ Challenge:** Antismokers love to say bans will be good for business, even bar and casino business, and that overall, "no one will be hurt" by them. My challenge here hits right at their pocketbooks. All they have to do to quell most of their ban opposition is promise to cover the business losses (which they claim won't happen) out of their own pockets, both personal and organizational. After all, if they're telling the truth, they'll have nothing to lose. But if they're lying... well, then they'll run away faster than a little girl from a pack of tarantulas.*

(2) **The Vote Challenge:** Antismokers love to say they're pushing bans to save the poor workers who have no choice or voice in the matter. I say, "Fine. If the workers want the bans, then they can have them. Give them that choice and voice and let them *vote* on whether they want a ban, either in their own establishment or city/county/ statewide." The result? Never once has any antismoking group thrown a ban decision to the workers they claim to represent; they *know* that most workers in the hospitality industry would have overwhelmingly voted against smoking bans.

(3) **The Stiletto Challenge:** After writing *Dissecting Antismokers' Brains,* I realized I needed something smaller and sharper to reach out to people who were not about to lay out $22 for a 400-page book and then spend hours reading

---

* To no big surprise, they *always* run!

it. I developed a short, large-print, openly one-sided (although both accurate *and* honest), and somewhat bombastic booklet that could be easily bound in cheap plastic term-paper covers and read quickly in a bar or smoking-pit where time might be limited or the light might be dim. Its formal title is *The Lies Behind The Smoking Ban* and its nickname is *The Stiletto* – a small, cheap, easily manufactured and carried, but sharp and deadly weapon perfect for self-defense. I'll usually just post a link to the site where *The Stiletto* is offered, invite attacks, and then stop back to defend it if any attacks materialize. After hundreds of offers, I've only had two Antismokers try to seriously attack its content – and neither of them did very well. *The Stiletto* is freely available for reading, downloading, and printing at: http://kuneman.smokersclub.com/PASAN/StilettoGenv5h. pdf and I'm quite happy to help anyone who'd like it customized for their particular fight.

(4) **The "Name Three Studies" Challenge:** This challenge was outlined at the end of *The Slabs*. It originally came from two sources: (A) Dave Hitt's idea of challenging Antismokers to name three people who'd died from secondhand smoke,[386] and (B) the common attack on Free Choicers that we "cherry-pick" weak studies to publicly attack. Basically I ask my opponents to pick the best three studies that they can find and defend that show significant harm to ordinary people from the durations and levels of smoke commonly found outdoors or in any decently ventilated Free Choice business today. As noted earlier, this challenge has never been adequately met at all. Despite specific requests to avoid such things, opponents offer statements by important people, generalized reports, advocacy fact sheets and websites, opinion articles, lists of irrelevant links, and kitchen sinks they've dragged up from their basements; but never, ever, have they been able to come up with, much less defend, even a single study that meets that one simple set of criteria.

(5) **The "Ban Lift" Challenge:** This is one that I use whenever I run across a news story celebrating a ban anniversary or gushing about how wonderful, successful, and universally loved a ban has been. The concept is simple. If the Antismokers are telling the truth, then they could make a strong statement to the world and advance their cause enormously by taking one easy action: simply repeal their supposedly successful ban and let the world see how all those happy bars and restaurants and other businesses keep their own bans in place without the law! Just as with the other challenges, not a single taker has ever accepted this challenge – despite the tremendous propaganda coup that such an action could hand their movement. Or it *would* be a tremendous propaganda coup if their message was actually truthful.

Hate

As I noted in the introduction to *The Trenches*, I'd originally intended this entire section to be a collection of examples of the back and forth repartee on the Internet's discussion boards. I'd also intended to use it to show the increasingly dirty feet of the antismoking movement: the hate that is slowly building up toward the sort of things fictionalized in the story of *TobakkoNacht!*

I've dropped the repartee aspect, but I still think it's a good idea to share at least some sort of hardcopy record showing the sort of hate that has been manifested online. This sort of stuff will generally be filtered out by well-meaning censors at newspapers or on TV, but it's important that the public see it and recognize the threat that it represents – not just to smokers, but to any minority that might fall out of favor in the future. All too often, people think that just because we're not seeing or hearing too many expressions of sexism, racism, or religious intolerance in public today, the human tendency to find groups of "others" to hate has disappeared.

It has not disappeared; it's simply been sublimated into targeting a group that even the most ardent of feminists, gay activists, liberals, and other such "progressive" types find completely acceptable as an object of hate: smokers. Are such verbal expressions of hate relatively harmless though? No. They form the foundation upon which acts of hate, those such as the beatings, killings, tortures, and punishments described in the Hate section of *Brains* and which have continued and multiplied in the years since, are built on. It creates the atmosphere that lets the haters convince themselves that their hate is socially acceptable or even commendable, and goads the psychologically vulnerable into eventually acting out that hate in harmful or even deadly ways.

Here then are some examples of such vitriol. I had originally planned to just share a few sentences typifying the material I've seen in the dirtier trenches of the Internet. As I went through my saved files, however, I'd actually gathered over the years and further realized that no sampling of five, or ten, or even twenty sentences could really transmit its reality. I was torn between wanting to share all of it and wanting to share none of it because it was simply too revolting. I soon recognized that there was no question of sharing it all as the files ran upwards of 30,000 words – almost seventy pages' worth in the general style of this book.

What I have decided to do is share just about ten percent of those samples and condense the style to a solid block of four pages in a reduced font, putting one piece of invective right after another, separated only by slash marks. In total, these words represent the feelings of well over a hundred different individuals. If I wanted to spend the time on it, and if there was any reason to do so, I could probably collect material like this authored by well over a thousand separate Internet users.

The ironic thing is that many of these Internet posters in other postings done in response to other news stories on other topics showed themselves as being "good liberals," the political type who normally would violently object to the idea of this sort of hate being expressed against blacks, or Jews, or Muslims, or women. When it comes to smoking and smokers however, they simply excuse their attitudes with such comments as "Smoking is a choice." (And, as noted earlier, they'll then immediately turn around to rant about protecting twenty-year-old "children" from "a lifelong addiction more powerful than heroin or cocaine.")

I considered trying to reference each quoted sentence or phrase and decided against it because there are simply so many and because most of them were posted under anonymous handles on message boards without permanency. If you've gotten this far in reading *TobakkoNacht – The Antismoking Endgame*, I hope you've developed enough trust in my statements to accept that each and every one of these is shown exactly as written for the public eye on the Internet. The excerpts are verbatim as originally produced, including their various typos.

Given the extent to which I've seen such expressions of antipathy grow in frequency over the last ten years, I'm sure anyone with net access can pretty quickly find their own set of examples if they have any doubts about the origins of what follows. For those without Internet access, these examples may serve as an eye-opener, although those who've already spent a lot of time on the net may just want to skip over the nastiness of the next few pages. It's only a small sampling of what's out there, but it's enough to be disturbing. The next four pages, if read aloud in segments, enunciating clearly and forcefully with perhaps just a slight pause between each thought, can give an audience the classic Orwellian daily dose of "Two Minutes Of Hate" for the better part of a month.

I suggest you read the next few pages, and then go back and reread *The Flame* in the opening story through a new set of spectacles.

All you smokers are idiots. Die. / If smoking were not so horrendously evil no one would bother to think about it. But when they smoke in my presence they should be stopped and if possible destroyed. / (Question)What would you do about the 50,000,000 smokers who the ban would drive onto the black market? (Answer) Shoot every last one of them / I vote to kill all smokers....perhaps then, and only then, will the putridness of their disgusting habits be truly understood. / a few tobacco stores may find them selves burnt to the ground – a few tobacco farmers may be found hanging in a few trees, a few smokers may find themselves on the other side of a baseball bat / you are such pathetically gutless spineless little whiners that you confuse addiction with freedom. How bloody stupid can people be? It has been argued before that people that stupid have no right to live / smoking around helpless children is probably the worst form of child abuse next to murdering them. you should be locked up. Smoking around children is child abuse, pure and simple. / At the Orlando Airport you should see the segregated small glass rooms that the Addicted Smoker Losers must sequester themselves into, by law, to fix their wretched addiction. Like dirty monkeys in a cage on display, their faces sullen and depressed with the pathos of their junkie addiction. Its humorous to watch these pathetic addicted simians, eyes sullen by despair, grotesquely overweight and pale corpses, wrinkled ugly and worn skin, all caged together in a stinky, smelly smoky filled chamber or wretch where they belong ha ha ha, suck harder and faster now you wretched whores! Even these putrid human scum can't stand the stink and filth of their own repugnant drug addiction waiting until the last possible second before stepping inside to light the putrid drug delivery device they had placed at the ready in their nicotine stained, foul smelling rotting suck-holes. / if Democracy allows poeple to throw away their lives by smoking then democracy isn't a good thing. / Every day you wretched drug addicts are further being restricted and taxed so that very soon, the only place you'll be able to fix your pathetic tobacco addiction is in the privacy of that festering hole you call home. How apropos! Suck harder and faster now you maggoty addicts / lick ashtray, lick ashtray, lick ashtray, lick ashtray, lick ashtray, lick ashtray, WE DON'T WANNA EAT AROUND YOU!!!! WE DON'T WANNA SMELL YOU, STINKERS, lick ashtray, lick ashtray, lick ashtray, lick ashtray, lick ashtray lick ashtray lick ashtray / Place a bounty on anyone seen in a Joe Camel or Marlboro T-shirt, baseball cap, etc. They may be shot on sight and a $1,000 reward will be given. / you want to force your filthy habit on others / Allowing people to smoke in public places is akin to allowing people to defecate in public pools. / boo hoo hoo for the worthless drug addicts / a restaurant that permits smoking was operating the tobacco equivelent of a crack-house, for heaven's sake. He was making his living by providing a place for drug addicts to gather. Now these pests are where they belong when smoking / they should be rounded up into concentration camps so they can smoke themselves straight to hell / You smell, your clothes smell, your cars smell, your homes smell, your breath reaks, your teeth are yellow and your lungs are black!!! / You filth. I'll rejoice the day the doctor sticks a traech tube in your throat and you scream from the pain and agony of chemo!! / you're just a dumb addict / you should be locked up. Smoking around children is child abuse, pure and simple. / someone who denies the risk they are taking, does not deserve my help. Nor the help of any physician, nurse, hospital, etc. / It is bad enough that we have to put up with your nasty habit but to then to heap on the offense by not even cleaning up after yourself is too much. Nothing a bullet wouldn't fix that is. / Smoker's have no idea how bad they reek since they can't smell their own filth. I bet a pig that is rolling around in its own filth thinks it smells just fine. / To discourage smoking ... more emphasis should be placed on the uglyness of Smokers Face stinking tobacco breath, premature baldness, impotence, the 400% increase in AIDS suseptibility / smokers will continue to be harassed, taxed, restricted and humiliated back into the grotesque stench-pits they call home by a society that has been too tolerant for far too long. / charge the tobacco company CEO's with GENOCIDE. They should be executed along with all the politicians who took money from them and collaborated. / I demand that this nicotine addiction not be allowed to be spread to another generation by the outlaw tobacco industry. I demand that smokers go quietly into their private, dark, closed dens and smoke themselves to death as the last generation hooked to this hideous, worthless practice. / Eventually the smokers will have to remain say 200 ft from any public building not just outside the door. Then you'll really look pathetic smoking in the rain because you're so weak and have to smoke regardless of where you are forced to. I'll look outside and laugh at you. / Get a life JUNKIE, addict, YOU are a NICOTINE JUNKIE!! / Smokers violate me. it should be against the law to exhale - with a charge of attempted murder. Moreover, smoking parents should be arrested for child endangerment and attempted murder if caught smoking around their kids.... We have no desire to be around you, your smelly clothes, yellow teeth, bad breath, wrinkly skin, chronic coughing and nic fits. Smokers in my opinion are mentally ill criminals. They should quit or be put down. / Pubs can certainly survive without smokers. I hope the cold winter kills a few more off in fact / I want smoking n the presence of a minor child to be a CRIMINAL charge. My daughter, a social worker, tells me that most child abusers are smokers which stands to reasons since smoking Is the worst form of child abuse. Smokers should be shot and killed on sight! / they are ignorant, abusive and don't care who they harm as long as getting high is on for their selfish selves. / screw the smokers!!! ban it everywhere / BAN SMOKING, it's sick, it's disgusting, it stinks like sh... and I have to breathe that in because some selfish d-bag 20 ft away is lighting up? / Smokers need to have the words, STUPID IDIOT across their foreheads! stop forcing others to breathe the poisonous fumes, I DO NOT want to smell it or be FORCED to breathe it ANYWHERE. / You are second-class citizens. If you don't like it, move. I don't want you here anyway. / Smokers are idiots. ... they are a drag on our health care system. We should do them a favor and give them a quick clean bullet through the head. / What's the ETA on the global smoking ban? I'd like to know when we can get rid of smokers. Hell, I'll even go ahead and say smokers should be labeled terrorists since they terrorize my lungs and threaten my health. / I like watching recent American TV police shows where the guy who smokes gets shot. That'll teach him. / Smokers are generally selfish, childish bastards / Have you seen them, they drop thir cigarette stubs and walk on like baboons discarding banana skins. Provide vats of slime for them to stand in when they smoke then just dismiss them for time wasting / Why is it that you are so opposed to murdering children with baseball bats, but not opposed to killing them with secondhand smoke? One is OK and not the other? / Crusty leather-skinned tar-ridden ash-tray mouthed, dogs-bottom-mouth, chimney-faced horrors the lot of them. / It's about time the govt started to get serious about getting rid of smokers. / enjoy your emphysema clowns, and keep your smoke to your own festering, brown walled, putrid holes. Suck it in suckers! / They are the worst. There is nothing worse than walking to class at nine in the morning with a smoker in front of you blowing smoke all over you. I want all smoker dead, but especially morning smokers and any one who smokes on campus. DIE!! / smokers are the lepers of our society and are treated as such. / I hate smoking and I hate smokers, the ban couldn't come quick enough for me. Best thing to happen for years. / Smokers scum of the Earth, a cull next. / Smoking is disgusting and dirty. And so are smokers. I think I'd prefer to have the lepers. / I have always looked down at the filth or brown fingered,brown teethed lower classes that smoke. / I hate smokers and I'm sure all other non-smokers hate them, too. I've hated smokers for many years and I am almost positive that one day, I will successfully kill someone who smokes. I encourage any non-smokers who are reading this to go out and kick the shit out of smokers. / The issue is whether you quit on your own steam or you are forced. And you _will_ be forced. And you will be forced by those who despise you with a ferocity that you simply can't imagine. The smoke _will_ stop. You are shit. You are puke. We will dance on your grave. You will die. / If a person is caught smoking, he or she should be shot on sight. The world would be a better place! / SMOKERS,

PLEASE die from diseases from cigarettes sooner rather than later, so there will be less of you around, stinking up every place you go. / I hate it when I'm waiting for the bus and someone decides to arrive and then have a smoke right next to you.... they don't just kill themselves but everyone around them. they might as well bring a knife a stabe people cos its the same affect!!!! / I should not have to pay for those weak-willed people who are too lazy and stupid to stop smoking let them die from their self inflicted diseases! / Bad smells, bad breath, choking air, horribly ugly smelling cloths, discoloured fingers and teeth, possibly unhealthy babies, stinking houses and homes, ash pit smelling cars, discoloured skin. / ignorant smoking hillbillies / they are huddled together like penguins, dragging away surrounded in smoke as though the last fag was going out of production, second lung smoke ugh, give me a bucket. / Smokers who walk around residential streets are anoying killers too as they allow the deadly poison onto everyone they pass and into homes ... Smokers should be give a lower priority at public hospitals with higher priority being given to people with non self inflicted illnesses. / I love how all the dirty, smelly, selfish, self-centred, bitter smokers make comparisons between their smoke and the fumes from vehicles. Go away and cough up some black phlegm and stop talking / Okay, someone should just say it...we should remove all smokers from society. Smokers are bad for society as a whole. Why not move all of them to a desginated State and let them live and die together and save the rest of us from the evils of smoking. I'm sure a smoker would like to live with others of their own kind. After all they ARE different than non-smokers, they pose a risk to the health of EVERYONE around them, that are bad for us normal people. / I would like to see a study regarding medical personnel that smoke and then have direct contact with the patient. I see people at work all the time that violate the sterile concept by going outside to smoke and then they return to the operating room area in the very same scrubs that they were wearing outside the hospital while they were smoking, yuk. The same person that is standing over you while you are cut wide open has these same carcinogens all over their body, clothes and this stuff is dropping into your body. / How can a non-smoker live Free as you say when they are constantly being killed each day by second and third hand smoke! / Have fun at your senior prom with your leather catchers mitt face and tiny dry titties / So, I guess we're to just roll over and let you cancerous, diseased, psychologically damaged, pieces of dirt screw up the air / The best part about smokers is that they're killing themselves off before the rest of us. / Smokers are the lowest form of life, I welcome anything which attempts to reduce their numbers. taxing them to death is a better way of ridding us of these smelly people. / people who smoke are mentally backward and I would never employ a smoker. / I hope they are unable to reproduce. They and their kind aren't needed, or wanted in todays world. / The sad, raddled addicts who are hooked on smoking would crawl in the gutters and recycle fag ends to get their fix, charge anyone who exposes their children to smoke with child abuse, and refuse surgery on anyone with a smoking related disease until they have given up. / perhaps we could put something in the water (or more rationally the cancer sticks / coffin nails) that made smokers violently sick every time the took a puff (or in every twentieth fag)! / compel manufacturers to put parts of diseased cancerous lungs randomly in every tenth packet rather than their evil product! Then show adds of smokers opening such packets in situations where they are trying to be sophisticated. / The majority of smokers belong to the underclass. Fact. / drive this pestilence underground and stamp it out / smokers are like gays. they think everything is their right. screw other people, it's all about me. everytime i smell a cigarette or cigar when i am outside i just want to punch you worthless, selfish prickkks in the face. / Breaking the smoking ban should carry the death penalty. / Smokers are evil, disgusting, vile creatures and should be exterminated like curs. / I hate smokers, should have their fingers broken as well. / Shoot a tobacco seller in the belly ... God grant me the chance! / All you dirtbag fat, smelly smokers polluting my air get cancer and f*cking die a quick, painful death. / Smokers are weak minded defectives. There's also plenty of fat smokers too. If I were consul and you smoked in public and/or your smoke was anywhere near a normal person you would be badly beaten. / Only losers smoke. It doesn't matter whether they are thin of fat, they are all losers. / just shoot the lot of them. / You are a disgusting human being. I hope you never have children and poison them with your cancer sticks. / smoker workers touch food with poison that we eat / your vice covers me in SMOKE and STENCH how about my URINE covers you / You are inconsiderate people with a filthy habit and you force your stench on all those around you. I HATE SMOKERS. Scum / I like to punch smokers in the face. I detest smoking and smokers. I have always wondered why I can't just go take a piss on a smoker. After all they are harming my health so why can't I just go piss on them? / Let them pay their taxes, let them die young sounds like win win :) / Better still - just shoot smokers and put them out of their misery. / I personal don't care if you kill your self and die slowly of cancer BUT don't blow your carp in the air I breath, and stink up the elevator I have to travel in. / persecute smokers, persecute big tobacco I don't want filthy smokers ruining the air I breath. That is what should be done the only way to beat an enemy is to destroy it with prejudice. No half hearted measures they don't work. If it pollutes the air I or anybody else breathes, you don't have any rights. Most of you are filthy butt dropping menaces and you stink / I just hope all you deluded smokers who do get lung, throat, or mouth cancer to name a few items apart from the big killer heart attacks all brought on by smoking, remember how you defended smoking when you are asking for more morphine after half your face has been amputated a lung cut out or your larynx have been removed. / You lot of addicts and half brains have no right / It's a pathetic sight to see these revolting creatures huddled outside offices in the horizontal rain. / I don't want smokers banned, I want them dead. Cremation would be fitting. / Shoot all smokers..quick solution! / Smokers disgust me. Me and my friends do see them as second class citizens. We all look down our noses at them and their disgusting and filthy habbit. We make sure they know it too. / I would shoot a tobacco seller in the belly as happily as I would a heroin dealer / Smokers are disgusting and dirty outcasts. / I suspect lepers have better complexions than smokers! / Children and the unborn have the right to be uncontaminated by selfish smokers. SMOKERS STINK / Smoking *is* disgusting, it *is* dirty and the sooner the smokers drop dead and wither away, the better. / The only good thing about smoking is that bthiose who practice this dirty and disgusting habit often die of it. / ban the rat poison cancer sticks / It's a no brainer. If you smoke, you are stupid. If you are stupid, you are probably low-income or no income beyond benfits. Smokers are not lepers. I feel sorry for people with leprosy. Smokers are fools./ Smoking is disgusting and dirty, we don't need campaigns to tell us that, smokers smell and are dirty, and are contaminating the rest of us. / ban the rat poison cancer sticks from effecting the non smokers / So why do the poor smoke so much? Could it be because they are stupid, and could that be why they are also poor? / Smokers deserve to be treated like lepers. It`s a filthy, stinking disgusting habit. It makes their clothes, hair and breath stink. It makes their skin wrinkled and grey. / disgusting, dirty and smelly people - why should we allow them to contaminate the rest of us / Smoking around children SHOULD BE seen as abhorent as paedophilia. / We could always just change the law to allow people to legally shoot dead anyone caught with a cig between their lips outside the four walls of their home... Probably won't happen. I can dream though. God, I'd have a field day if that ever came to pass... / smokers are nothing but vermin to the healthy, responsible society. I sincerely hope that you will all die a painful death. i should actually hope that it is slow, but in your case, the sooner and the quicker the better / This will finish in death, death for a cigarette lit in the wrong place, at the wrong time in front of the wrong person. / I no longer allow smokers in my house, period. / smokers' morality is in the toilet. It is a disgusting act. They ... continue to force their disgusting act onto others. It is the moral equivalent of rape. / Screw em / all smokers are garbage of society. / smoker morons / They'll just have to die.

Let me supplement that collection with an extended rant from a truly unbalanced Antismoker who repeated this general theme, with minor variations, many times over on various internet sites during the 1990s:

Most sysops realize that you are nothing more than a wimpering whining little shit that can't keep his own diapers clean. They aren't stupid enough to hop just because you want them to. All they have to be told is that you are an advocate of smoking and they immediately understand that you must be either a moron or a murderer and most likely both. But they do care about the smoke. And they do care about irresponsible people that are poisoning the planet. We will make the choice for you. We will take the source of your antisocial attitude and behaviour away from you. And then we will throw you in a very small cage.* Laws are being passed. Irresponsible selfish little drug addicts like yourself should be caged indefinitely. And maybe in some states you will even be executed. Hopefully it will be televised although your furiously sucking on that last cigarette may fuzz up the video somewhat. Hurry up now and go do it. Stop talking about it. There's nothing more to say that hasn't already been said. It's like kicking a dead smoker.

Finally, I'd like to add just a touch of humor for anyone who's actually made it through the entirety of the last four pages. One recent poster became rather upset at the way I was countering every argument being put up by the other side on a particular news board. They thought they'd put me in my place with the following note:

Michael J McFadden is a blatant homosexual bed-wetting bastard. Like most New Yorkers, I am sure he talks like he has a mouth full of Linda's dirty underwear. Just like most New Yorkers, he is loud, obnoxious, self centered and clueless. I am sure fat, and smelly are also appropriate.

Aside from the fact that I actually was born in New York, it's amazing how absolutely completely the poster missed on the rest of it. Additionally, I have to admit that it was rather pleasant having someone call me "fat." At about 120 pounds on a heavy day, I take that as a compliment!

Remember: this division and hate and anger and harm are not natural phenomena that simply popped into existence on their own.

---

* Despite the similarity of the "cage" theme, this poster is not the same as the one who took pleasure in looking at the "caged" airport smokers.

Speak to almost any smoker, or even non-smoker, over the age of 50 or 60 and ask them if they heard stuff like this back when they were in their teens or 20s. The generation of this new and destructive ripping of our social fabric was a known and accepted consequence of the antismoking movement's effort to create denormalization of smoking.

The hate campaign didn't start as a hate campaign; successful ones never do. You can see the early seeds, though, in such things as this 1996 request from the Tobacco Control Section of the California DoH for a comprehensive media campaign. The request explained, "(Our campaign) has the goal of reducing tobacco use in California by promoting a social norm that does not accept tobacco."

The movers and shakers in the general antismoking movement have come to feel that almost any amount of collateral damage caused by their efforts to create such a "social norm" are worth the cost to civil society. I disagree, and I think many of my readers here will disagree also as they consider the world that this campaign has now created. The end goal of a better public health may be a laudable one, but the means being used to achieve it are imposing a harm far beyond anything that's justified by that end goal.

You'll notice that roughly a quarter of the hate postings on the four condensed pages make reference to drugs, addicts, or addiction. This tool only became useful to the antismoking community in the 1980s after the formal definition of the words were changed so that they would better include tobacco, with that change being extensively promoted by the US Surgeon General.[387, 388] The change in terminology had two distinct advantages for the antismoking movement: not only did it tinge smokers with the nastiness of true illegal drug addictions and their associated crime, violence, poverty, and very early death; but it also provided a much stronger springboard for using the "Save The Children!" propaganda tool.

While it had always been seen as undesirable for teenagers to smoke, it was also widely accepted as quite normal that many of them would try smoking at some point, with a significant minority taking it up regularly in their mid-teens regardless of tobacco purchase laws and enforcement efforts. The change in definition suddenly morphed the image of those teens from being rebellious youths into being helpless, lifelong victims of an evil industry intent on "hooking them while they're young." The claim was made that 80, 85, or even 90% of smokers began smoking as children, while rarely noting that the definition included even a single puff on a cigar taken on the eve of an eighteenth birthday. And then, when the push started for college campus bans, Antismokers suddenly started moving the goalpost for the word "children" to include those up to ages twenty-one or even twenty-three years old, while talking about 30 or 40% of smokers picking up the habit during their college years.

So, what has been the end result of this decades' long hate campaign (or, euphemistically, this "denormalization" campaign) so far? Aside from the emails and conversations I could share from people who've felt rejected from their families, had their businesses shut down, or been threatened with eviction, there have been a good number of outright "hate crimes" of various kinds aimed at smokers. When I wrote about hate in *Dissecting Antismokers' Brains* I cited two instances of such crimes, and several people criticized me for picking out uniquely extreme cases. It's less than ten years later, but I feel I could now fill page after page with examples – and it's getting worse. *The Flame* of the opening story is still just a story, but there have been a lot of hints of it in the news of random killings, beatings, and outright punishment-by-torture of spouses, children, friends, roommates, or even passing strangers by those unhappy with their smoking activities.

I'll spare you another horror show, but keep the concept in mind as you read the daily paper or watch the talking heads on the morning news-talk productions. The "rare" instances you might hear about locally are not as isolated as you might think, and the ones that make the papers are probably far outnumbered by the ones that don't or the ones where the smoking factor simply went unnoticed and was brushed off as unimportant.

A final heart-warming hate posting from a paramedic who oozes the milk of human kindness as he cares for the sick:

> In one town in ontario infested with the obdurate elderly, one lady had a hole cut in her oxygen mask so she could smoke. She demanded that I the paramedic allow her to smoke in the ambulance. One of the great joys in my life was lying to her, and then not allowing her to smoke. Oh she was furious. Tough toodles.[389]

The next time you see someone make a negative remark to or about a smoker, think how you might feel if you were witnessing the same remark made with regard to a black, a gay, a Jew, or a cripple. And then think a bit further about how such remarks and the acceptance of them have grown over the past twenty years... and where that points for the next twenty. Maybe, even if you're not a smoker yourself, those thoughts will encourage you to stand up for the right of smokers to be treated like human beings.

Censorship

I've always been strongly opposed to censorship in almost any form. While I might approve of blocking the publication of step-by-step instructions on how to make a nuclear bomb out of ordinary kitchen ingredients, or a detailed guide on how to assassinate world leaders, or a handbook on kidnapping children, or a few other such things, for the most part I believe strongly that even the vilest expressions of hate and pornography serve to alert the general population of what exists and needs to be guarded against. If a Fourth Reich is building a strong following out there, I'd like some advance warning rather than have it develop unseen in a censored underground.

Censorship in print has always been difficult to achieve. The Tom Paines, David Dukes, Abbie Hoffmans, Saul Alinskys, and Salmon Rushdies out there have survived refusals by traditional publishers and bookstores, book burnings, and even murder threats and assassination attempts. *1984*'s Winston Smith[390] has never really managed to keep up well with the spread of the printed word; it's hard to stop its production, and once it's out there, it's effectively impossible to track down all the copies and rewrite them to the desired content.

The Internet is similar to some degree in that things written on the net can quickly spread beyond recall, even when the original source is thoroughly erased, but those initial erasures and the power of Internet sources to refuse a voice to those they disagree with while simultaneously seeming to open a platform for all to speak equally have always grated on my nerves.

While books can't be "burned" on the Internet, original web pages can certainly be altered or erased. If you remember the story of Helena's PowerPoint graph back in *Slab II*, you've seen how such a thing works and also the primary defense against it: the Internet's archival engine, The Wayback Machine. Unfortunately, the Wayback

Machine has some limitations. For one thing, unless a page is sitting out there on the net for six months or so, there's no guarantee that Wayback's web-crawler will grab a snapshot of it. For another, for some unknown reason – perhaps simply to avoid fighting expensive lawsuits – the Wayback Machine moderators seems to allow those who don't like their history exposed to erase pages from their past beyond retrieval. This option isn't widely used and there are probably strictures upon it, but it exists, and I believe I have seen several examples of its use. Finally, Wayback stores only fundamental page content seen at a particular random moment; the content most often censored, the pages of comments left by readers, is not preserved.

Fortunately, there is now another mechanism available to net users in a web offering at http://iCyte.com where a user can "store" any current page they are viewing with all of its text, graphics, ads, comments, and, I believe, all relevant computer coding in the background. The user can then share that documented version of the page in the future with others – either from their own computer or exactly as it was saved on iCyte's online database itself. It's similar to a "screen capture," but has the double additional advantages of saving an entire page (which may comprise a dozen or more screens of material) at one time and also automatically giving the saved page a permanent home on the Web for easy referencing and sharing with others.[*]

iCyte is particularly handy for quick and concrete exposure of a particularly nasty form of web censorship known as "shadow-banning." Shadow-banning was developed in the 1990s as a defense against spammers who would post ads or other offensive material and then stop back and repost them if they were removed. When a posting is shadow-banned, the original poster in question will stop back and see their work sitting there in all its glory and feel no need to put it up there again. Meanwhile, no one else on the Internet who happens to visit that page will see even a single word of the offensive posting or

---

[*] You can see a quick example of the usefulness of iCyte in the block quote from the paramedic just two pages ago. At some point a week or so after that comment was posted it was removed. If you check the cited iCyte link though you can see it as it appeared originally and also click on the "View Current Page" option to see how it was removed.

any obvious indication that anything was even erased. A news story about smoking bans followed by a dozen supportive posts may have actually had a hundred in disagreement as well... all of which have been silently rendered invisible to all but their individual writers who would have no clue about the censorship going on in the background and no idea that a hundred other posters had written to agree with them!

The people who developed this tool, one of whom I've had some extensive conversations with over the last few years, never intended for it to be used as a form of censorship of ideas – but sadly, that is what it has become. Fortunately, programs like iCyte now allow users to publicly verify and effectively protest that censorship.

The most egregious example of the use of shadow-banning that I'm aware of is found on the Internet conglomerate news posting boards of Topix.com. If you go to an active and contentious Topix news thread, you may notice that while it seems to contain only 10,000 posts, the last post to it is numbered 12,277. If you then go back along the thread history, you'll find that there are "missing numbers" scattered throughout. You'll find areas where the small numbers next to the posts might look like this: 517, 518, 521, 522, 524, 526 ... etc. What has happened is that the group moderator has banned or shadow-banned about 20% of readers' postings, including posts numbered 519, 520, 523, and 525. There is no notation or blank spot pointing out the deletions and most readers will never notice the censorship. And if you were the author of just posts 520 and 523 and they had been shadow-banned, they would still be quite visible to you (although 519 and 525 would not). You'd simply be left wondering why your beautifully worded arguments and carefully researched information and references were receiving no responses from either your opponents or allies.

I discovered this phenomenon back in the early 2000s on Topix when suddenly, after several years of spirited debate on several smoking-related threads, I found my comments being totally ignored. Even the regular "trolls" on the boards who would normally follow every posting of mine with cracks like "McPrick just won't shut up." seemed to have nothing to say. It wasn't until I signed on one day from another name just to check on an old email box that I suddenly realized my last few weeks of posts were missing! Returning to my main screen name miraculously made them reappear; but clearly they were visible only to someone with my unique ID and password. Complaints went unanswered and then several weeks later I discovered that my entire history of hundreds of postings had *all* been relegated to the Internet's dustbin and seemed to be unavailable even to me.

Sadly, at that point in time there was no such thing as iCyte, so aside from a few examples of text sitting on my hard drive and lost-in-the-sands-of-time instances of un-erased responses to those burned postings, my writings are gone forever and were deleted without warning or notification. I have rarely posted on Topix since.

Topix is not alone in this. I have noted shadow-banning used by several respected news websites. One case that I was able to document with the aid of iCyte involved a posting I had made to the *Argus News Leader*. I was able to send that documentation in to them along with a formal complaint noting that my posting had not violated their Terms Of Service guidelines. However, just as with Topix, no explanations, excuses, or corrections were forthcoming. The entire conversation thread has since been deleted, so the matter at this point is somewhat immaterial, but nonetheless, the action was reprehensible and is probably being repeated against other, more current, posters.

The censorship can even extend beyond one's posts. The more responsible members of the internet community often use their real names when posting, and if they have a potential conflict of interest they'll note it explicitly or link their signature to their homepage so people can fairly evaluate how their background interests might influence their stand. Throughout most of 2012 , I was quite active on a fairly well-known blog that specialized in questioning global-warming science. The blog often featured fairly technical and scientific arguments that had strong parallels to the sort of argumentation and use and misuse of statistics in the smoking arena. My posting history there was active and appreciated by a number of the blog participants, but then, early in 2013, I got an email from the main moderator stating that my future posts would be deleted unless I removed the interest-identifier attached to my name (It led to a free booklet titled *The Lies Behind The Smoking Bans*.) In the emails that followed, I discovered that the moderator had been going back over my previous comments and eliminating that identifier without permission or notification. Given the accusations I've had to deal with over the years for supposedly hiding my "special interest," this was a very serious matter. I was unwilling to delete my identifier and am no longer permitted to post on climate politics... simply because of my stance on smoking.

As things stand today, Internet posters have no defense against such censorship. They are not covered under the First Amendment, since it is not the government that is censoring them; private parties clearly and properly have rights over what material they choose to publish. My quibble is not so much over the censorship itself as it is with the fact that some of those private parties project a false appearance of free speech when in reality it does not exist. Legally they may be on safe ground, but morally, they stink to high heaven.

The First Amendment may come into play in the future, though. Antismoking advocates are greedily eying the concept of using .xxx domain names – supposedly to be reserved for marking websites dealing in outright pornography – to put Free Choice blogs and websites off-limits to those under twenty-one years of age. There have already been instances in the UK where outspoken but otherwise inoffensive websites that speak out too strongly against the bans have found themselves suddenly rendered invisible on library search engines. If websites espousing undesirable political content begin to get shut behind the same doors as the ones with pictures of Farmer Gray playing naughty games with his barnyard animals, we will all be poorer for it and the world will be a more dangerous place.

Join The Party!

With the above thoughts and warnings in mind, sign on to the Internet, fire up a Google search screen, select "News," and type "smoking bans" or "secondhand smoke" or "cigarette taxes" into the search box. Welcome to trench warfare. Anyone who's spent any time at all on Internet message boards knows that they're not for the thin-skinned or the fainthearted. People can sit safely behind their keyboards and the anonymity of a fake name and let bile and hate and invective loose on others in ways that they would never dream of doing in person.

When you're done with *TobakkoNacht – The Antismoking Endgame*, take your new weapons, pull on your hip boots, and wade on in. Be careful though: it can get pretty nasty out there!

Launched O'er The Ramparts

Aside from letters to newspapers, occasional articles or satires, and the thrill and grind of trench warfare in the flame pits of the Internet, there's another battlefront where one's efforts can have a significant impact in the War On Smokers (or as I might prefer to phrase it, the War Against Antismokers). That battlefront consists of official communications to government officials and political bodies involved with passing restrictive regulations.

O'er The Ramparts offers three examples of such communications. The first consists of testimony I offered to Philadelphia's City Council in May of 2000, the second was launched over the ocean to the British House of Lords and entreated them to deny the Welsh Assembly Government the power to unilaterally mandate smoking bans in its own sector of the UK, and the third was sent Findlay Ohio's Department of Health (FDoH). The first arrow helped to delay Philadelphia's smoking ban for over five years, the second sank in the British Parliament with barely a ripple, and the third set off a ripple effect that not only helped scuttle their ban but evidently resulted in the resignation of a third of the FDoH! (OK, the FDoH only had three members at the time, but hey, saying I forced a third of them to resign makes for a good stratistic if I leave out the details, no?)

So here are three flaming arrows shot o'er the ramparts of castles in the midst of battle. The second one landed in a pond somewhere, but the other two scored nice hits on munitions bunkers!

Before the year 2000, I had never spoken publicly in front of an audience on the smoking issue. That February I traveled to New York City and gave an impromptu five-minute presentation to Vincent Robles and other City Councilors at a hearing purportedly held to evaluate the current restaurant ban, but actually used by Antismokers to begin pushing the idea of extending the ban to bars.

With the help of Audrey Silk, the outspoken Brooklyn policewoman who founded NYC CLASH (Citizens Lobbying Against Smoker Harassment), and other outspoken Free Choice supporters, the City Council heard, in no uncertain terms, that there was no need to extend the current restaurant ban, and that even that ban itself should be revisited for amendment. Every speaker, from the high and mighty James Repace who had been flown in from who knows where, down to a waitress from a bar in the neighborhood, was given just five minutes to make their presentation, and Councilman Robles saw that they all got a fair and equal hearing.*

I expected the Philadelphia hearing to go the same way. I was sadly surprised to find that antismoking Councilman Michael Nutter and his supporters had the first day's hearing structure set up to give the antismoking speakers almost unlimited rein throughout the opening hours of testimony. Their presentations were so overwhelming, so filled with official-sounding facts and figures and claims, that I ended up more or less slinking out of the door at the back of the hearing room without having spoken once I knew the hearing was to be continued on another day. I honestly had no idea what I could say in the five to ten minutes that ordinary citizens (mainly against the ban) were being allowed that would counter the avalanche of material presented by the other side.

I'd learned from my New York experience that I was not yet a skilled and confident extemporaneous speaker. So I sat down with my notes to prepare a formal, written testimony for the Council's next session. My six minute testimony had to be peremptorily cut to about three minutes at the sudden dictate of a Council that had no desire to listen to lengthy presentations from those against the ban, so I improvised by eliminating most of the middle of my testimony. What was left was effective however. I got one of the few rounds of outright applause from the chamber's audience as I slowly and forcefully read the introduction and ending of my testimony into the public record.

---

* Although he did bend the rules a bit when Antismokers paraded their kids up to the microphone to read testimonies obviously prepared by the parents. The antismoking tactic of abusing our love for children worked, as it almost always does, and Councilman Robles soon had them up on his knee, praising them for their civic concern while giving them City Hall mementos to take home.

Testimony To Philadelphia City Council
May 31, 2000

Councilpeople, thank you for the opportunity to speak today. My name is Michael McFadden and I'm a long term resident of West Philadelphia. I have no connection to any bars, restaurants, or tobacco companies other than being a good customer.

Last Wednesday you heard from Antismoking advocates that Environmental Tobacco Smoke had been added to the National Toxicology Register as a carcinogen. What you DIDN'T hear was that alcohol and sunlight were ALSO added at the same time.

Antismoking advocates consistently ignore the clear and simple fact that there never has been and very likely never will be shown any threat at all from the microscopic exposures to smoke that might exist in well-designed nonsmoking sections. If you refer to the testimony that began today's session, you'll note that the Advisor to the Surgeon General herself carefully referred only to "smoky" and "smoke-filled" rooms in most of her testimony. At no point did she indicate that well ventilated smoking and nonsmoking sections posed a threat to the general population.

There's no need for glass partitions or airlock doors: smoke does NOT "travel where it wants". Even a mild movement of air pushes smoke quite effectively in the direction of the air movement. Passing a law prohibiting smoking sections in restaurants on the basis of cancer risk is like passing a law prohibiting windows in restaurants where sunlight might come in and give cancer to innocent diners.

Most studies that have looked at the long term health effects of smoking upon nonsmokers have been based on situations where nonsmokers lived and worked closely with smokers, often in poorly ventilated conditions, every day of the week, over periods of up to 40 or 50 years. Even at THOSE extreme exposures only about one study out of 10 has consistently found any statistically significant link between secondary smoke and even small increases in diseases like lung cancer. The claim made here last Wednesday that these studies were "unanimous" in their findings is clearly false.

Some studies have even come up with NEGATIVE correlations ... (Text of WHO Abstract, list of 124 study results, and critical examination of two previous pro-ban testimonies submitted for the record.) ... *

A *lot* of the testimony heard from Antismoking advocacy groups at these hearings was like this: basic touches of fact, expanded, tortured, and twisted into visions of death and destruction far beyond what reputable scientists not connected with such groups actually subscribe to.

Councilpeople, I do not believe we are here today because of a real concern about the health of nonsmokers. We're here because smoking bans are seen as one of the most effective weapons in the arsenal of social engineering when it comes to reducing smoking and getting smokers to quit smoking. At least three of last week's Anti-smoking advocates spoke of that in their testimony.

I'm here today Councilpeople because this is NOT George Orwell's 1984. Our government should NOT be in the business of making laws designed to pressure citizens into thinking in proper ways or conforming to a politically correct healthy lifestyle.

Education about the dangers of smoking is fine. Social engineering and behavior modification is not. That's NOT what government in the United States is supposed to be about, and by voting for proposals like this one I believe you will actually be hurting our country and its people a LOT more than you'll be helping.

Thank you.

---

* The ellipse marks (...) used several times here and in the following documents indicate deletions to avoid repetitive material, material not verbally presented, or simply to move the reading along a bit faster. The original materials were up to 30 or 40% longer.

In the summer of 2005, I got an email from across the ocean asking for assistance. A John Gray in Wales had learned that the Welsh Assembly Government (the WAG) had produced a Report justifying its desire to institute a wide ranging smoking ban. The arguments in that Report were also going to be used the following year in the House of Commons and the House of Lords in relation to the English ban. Mr. Gray was quite distressed when he heard that news and, having recently read *Dissecting Antismokers' Brains*, he thought that perhaps I could help.

I had some significant doubts about what I could do. I certainly didn't have the funds to fly across the ocean to give a presentation, I had no knowledge at all of what was being proposed, and I was completely unfamiliar with how the British Parliament was set up or operated. My attempt to duck out met with failure though: Mr. Gray was both persistent and persuasive, and he sent me the detailed WAG Report itself.

Once I read that request I was hooked. My understanding of the politics involved was a little confused as it seemed to me that the Welsh were actually asking for permission to institute their ban, but the analysis of the document was accurate nonetheless. It was *so* filled with misinformation, propaganda, and misleading claims that I couldn't have stopped my fingers from typing a rebuttal if I'd tried. The rebuttal was lengthy, going point by point through the similarly lengthy document from the WAG, but I believe it succeeded in thoroughly skewering their claims that there was an urgent public health crisis necessitating their imposition of a smoking ban.

I think you'll agree that the strength of the case made against the Welsh argument was sufficient enough that it should have brought the process to a grinding halt in both Wales and in England – as copies of it were formally sent to a good number of members of the House of Lords. Instead it sank with barely a trace. I received two somewhat positive email responses from Lords, and one negative, but quite interesting response from a Lady.

The Lady was Lady Elaine Murphy. Lady Murphy informed me that I had misunderstood the intent behind their ban, saying, "The aim is reduce the public acceptability of smoking and the culture which surrounds it."[391] That acknowledgement, back in 2005, was actually the first official public recognition of the denormalization strategy that I had seen from a major political figure. At that point in time organized antismoking forces almost universally denied any real goals for their bans beyond those of "protecting the workers" (and the children, of course). Evidently Lady Murphy hadn't gotten the memo.

Critique of the Committee's Report on Smoking in Public Places

Submitted By

Michael J. McFadden
August 30th, 2005

Cover Letter

Dear Sirs and Madams,

I ask you to read and carefully consider the contents of this document, not just in making a decision about the recommendation for Welsh power to pass a smoking ban, but in making the wider decisions about governmentally imposed smoking bans in general.

My criticisms will be very harsh, and I believe quite deservedly so. I am acting on behalf of no lobbying group, and have no funding from Big Tobacco or Big Hospitality. I am concerning myself with the affairs of the United Kingdom because your decisions over there will ultimately affect the situation here in America and in my home town of Philadelphia, just as our decisions are affecting you.

My critique will not be exhaustive but will instead concentrate on the introduction and health aspects sections of the [Welsh] Report of the Committee on Smoking in Public Places. I will pick selected passages from its body in order to demonstrate why the Report should be discarded and a new one commissioned under a different group.

I do not have a medical background myself, but am well educated and well-grounded in my readings in the area of concern. [biographical information snipped here]… The content of my work has always been careful and exacting in its statement of facts and has never been criticized for such statement. The great bulk of the argument against exposure to secondhand smoke in public places comes from epidemiological research which is far more statistical than medical in nature: thus my lack of an M.D. should not be treated as an excuse to dismiss my observations and arguments.

I hope you will seriously read and consider my presentation: your decision in this matter will [reach] far beyond simply granting … Wales the power to ban smoking in workplaces.

Sincerely,

Michael J. McFadden,
Philadelphia, Pennsylvania,
USA

Chair's Foreword

I will start by noting that the material in the very first paragraph of the "Chair's Foreword" exposes the bias that I found evident throughout the Report. At the end of that paragraph it is noted that the Welsh Assembly Government "seeks the power to ensure that we can have clean air indoors."

Actually, what the Assembly is seeking has nothing to do with seeking the power to "ensure... clean air indoors." Clean indoor air is already quite well ensured by laws regulating all the pollutants of concern, whether we are speaking of carbon monoxide, nicotine, particulates, or arsenic. What the Assembly is seeking is simply the power to regulate smoking. Whether the air in a pub is "clean" or not after such a ban is not addressed in this Report as a concern. The bias of the document is evident in that the Committee starts out with the assumption that a venue that allows smoking will perforce have air that is "unclean," something that is obviously not a necessary follow-on given the capabilities of modern filtration and ventilation systems.

Antismoking extremists will assert that any practical degree of ventilation will be inadequate to remove all the elements of smoke from the air. To some small extent they are being truthful. It is also truthful that no practical degree of ventilation will remove all traces of fumes of the highly volatile Class A Carcinogen ethyl alcohol from the air of any venue that serves alcoholic drinks.

The question is not whether such zero-levels can be achieved but rather whether they are necessary. In the case of alcohol fumes few would argue that a zero level is necessary despite the carcinogenic classification of the substance. That would be true for tobacco smoke as well if not for the political and social engineering push by extremists for its outright banning.

The relevant point is simply as stated in my second paragraph: the deliberate confusing of the terms "clean air" and "smoking ban" is full evidence of the bias that will follow in the body of the Report.

I will demonstrate that the Committee not only failed to present a sound case for the necessity of such bans, but it even failed miserably in presenting a competent review of the available factual and scientific materials at hand.

Section 1: Introduction

Section 1.1

This section notes a previous vote that affirmed *"the well-documented and proven life-threatening dangers"* of secondhand smoke. As will be shown in Appendix A such dangers are far from being "proven."* Only about 15% of the studies in that Appendix were able to pass even the most basic of simple scientific tests for correlation, that of statistical significance, to say nothing of "proving" causality of effect. Again, being so near the beginning of this Report such a statement foreshadows the bias evident in its body.

Section 1.3

The Committee was instructed to "consider current evidence on.... The economic impact of restrictions on smoking in public places" and to "consider the experiences in other countries where a ban has been introduced." The 160 specific real-world citations in Appendix B† show clearly that the Committee failed to fulfill this directive. There is no mention in its report of the significant body of the data exemplified by those citations, nor is there mention of the research I carried out with David Kuneman and published two months before the Report. Incredible statements such as "There was a trend away from drinking in clubs and pubs" as an excuse for the economic downturn after the introduction of Ireland's smoking ban were taken at face value rather than being questioned. Why do you **think** there would such a "trend" after they threw their smokers out on the street?

---

* The Appendices will not be reproduced here, however they are freely available online at: http://nycclash.com/Philly.html#ETSTable and http://smokersclub.com/banloss3.htm. Appendix A also appeared in *Dissecting Antismokers' Brains*.

† Appendix A was a listing of summary results of over a hundred epidemiological studies, cross-checked against all available sources. Appendix B was a listing of actual, real world ban losses and closures experienced by hundreds of pubs after the imposition of bans elsewhere.

Section 1.4

Here is the first mention of the importance of "the impact of a ban in reducing the prevalence of smoking."

This is a far cry from the Committee's official charge to examine the effects of secondhand tobacco smoke and should not have been included here. I do not believe a directive for a social engineering plan was, nor should it have been, in the original scope of the report.

Section 2: Summary of Conclusions and Recommendations

The Committee concluded that:

A) "There is overwhelming evidence that environmental tobacco smoke is damaging to health."

This conclusion is meaningless without reference to concentrations and exposure levels. There is overwhelming evidence that carbon dioxide can kill a person, but exposures to it in the amounts usually encountered from being in a room with other people exhaling it are not damaging. There is also overwhelming scientific consensus that the ultraviolet radiation in sunlight is a dangerous carcinogen, yet going out briefly to grab the morning paper would not be considered "damaging to health" by any sane definitions of the words.

The Committee's conclusion contains an unstated premise that is of prime importance in foreshadowing their ultimate recommendation. What they are actually trying to say, but know they have no grounds for saying openly, is that they have concluded "that **any trace exposure, no matter how microscopic and undetectable** to environmental tobacco smoke is damaging to health." Such a statement as an opening conclusion would have given away the bias of the conclusions and recommendations to come and thus those words were left out – although they were fully intended to be "understood."

This conclusion also blithely ignores the fact that of the 130 studies detailed in Appendix A, only 15% passed the bare minimum scientific test of statistical significance, to say nothing of the more stringent standards required for a true determination of causation. ...

It is clear that the evidence is **far** from being "overwhelming" and that the Committee's conclusion on this point needs to be revisited.

The Committee also concluded that:

B) "Ventilation equipment is not capable of removing the majority of health damaging particulates from the atmosphere."

Wait a moment. Has anyone on the Committee completed high school math and physics? If I have 10 grams of Bubonic Plague dust in the air of a room, and I swirl 51% of that air/dust out a window, replacing it with fresh air, then I have clearly "removed the majority" of Bubonic Plague dust. What laws of physics are the Committee referring to that would claim that a bar providing a normal 15 air changes/hour (ach) could not remove "the majority" of particulates from the air? 15 ach, even without added filtering, would remove well over 99% of particulates. High quality filtration could improve air quality a level more particulate-free than the "fresh air" outside and certainly cleaner than the air in a many "smoke-free" restaurants. This particular claim of the Committee is patent nonsense that should never have been allowed in an official government document. The people responsible for this claim should be fired and any funds spent to support or pay them during its production should be returned to the taxpayers.

The Committee also concluded that:

C) "There is no evidence that the introduction of a ban would have an overall negative impact on the economy."

There is an enormous difference between saying "There is **inadequate** evidence" and "there is **no** evidence." Since there clearly **is** evidence of such impact (Once again, see Appendix B)..., one can only conclude that the Committee is grossly negligent and incompetent or that it is outright lying. In either event the remedy outlined at the end of the previous section's critique should again be implemented.

The Committee also concluded that:

D) "The majority of the public who do not smoke should be able to go to their place of work... without risk to the health."

We will ignore the prejudicial qualities of referring to concern for the "majority who do not smoke" other than to note the conclusion would be equally valid if it concerned a "minority who do not smoke."

But we will not ignore the final four words of this conclusion, "... without risk to the health." As an absolute statement this is arrant nonsense.

If I am a working as a counter person, I am constantly at risk of catching and dying from influenza brought in by customers I am serving. This could be prevented by airlock arrangements similar to those proposed for smoking areas but does the Committee demand such?

As a bar-bouncer, I am constantly at risk of being stabbed by drunken patrons. Does the Committee insist I be required to wear a Kevlar vest while on duty?

As a waiter assigned to an outdoor patio, I am forced to work in a carcinogenic environment that may result in deadly melanomas despite any amount of sunscreen I put on. Does the Committee intend to ban patio dining?

All three examples deal with relatively low levels of risk, but all three deal with risks that are probably greater and certainly far less disputable than the risk posed by low levels of exposure to ETS. And yet the Welsh Assembly Government is doing nothing about them.

The Committee's first responsible action, after noting the complete absence of any compelling scientific studies showing a risk from ETS levels in any decently ventilated business, should have been to dissolve itself and save the taxpayers' money. The fact that it did not do so once again indicates a need for the remedy heretofore proposed.

Section 3: The Health Evidence:

The Committee leans heavily upon "six key documents."

Section 3.3

The first of these, from the California EPA, was produced by an organization that has been very heavily influenced by some of the most radical elements of the US antismoking movement. Any EPA employee who disagreed with the concept that secondhand smoke is one of the most deadly substances ever known to humankind would quickly be shown the back door. This report was no more produced by a "neutral body of scientists" than one emanating straight from the labs of BAT or RJR. ...

Section 3.4

This section, based on SCOTH's 1998 report, states that [ETS] "is a cause of ischaemic heart diseases" and then goes on to speculate that **if** estimates of risk were validated then it would be a "substantial public health hazard." Note that it does not state that there **is** such a hazard, only speculating that **if** there was one it **could** be substantial. This is no more valid as a basis for legislative action than if they produced a report stating that **if** there was a health hazard from candle wick fumes that it **could** be substantial.

It goes on to speak of parental smoking in front of their children, something that really is not within the purview of powers that I believe Wales is seeking, unless it intends to monitor people's homes.

The Report then examines sudden infant death syndrome, noting an "association" but carefully declining to state outright that the association is causal because the members of the Committee know full well that such has not been proven. Nevertheless, they dance around this by simply saying the association has been "judged" to be causal.[*]

And finally, it speaks of middle ear disease being linked to (not "caused by") parental smoking, which is again outside the purview of any laws being considered openly by the Assembly at this time.

This section has nothing at all to do with what the Welsh Assembly Government is actually seeking the power to do: namely ban by law the smoking of tobacco products in pubs and restaurants. ... Their inclusion in the body of "The Evidence" once again points up the biased nature of the Report in that it here clearly seeks to use the reader's ingrained love of children and biologically hard-wired instinct to protect children in order to advance its goal of regulating the public smoking behavior of adults in largely adult venues. ...

Section 3.7

This section speaks of maternal smoking during pregnancy. Is the Welsh Assembly Government seeking the power to prosecute women who smoke during pregnancy? If not, then this has no proper place within this document. If so, then it should be made clear to the public that this is the power the Assembly is seeking. ...

Section 3.8

... Again, as in previous sections, there is an emphasis on protecting infants and young children who would primarily be exposed to secondhand smoke at home. The extent of this emphasis in this document and its implications for the sort of governmental control powers being sought is highly disturbing. I doubt the general public would approve of such powers if it was aware they were being sought.

Section 3.9

[This] is an advocacy piece titled "Towards Smoke-Free Public Places." produced with a group whose mission statement stresses "**promoting** tobacco control" and "**supporting** ... tobacco control activities." ...

---

[*] And, while I didn't note it at the time, there was also no explanation of how bans in pubs would reduce deaths among infants who rarely drink in pubs.

[T]his document simply summarizes the previous information contained in the EPA and SCOTH documents. If seventeen more such documents were produced by various bodies, each summarizing the results of the previous ones, we would then have **twenty** documents supporting the eventual recommendations of the Committee. This is not how science is generally conducted.

This section again heavily references children and twangs the heart strings with images of pregnant women and ends with a particularly telling statement: "There is no safe level of exposure to tobacco smoke." This statement has become a veritable mantra among Antismoking Lobby groups in the last five to ten years.

Scientifically of course it's nonsensical. It is similar to the aforementioned risk of getting the morning paper. Technically, in some arcane way, it is a risk: sunlight causes skin cancer and during your brief outing you will, even if you slather yourself with SPF 100 sunscreen, be exposed to sunlight. There is no safe level of such exposure: sunlight is a carcinogen. Alcohol faces a similar fate on our scales: if you are in a restaurant and a couple sitting fifty feet away from you toast each other with champagne it could be argued that you are being exposed to a "dangerous" level of the highly volatile carcinogenic element, ethyl alcohol.

Are the words "dangerous" and "no safe level of exposure" properly used when describing such exposures to alcohol and sunlight? Of course not. No sane person would claim so. But many sane people have come to accept the same sorts of statements about secondhand smoke simply due to its social opprobrium and the constant repetition of such claims ...

Section 3.12

This section cites and rests entirely upon the "Great Helena Heart Miracle" study. [The bulk of this section critique is deleted here as the points made were quite similar to those presented in *Slab II*.] ...

"The Great Helena Heart Miracle" would be more accurately called "The Great Helena Heart Fraud" when it is cited as evidence of the danger of secondhand smoke. To cite the Helena study as the basis for justification to pass laws affecting the lives and livelihoods of millions of people is disgraceful. The Committee should be ashamed.

Section 3.14

Section 3.14 [asserts] that concentrations of salivary cotinine are associated with risks of cancer and heart disease.

Tracing the references for this assertion leads through a chain that concludes at two end points: Kawachi's 1999 study in *Environmental Health Perspectives* v. 107 which was simply a review of other studies and did not focus specifically on cotinine exposures; and Lubin's study in the same volume which did not so much as mention the term "cotinine."

The Committee's assertion that these studies strongly support its contentions about cotinine levels seems somewhat misguided.

Section 3.15

This section cites one non-published study by a smoking cessation worker who, not surprisingly, concludes smoking bans save lives.

Section 3.16

This section [states] that smoke-banned pubs have had air particulate levels reduced by 53% and breath carbon monoxide in bar workers reduced by 45%. Those facts may be true, but they are not as "important" as they might appear.

The reduction in particulates simply means there's less cigarette smoke in the air after smoking is banned. I am sure many non-scientists could have told the Committee that without the need for a study.

The reduction in breath carbon monoxide could be meaningful, but there's pointedly no discussion whether pre- or post- ban levels were at a level normally judged to be harmful. The fact that this was not discussed would indicate that they were **not** harmful ... [t]hus their reduction [also] has no bearing on whether the Welsh Assembly Government should be seeking the power to ban smoking in pubs.

Section 3.20

To its credit, the Committee did not (simply seek) to discredit the Enstrom/Kabat Study simply because part of its funding was provided by the tobacco industry. However, the disrespect for the value of the study is shown when the Committee here emphasizes the "small sample size" of Enstrom/Kabat. Oddly enough the Committee somehow missed mentioning concerns about the "small sample size" of the Helena study earlier. It should be noted that the Enstrom/Kabat Study was roughly 1,000 times the size of the previously referenced Helena.

Odd that the Committee didn't notice that in its references to the two.

Section 3.22 (3.22 – 3.27: Conclusion)

Despite all the various difficulties in the data and evidence pointed out in the preceding pages the Committee concluded that secondhand smoke (again, at totally unspecified levels) is a "significant risk to the health of non-smokers."   Thus, pursuant to their conclusion, the statement that a smoker on the street a block away from a nonsmoker is presenting that nonsmoker with a "significant risk" to their health would be valid. If this seems nonsensical it is by no means more fundamentally nonsensical than saying that there cannot exist reasonable levels of ventilation and filtration to remove such risk indoors… and yet such a conclusion is exactly what the Committee is preparing itself to arrive at in Section 3.28.

Section 3.26 – 3.28

These sections have many problems, almost all stemming from the ruse of casting the problem of secondhand smoke exposure and its risks in absolute terms rather than the more appropriate relative terms. The BMA advised the Committee that ventilation and filtration "does not remove the fine particulate matter" from the air. Obviously, to anyone with an understanding of physics, ventilation and filtration **does** remove at least some of that matter. The air cleaner testimony from the industry claimed their product removed 99.997% of such matter. A simple window fan, if given enough power, could do even better although it might ruffle a few feathers here and there. … but ASH Wales and Professor Hastings contended that tornadic winds – winds approaching 300 miles per hour – would be needed to do the job. According to their testimony, if you were sitting in the middle of a hurricane and a smoker were nearby, your focus of concern should perhaps be on wisps of smoke fighting their way through the winds toward you, rather than upon the lorry that was about to be blown onto your lap. …

The Committee, in taking those claims as the basis for its final recommendation that ventilation cannot be an effective or feasible solution, has ... laid waste to its own claim to be an impartial body presenting scientific evidence to the National Assembly for Wales and the Government at Westminster.

The Economic Impact

The Conclusion notes that "there is no credible evidence of an overall negative impact on the hospitality industry or the wider economy." ... [T]he Committee has again failed in its task, seeking instead to simply proclaim its foregone conclusions. To look at Appendix B and declare there is not a shred of credible evidence is disgraceful. Once again the Committee should be ashamed.

The cavalier attitude with which it approached the economic question can be seen in the final sentence of its Conclusion where it notes businesses may have "some difficulty in adapting to the changes and **opportunities** a smoking ban would bring." Opportunities? Of the entire list of 160 businesses and personal statements listed in Appendix B somehow I missed seeing a single one that was having "difficulties" with the "opportunities" the ban was giving them.

Perhaps the Committee wears better eyeglasses than I do?

Committee Recommendations

Finally we come to the heart of the matter [as noted in section 1]: ... The belief that such bans will *"reduce the prevalence of smoking."*

That is a social engineering goal, and simply stated as such would be unacceptable to most free thinking people. Thus we now see the real reason for all the cloaks and daggers and folderol and muzzamarole of the preceding sections. It was all simply an attempt to justify this final acceptance of a ban designed to artificially pressure a population into adopting health habits it would not otherwise want to adopt.

Such a social engineering goal could not be achieved if smoking was still allowed in separated but comfortable carriages of trains, or in enjoyable smoking sections at restaurants, pubs, and clubs, and thus such things cannot be stated as acceptable. And so, thus pursuant, the science demanding such restrictions must be created, far beyond the bounds ever conceived of in the era of Soviet Lysenkoism, and total public bans must perforce be the only and the final solution.

Final, that is, for now. Let us not forget that the Welsh Assembly Government is seeking the powers that will next allow it to move into our private homes to "protect the children," or perhaps our next door neighbor. Whether you are aware of it or not, children in America are already being taken from smoking parents in child custody disputes and neighbors are being forbidden to smoke on their own properties. Public smoking bans in bars and restaurants, just as all the little bans that came before, are nothing more than a pit stop in the larger scheme of things.

Overall Conclusion and Recommendation

I submit this with a significant degree of sadness and regret. I am a strong believer in small government, locally responsive to and administered by its people, and I feel that this ultimately is what the Welsh are striving for.

And yet I am recommending that the Welsh Assembly Government be not granted the authority, as the Committee put it, "to ensure that we can have clean air indoors."

My recommendation started and ends with that particular statement from the Chair of the Committee. The purpose of those creating this Report was not to determine whether a smoking ban was a good thing, a bad thing, a necessary or unnecessary thing, a thing that would hurt businesses or help them.... The purpose of those creating this Report was to create an excuse that would allow them to ban smoking and to use the lobbying power more freely available at the local level than would be possible for the entire UK in order to achieve that ban.

Here in the United States we are facing similar battles. The Antismoking Lobby has decided that it's easier to muscle smoking bans through in small communities first and then use their cries of economic pain as a base to appeal for a "level playing field," falsely assuring one and all that such a field will solve the problem and guarantee wealth, happiness and healthiness for everyone. ...

In 1975 your own Sir George Godber chaired an international conference on smoking and health which reached the conclusion that in order to be successful at widespread smoking bans, it would first be essential "**to foster an atmosphere in which it was perceived that active smokers would injure those around them, particularly women, infants, and young children.**"

Sir George's recommendations have been taken up with a vengeance over the past thirty years and now the ravens are coming home to roost. The question is whether you want to promote the conditions that will allow a small group of extremists to foist their rule upon a larger and for the most part unwilling society, or will you, as the government, seek to stop them in their march?

For do not be fooled: it **is** a march, and it will not stop and set up permanent camp once workplace smoking bans are in place. They will storm your beaches as they are already doing in the United States, and the parks, and the private clubs, and eventually the private homes.[392*]

When the Demon Weed has been stopped don't be surprised when Demon Rum comes under attack and don't expect the attackers to employ the primitive tools of 1920's American Prohibitionism. Moral Crusaders of all types have been watching and learning from the successes of the Antismoking Crusade, and that knowledge and the strength of their past victories will be used to telling effect in order to make sure that the corner pubs close and all citizens live uprightly.

---

* When I wrote this missive to the Brits, I was unaware that just a few years earlier one of their chief campaigners against smoking, Clive Bates of the UK's Action on Smoking and Health, had come out with a statement against "scaremongering by tobacco industry front groups." According to Mr. Bates at that time in 1998, ""No-one is seriously talking about a complete ban on smoking in pubs and restaurants."

The Battle Of Findlay

The City of Findlay Ohio began consideration of a smoking ban in early 2004. Two local citizens, Larry Hershey and Frazier Webb, compiled a number of arguments and writings into an argument against such a ban. Among those writings were several of mine that they had found on the Internet. The Findlay Department of Health (FDoH) made an official response to the submission, supposedly "answering" each objection clearly and completely.

Upon receipt of that response Mr. Hershey sent me an email and asked if I would look over what had been said and offer some suggestions as to how they might respond in their fight. I looked at what they sent and found that, in reality, the "answers" from the Department of Health officials were neither clear nor complete and consisted largely of *ad hominem* attacks and unsupported assertions.

My annoyance with the high-handed way they had treated serious arguments combined with some anger over their imputations that I and NYC CLASH were probably getting payola from Big Tobacco. If you have read *Dissecting Antismokers' Brains* you'll probably remember the lines I chose to open the book:

> *I am not now, nor have I ever, been a member of*
> *the Communist Party. I am also not now, nor*
> *have I ever, been affiliated with Big Tobacco or*
> *with their stocks, nor do I have any plans to be.*

As will be discussed more deeply later in *Fighting The Endgame*, Big Tobacco funding is a very touchy subject among those of us fighting for Free Choice. We sorely need funding from almost anyplace we can get it, but touching even a cent from Big Tobacco has been made almost impossible due to the successful tactics of our opposition. Once they can claim that you're "in Big Tobacco's pocket," your credibility is destroyed.

So, in my anger and annoyance, I did far more than simply suggest a path of recourse for Messrs. Hershey and Webb. I sat down and wrote the following few pages and sent them to the FDoH as well as to local Findlay media and government officials. While no FDoH personnel or city officials ever made any attempt to respond to or even acknowledge my arguments officially, my efforts helped produce a very desirable outcome – as will be seen in the Postscript.

General Response to Findlay DOH Rebuttals

I would like to start off by saying that I appreciate the effort the FDoH put into responding to the criticisms of the proposed smoking ban. I am however somewhat distressed because I am going to be rather harshly critical of that response. While I reserve my most specific criticisms for that portion of the response directed toward my own writings, I noted a disturbing pattern throughout the entire body of responses.

In almost every case it seemed that the FDoH chose to respond to scientific arguments, studies, and facts with *ad hominem* arguments or information from press releases rather than on a more professional level. There is nothing wrong with pointing out the possible background motivations of those presenting evidence to you, but it certainly does not serve as a substitute for a proper response to those challenges themselves.

To illustrate with one of the more extreme examples: Topic # 12 dealt with a detailed analysis of the study done on air quality in smoking workplaces by Roger Jenkins of ORNL. The results found in this study, showing levels of pollution even in a smoky bar to be far below what OSHA considers to be a problem even for full 8-hour-a-day workers, are quite valid. They actually mirror the results found by the Antismoking extremist, James Repace, in his 2003/04 study.

However, instead of examining those figures and arguments the FDoH makes eight separate points in refutation and **all** eight of those points are aimed at trying to tie the Oak Ridge National Laboratories and Dr. Jenkins to Big Tobacco. It was more like something I'd expect from the Catholic Church inquisiting Galileo than I'd expect from a Department of Health defending an official policy stance.

Even an elementary school debater knows that if you actually have a good argument to use against your opponent, you don't ignore it in order to simply attack their character. If you have a valid character attack you use it **in addition** to reasoned refutation of their position, not in substitution for it.

Response to Findlay DOH Rebuttal of McFadden Docs

I would like to start off my response to the FDoH rebuttal of my testimony and research by saying that I believe it was both poorly and prejudicially presented overall. In my view it represents a lack of professionalism disappointing in an official public body.

I will respond, point-by-point, skipping none, despite the fact that their rebuttal ignored the vast majority of the many and well-researched points made in my own writing. The FDoH instead chose to attack just two of the points I had made while adding two pieces of pure *ad hominem* attack.

Point One: Sunlight and Alcohol

This part of the rebuttal simply claims that I fail to note that drinking and sunbathing do not harm those around the drinker or tanner. It does not address at all the points I **did** make about the relative lack of harm of reasonable levels of secondhand smoke exposure, nor does it address the fact that, since the smoking ban is primarily being implemented to protect the health of workers, my concerns about workers being forced to work in outdoor patio situations would be quite valid. Just as there is supposedly "no safe level of exposure to the Class A Carcinogens" in secondhand smoke, in exactly the same sense there is "no safe level of exposure" to the Class A Carcinogen of ultraviolet radiation present in sunlight. Sunscreen is no more a protection for these workers than filtration and ventilation systems are to workers working in a smoking environment.

Would it be silly to ban restaurants from offering outdoor dining on such a basis? Of course it would be. It is similarly silly to ban them from offering decently ventilated smoking accommodations for their workers and customers.

As for drinking, the FDoH seems to forget that since they argue that there is "no safe level of exposure" to Class A Carcinogens, and since the very volatile liquid known as ethyl alcohol has been declared to be such a carcinogen, that there is obviously, by their reasoning, "no safe level of exposure" to the invisible and unnoticeable alcohol fumes that permeate the air of establishments that serve alcoholic drinks.

Would it be silly to ban alcohol service in bars and restaurants on that basis? I believe it would be, despite … the fact that a single drink may put as much as 2,000 times the absolute amount of Class A Carcinogen into the air as a cigarette. (Figure based on measured evaporation rates of grain alcohol in a martini glass of over one gram per hour, and the total weight of the six identified Class A Carcinogens in the smoke of a cigarette as being roughly one half of a single milligram.)

The FDoH has no solid grounds for ignoring the minute threats posed by such things while seeking to impose laws banning what may well be the even more minute threat, if any, posed by secondhand smoke in a decently ventilated environment.

To address FDoH's final and related point on this subject: sunlight is not a "necessity for one's body." Again, a statement like this displays a disappointing lack of professional expertise: medical research has never found any direct physical benefit from sunlight other than as a mild topical disinfectant and a promoter of Vitamin D production. Since other disinfectants are readily available and Vitamin D is quite abundant in the standard American diet, the FDoH's statement about it being "a necessity" is something more along the lines of homemade folk wisdom than science.

In summation, it is quite possible that the health of Findlay's workers would be better served by prohibiting outdoor dining and banning the service of alcoholic beverages than by prohibiting smoking. In reality of course, the health and the freedom of Findlay's people as a whole is best served by banning none of these things: let individuals decide for themselves what conditions they wish to work in when prohibitions involve ridiculously small possible risks. The zero-tolerance insistence of the Antismokers is simple insanity, rooted purely in their drive to do anything, and say anything, that will force people to smoke less.

Point Two: CLASH

The FDoH claims that a group called CLASH "sponsored" my writings, thereby providing a base to launch an ad hominem attack on me by implying that I was somehow paid by this alleged "tobacco connected" group to say what I said. I had never even MET anyone from CLASH until the day I went up to New York to give my testimony there.

This totally and completely untrue accusation is all too typical of the "disproof by slander" approach of Antismoking groups when they are faced with scientific arguments and evidence that contradicts their positions. The only things stopping me from considering a lawsuit against the town of Findlay over this are the facts that, one, the unexpectedly irresponsible statements of the FDoH should not be the responsibility of the entire town in this one instance, and two, despite the intentions of the statement, I am actually proud of the fact that the independent activists of CLASH, a group with NO connection to Big Tobacco, chose to reproduce my writings on their website so others could see and benefit from them. ...

Audrey Silk, the founder of CLASH and a wonderful and very hard-working member of New York City's police force for twenty years, has already written the FDoH requesting a correction and public apology. The citizens of Findlay should read what she has to say and be ashamed that their Department of Health, rather than arguing the science that they claim is so solidly behind them in pushing for a smoking ban, should instead choose to attack such a person.

Point Three: Homemade Conclusions & Lack of Training

Given that virtually every one of the other pieces of evidence presented to the FDoH by Messrs. Hershey and Webb was attacked by *ad hominem* arguments seeking to discredit their findings by linking their sources in some way to Big Tobacco, it is not too surprising that they ran into some frustration in examining my own case. I have never applied for grants from either Antismoking funds **or** from Big Tobacco. I fight these battles on my own time and out of my own pocket.

So what does the FDoH do? They accuse me of presenting "homemade conclusions" since I am **not** getting paid! Since my views would obviously never get funding from Antismoking sources they are performing a Catch-22 and trying to discredit me because I am **not** getting money from Big Tobacco. Truly incredible.

They also criticize my conclusions because I supposedly do not have "the medical training required either to interpret or put into perspective such numbers with any amount of certainty." My doctoral training in statistics and propaganda analysis at the University of Pennsylvania's Wharton School is evidently thought to be worthless by the FDoH. I will freely grant that I left Penn before finishing my doctorate, and I will even grant further that over twenty years of time has severely blunted my statistical expertise, but to interpret the very basic conclusions represented by the Relative Risks and Confidence Intervals derived from epidemiologic analysis requires nothing beyond a high-school level of mathematical sophistication. I believe the FDoH is fully aware of this and attacked my data on this ground simply because they had little else they could attack.

While many prominent Antismokers have received truly enormous sums of money for using their time and staff to promote bans like the one proposed here, I do not get paid for my efforts nor do I have millions of dollars in grant money to hire staff to write and research papers for presentations. I am getting paid nothing and I have no staff. My work is, by definition and with pride, homemade. I have just finished writing a book that I hope will help those who are fighting bans, and if someday, I actually make any significant amount of money from it I'll be very pleasantly surprised.

The fact that I am able to mount any credible challenge at all to the opposing arguments should by itself give you pause for thought. ...

Point Four: Outdated Studies, Misinterpreted Libertarianism

Since the writings examined by the FDoH were completed largely before the year 2001, it seems disingenuous to criticize me for not referencing work done in 2002, 2003, and 2004. They then make a further accusation that I deliberately chose not to list sources that I thought were "questionably funded" or "never peer-reviewed." **All** of my sources were listed clearly and accurately, along with what I perceived to be their limitations and possible sources of bias. To say I did not do so seems to be a simple lie, easily checked by going back to the table I presented and examining it for yourself.

The FDoH then refers the reader to "unanimous medical agreement" found by "similar studies" done more recently and cites a list of such studies on another page as proof. What happens if you go to see those unanimous studies? You find a list of ten... *things?* (I honestly don't know what else to call them.)

"President George W. Bush" is being cited as a study similar to but more recent than the 130 studies whose results I so carefully listed and double-checked? "Philip Morris" is a study whose results the FDoH somehow believes? Four different magazines, without reference to dates, pages, authors, **or** results are "studies" that show unanimous agreement against the 130 actual published scientific studies presented by me?

Is this supposed to be a joke? I cite 130 studies, include verbatim excerpts from the conclusions of one of the largest and most important, provide statistical analysis of the results to clarify the numerical terms for those uncertain of them, and the FDoH says that the study "President George W. Bush" proves my conclusions wrong???

The FDoH then goes on to make another criticism, this one a criticism of my choice and use of the 1998 WHO study. I offered in my testimony the results and the verbatim scientific abstract of the study itself, taken directly from the medical journal where it was printed. In response the FDoH cites a **press release** that was put out with a large screaming headline warning the public, "Do Not Let Them Fool You! Passive Smoke Does Cause Lung Cancer!"

I offered an actual medical study. Findlay's Department of Health responds by pointing to a press release that reads more like a fanatic's street corner pamphlet than the reasoned output of a scientific body.

The material submitted under my name was written at a time when I understood much less about the Antismoking Crusade and its manipulation of truth than I do now. Nevertheless, I believe that the bulk of what was submitted is sound and certainly deserves a more well-reasoned and scientific "refutation" than a citation reading simply, "President George W. Bush."

A final note in conclusion: given the near universality of the approach of the FDoH in trying to discredit their opponents' science merely by pointing the finger of "conflicting interest," it seems only fair to question the FDoH itself as to conflict of interest. Has the FDoH, any of its members, or any Findlay activists working to promote a smoking ban ever received any grants, financing, expenses, or other consideration from any groups such as ASH, CTFK, ALF, RWJF, ANR, ANRF, or any pharmaceutical companies involved in the promotion of Nicotine Replacement Therapy products? And would such FDoH members and activists be willing to provide the same full disclosure of such personal investment and other financial connections to such entities as I do below? If any are not willing to do so, I think it would be only reasonable to ask that they step down from their positions and activities, at least insofar as a smoking ban is concerned.

And finally, has the FDoH allocated funds or have knowledge of funds being spent to promote the smoking ban, either directly (by such things as TV ads, posters, or mailings saying "Vote To Save Lives") or indirectly (by frightening the citizens of Findlay with information – i.e., propaganda – about secondhand smoke)? If so, in the interests of fairness, would the FDoH be willing to ensure that similar levels of funding be made available to those in opposition in order to guarantee a fair and representative process?

I look forward to your response and answers, and will be happy to respond to any further questions you may raise.

Sincerely,
Michael J. McFadden
Author of "Dissecting Antismokers' Brains"
http://www.Antibrains.com

{Footnotes for the scientific material/claims cited, as well as a detailed non-competing interest statement were appended.}

Postscript

Before the above communication was sent to the Findlay Department of Health and distributed to the local media and politicians in Findlay, the near-term imposition of a smoking ban seemed almost a sure thing. The hard work of Larry Hershey and Frazier Webb in distributing copies of this *Response* and other materials where they would have an impact resulted in Free Choice continuing to reign for several more years in the region. It may also have played a deciding role in one of the three FDoH members stepping down from his position after its receipt.

No formal response was ever made by the FDoH to this submission, nor was a satisfactory apology and retraction ever offered to NYC CLASH for their mischaracterization.

The Endgame

As this book was in its final stages of editing, formatting and index-ing, what I have referred to several times as the pseudo-journal, *Tobacco Control*, came out with its May 2013 issue, quite serendipi-tously titled *The Tobacco Endgame*.[393] The issue featured a cover photo of dozens of crushed cigarette butts with the words "THE END of tobacco?" superimposed over them. Its introduction described the publication as filled with writing by authors who are "some of the world's most brilliant tobacco control scholars, strategists and activ-ists, including those who originated the principal endgame concepts, [and who] are all highly regarded members of the tobacco control community who were invited to offer whatever thoughts they wished to share on whatever endgame issues struck them as most important or most intriguing."

Oddly enough, they didn't ask me for a contributing article. *Tsk.* An unintended oversight on their part, I'm sure. Despite the timing and the title, the issue didn't present much that required any sig-nificant change in anything I've written in *TobakkoNacht – The Antismoking Endgame*. The Antismokers are still divided as a community, with some supporting such things as e-cigs as a safer alte-rnative to smoking, some supporting plans for reducing smoking to an "acceptable" level (five or ten percent of the adult population who'd then continue supporting tobacco controllers' salaries through their taxes), some supporting step-by-step smoking prohibition through increased ages of legal purchase and use, and some support-ing a gradual reduction in nicotine levels until they're the equivalent of the USA NoNic brand in this book's opening story.

In any event, there was nothing new enough or significant enough to really require much change in these 500 pages – which says something all in itself about how irrelevant tobacco controllers have become in controlling the juggernaut they've set loose upon the world. Let us hope that those of us outside the control community have better luck.

The Endgame

"Qui desiderat pacem, praeparet bellum."
(Let him who desires peace, prepare for war.)

- Vegetius

As I noted in the *Introduction*, the October, 2010, issue of *Tobacco Control* dedicated its editorial concept to "new endgame ideas for tobacco control." The arguments, information, and ideas presented in this book have been meant to aid readers in their fight against that endgame. There's no single "magic bullet" that can be used to stop the massive engine of tobacco control and move it toward a saner path of proper education and harm reduction based upon people's desires and decisions about how to live and balance their immediate pleasures against their chances at longevity and gradual decline.

No single magic bullet, but hopefully, the total compendium of information, ideas, and inspiration of the last four hundred pages will come together to provide the strength that's needed for individual dedicated Free Choice activists and groups around the world to stop a movement that may have once had some valid idealistic base but has grown into being a negative and destructive force in our lives and our communities.

The Endgame looks to plot a future for Free Choice, just as a future is being plotted by those who would take that Choice away.

Funding The Endgame

Smokers and their friends never looked for conflict in this area. For the most part smokers respected requests not to smoke in areas where smoking bans were clearly reasonable (grammar schools, sensitive hospital wards, elevators, mass transit, fireworks factories and space capsules, etc.) but no matter how much they tried to appease the conquerors, the greed was never satisfied. As of 2013, even such things as private living spaces and public sidewalks have become fair game for attacks by antismoking organizations. So, despite the desire for peace, there's no avoiding a battle.

The playing field won't be level by a long shot – a few books, websites, blogs, and unpaid activists versus organizations controlling annual budgets in the billions of dollars isn't the fairest of odds – but, again as noted in the *Introduction*, the tobacco control industry's foundation is built on a bed of sand. Lies that are properly exposed end up costing the liars far more than any gains acquired through their use, and the Antismokers have told far too many lies over the years to escape unscathed.

Because they got caught with their pants down in the 1980s and 1990s, the public turned on Big Tobacco the same way it turned on Richard Nixon in the 1970s. Nevermore would their words be trusted. Indeed, the opposite was the case; people generally took anything they said and simply assumed that the opposite must be true. Because of this, Free Choice groups have largely kept a significant distance between themselves and Big Tobacco and have thus been cut off from the one solid source of financial support that might have been available for a professional and organized resistance.*

---

* The North American exceptions to this, the National Smokers Alliance in the US and *Mon Choix* in Canada, both started out with tobacco company funding but largely dissolved once that funding was pulled. The UK's FOREST organization is tobacco-funded and fights on under the leadership of Simon Clark, along with its sister group in Ireland, John Mallon's FOREST Eireann; but, just as with their past North American counterparts, the value of their efforts is at least partly discounted because of that industry association. In essence, any effort funded by Big Tobacco has to spend several dollars to get the same public effect that its opposition gets by spending a single dollar.

Is there a way that the world's smokers can get financial support for their fight through Big Tobacco or related commercial interests without the taint that goes along with such support? As we've noted, there are hundreds of millions of dollars poured straight into the pockets of the United States antismoking movement every year that come from Big Tobacco – courtesy of the MSA's "invisible tax" on smokers – collected by the industry and then given to the mob... er, I mean, the various state governments... to be properly laundered before being ladled out.

Unfortunately, the chances for Free Choice organizations to be included in that payout are simply nonexistent. The antismoking lobby screams to high holy heaven every time a dime of its desired funding is diverted even to more neutral community purposes. Touch a penny that they expect in their own pockets and they'll wade into battle waving a fantastical set of made-up guidelines in the air about how MSA money "should" be spent, while claiming state governments are killing our children when they fail to hand over the payola to meet those guidelines. No politician with any dreams of being re-elected in today's America would dare take money from tobacco control advocates and hand it over to their opponents. It would have been easier for Gus Hall, the perennial US Communist Party presidential candidate of the 1970s, to have had money transferred from defense spending to Communist Party meetings in the 1950s!

It's quite possible, however, that there is a way that tobacco money could be cleansed and used by true grassroots citizens' groups. Proper safeguards would have to be set up to require its continuation even if such groups began acting in ways contrary to industry interests (e.g., if they supported the promotion of independent small e-cig companies or encouraged "cottage industry" natural tobacco farming), and specific micro-control over allocating the money would have to be removed from the industry control entirely. The companies could be contractually held, provided they initially agreed, to continue that funding for a defined period of years unless outright misfeasance was shown; and they could be held to that agreement whether or not they liked how the money was spent.

The grant distribution of the money itself could be controlled through a volunteer board drawn from major grassroots Free Choice groups and would thus be completely free of any actual Big Tobacco influence. There would be no puppets jerking to pulls of hidden gears and strings. It wouldn't be a perfect solution – the money would still come from Big Tobacco – but it would be at least as free of corruptive influence as the money funneled through state- and lawsuit-based antismoking organizations funded through industry MSA payments.

Casino interests are also a potential source of support for Free Choice activities. Despite the ravings of some of California's loonier antismoking "researchers," casino owners know perfectly well how damaging smoking bans are to their bottom lines. State legislatures have also recognized this reality as they have passed bans exempting casinos. While politicians are ready to throw the independent small-pub owners to the wolves – knowing that the economic harm done to them will be buried by the larger ebb and flow of the hospitality industry – the carefully accounted for pot of gold represented by casino revenue is just too closely overseen for them to meddle with.

Go back and take another look at the charitable gambling figures graphed in Figure 2 and note the tens of millions of dollars' worth of damage done to charitable gambling by a smoking ban that didn't even include casinos – and then imagine the damage that would be done in a state dependent upon a huge casino industry. However, despite this inescapable evidence, Antismokers are still heavily pushing state legislatures with rhetoric about "the only workers in the state still forced to trade their health for a paycheck," and the casinos know that they're living on borrowed time unless something changes.[394] Studies arguing that bans don't hurt the casinos are so outright silly that even a few moments of thought reveals their weaknesses. A prime example in this area is a 2011 study done by a "Dr." Jenine Harris (doctorate in Public Health Studies/Biostatistics but writing on casino ban economic effects)* in *Tobacco Control.*[395,396] Unable to find any serious way to contend that the ban hadn't cost Illinois casinos money, Dr. Harris simply argued that the number of people walking in and out the front

---

* There's nothing wrong with a Ph.D. doing a study of course, but the confusion between public health, biostatistics, medical appearances, and casino economics is all too typical of antismoking research. For example, in mid-2013 a "Dr." Sara Summers spearheaded "Clean Air St. Joe's" in Missouri with numerous media appearances, but never once did I see it noted that her doctorate was in Education Leadership rather than medicine.

door turnstiles every day hadn't gone down after their ban. Of course a single smoker going in and out five times to smoke was thus magically transmuted into five different casino customers! Needless to say, the casino moguls were not impressed. They could have achieved even better results simply by holding fire drills once every hour or two to force *all* their customers to run in and out the turnstile doors!

If the casino industry supported the Free Choice movement, would that be their golden ticket? A guarantee that they'd never suffer from full bans on their gambling floors? No. Of course not. But it would improve their odds substantially. And they could get that improvement with an investment of far less than a single percent of the money they stand to lose if bans come into place. Again, as with the tobacco industry proposal above, safeguards would have to be arranged to ensure the independence and continuation of funding even if it was not always used in ways the funders might prefer. But, also again, that independence would be vital to the perception that the work done under such arrangements could be trusted.

Is such a partnership likely? Unfortunately, it may not be. I tried to work with the casinos in Atlantic City back in 2006 when the bars there were under attack. I met with one of the officers of the association representing the state's casinos and tried to make him understand the threat they faced. My warning went unheeded: he simply assured me that I didn't understand the fact that the casino industry in Atlantic City was "far too big" to ever be touched by the Antismokers. They learned their lesson a year or so later when they were hit with 75% floor bans and the obligation to throw their bar and restaurant patrons out if they dared light up after a meal or with a drink. The first year of Atlantic City's partial smoking ban marked the first time in its thirty-year gambling history when total casino profits and tax payments went down. And a few years later, when anti-smoking forces in New Jersey were powerful enough to force a full ban on the casinos, their profits took such a sudden and drastic drop that they convinced legislators to reverse the move almost instantly, although the door was left open for a future total ban.[397]

Maybe by now they have learned their lesson. The question then becomes simply whether they're willing to offer even relatively small levels of financial support to activist organizations not directly under their day-to-day control. The money would be cleaner than Big Tobacco funding, so it's definitely something that should be considered by Free Choice organizations in the future. It certainly could be sizable enough; if the casinos were willing to risk even that single percent of their potential losses, a bet with a hundred-to-one payoff, it would allow the organization of a larger force of grass roots opposition to the US anti-smoking movement than it has ever seen before.

How about funding from the wider hospitality industry? Restaurants and hotels probably haven't been hurt too much, but again, despite the stratistical machinations of advocate researchers arguing the contrary, it's pretty widely recognized that small bars are often badly hurt by smoking bans. One of the greatest failures of the early Free Choice movement in this century occurred at the end of a massive protest when the bar owners and patrons of New York City participated in a march and demonstration numbering in the thousands against Mayor Bloomberg's new ban law. The failure came when the leaders of the resistance, instead of exhorting the mom and pop owners to refuse forcible induction as citizen vigilante law enforcers, instructed the owners to go back to their bars and enforce the ban while the larger organization went to court to fight it. Without an active leadership uniting them to resist – indeed, with an active leadership telling them *not* to fight – the bar owners went back to their bars and threw the smokers out on the street "temporarily." We've all seen how well that worked out, eh?

Could the bar, tavern, and bar/restaurant owners in states and cities that are still free hold onto that freedom by working with or giving financial support to their local Free Choice groups? Perhaps, but the big problem that surfaces in such situations tends to be one of lack of organization. Independent bar owners are just that: independent. Free Choice advocates in a number of cities have tried to get them to work together, but without anything resembling a coherent union where all are required to contribute, most bar owners simply want to pass the buck onward to the owner across the street or down the road. If there's a real solution lying in this area for support with fighting smoking bans, it has rarely, if ever, surfaced.

Things may change in the future. A lot of bar smoking bans have remained in place throughout the world with the help of the proverbial wink-and-a-nod enforcement – situations where the regular nighttime crowds or even the regular quiet, but dependable, daytime visitors know they can get away with a smoke as long as there are no suspicious looking guests wandering in and nursing a wine spritzer while sniffing the air. They're surviving for the moment, but they know it won't last forever. Perhaps they'll realize the need to look outside their traditional boundaries in seeking aid in this fight. Time will tell.

Beyond Big Tobacco, casinos, and bars, there's a newly emerging area that Free Choice advocates might look to for support, Internet crowdfunding. For those unfamiliar with the term, take a quick trip to Kickstarter or Indiegogo on the net and you'll find hundreds of small entrepreneurs and activists seeking modest levels of start-up funding for all sorts of various projects. Some of them are purely commercial, perhaps someone trying to start up a market for "zombie rocks" in memory of the old craze for "pet rocks." Some are artistic, such as a comic book artist seeking to have his name and talents become better known to fans and companies as he shares his visions of a different world. And some are purely activist/advocate oriented as idealists seek funding to produce a summer fair project raising awareness of the need for breast examinations, or to cover the base funding for a concert benefiting construction of bicycle paths.[398,399]

While general support for organizations or open-ended support for advocates on a mission are not acceptable as Kickstarter projects, anything that's clearly defined, a project with a limited scope, goal, and endpoint, is a possible winner. A local Free Choice group could raise money for printing and mailing flyers and posters to all the bars in their city prior to a ban vote. A theater troupe could raise money for props to put on a production of *The Smoking Police Are Coming For YOU!* or a band could fund a studio recording of *Smoke Gets In Your Lies.* A larger group could seek funding for a national conference, perhaps not like the multi-million dollar galas hosted by antismoking funds, but at least enough to cover the cost of a hotel conference room in Atlantic City, Vegas, or The Hague for a single weekend of meetings.

The concept of crowdfunding, funding that clearly has no strings attached other than what's been stated in a project's initial proposal, is an idea that shows a promise of growth for all sorts of advocacy movements with specific projects in mind but which are unable, for whatever reason, to attract support from more traditional sources. And while some Kickstarter-type projects aim for goals in the hundreds of thousands of dollars, there are plenty that have been quite successful at simply raising the few hundred or few thousand that was needed for the birth or nurturing of an idea. A few hundred or thousand extra dollars could often mark the difference between success and failure for a local Free Choice group in the course of a particular battle.

All of these funding arrangements are possible, though some may be more likely than others, but realistically, for the most part at least, we need to plan for continuing our fight without such support. Every letter to the editor that is published, every Internet posting with sound information that gets read, and every individual standing up as a dissenting voice at a City Council hearing has an impact. Sometimes it might feel as though your voice has been ignored, but when it's added into the larger pot of growing resistance to wider bans and higher taxes, it counts. You may lose an individual battle, but the very fact that you helped make sure that it *was* a battle will make a difference in what comes down the line next time around.

We've begun to benefit to some extent from the fact that people are now more suspicious of the motives of Big Pharma than they used to be. Its constant advertising for questionable prescription drugs, as well as the revelations about its shady research, product-approval practices, and drug side effects have begun to erode the sheen of Holy Medicine that covered the cheap brass underneath.[400] We depend upon doctors and medical researchers for our lives, and in the past it was just too scary for most people to think that their saviors in time of need might share the same dollar-lust that drove Big Tobacco to its ruin. The excesses of the antismoking movement have combined with revelations from within the drug and health industries to destroy a good bit of the basis of that trust.[401, 402]

Free Choice advocates can achieve a lot by simply getting truthful information out to the public through such avenues as newspapers' letter columns, Internet blogs and comment boards, and the all-too-rare supportive articles in the press. It's also important to remember the power of radio. Talk show hosts are constantly looking for interesting and contentious topics that can be hotly debated by knowledgeable advocates, particularly if they think their listeners will walk away with a feeling of "Gee, *that* was interesting. I'd never heard those arguments put that way before. I'll have to listen to that show again!"

Unfortunately, for the most part, professional Antismokers refuse to meet Free Choice advocates for fairly-moderated debate in the public arena because the open challenging of their dogma on a level playing field is the last thing in the world they want. They depend upon the popular perception that they are the "good guys" and anyone questioning them must somehow secretly be connected to Big Tobacco and are thus the "bad guys." When inconvenient questioners actually do succeed in getting a moment at a community microphone, they are routinely brushed aside with such mantras as "The Science Is Settled" and "All Cognizant Authorities Agree." If that doesn't work, Antismokers will simply try to dismiss their challengers as Flat Earthers, Holocaust Deniers, generalized Conspiracy Theorists, or just outright cranks.

People are not stupid however. Even without the funding to get a high profile in the media, we have opportunities for our arguments to be heard as long as we're willing to put the effort in and be persistent. And once those arguments are heard, they're not easily dismissed. Lack of organized funding for a movement can be a strength as well as a weakness. Free Choice supporters don't need to have their messages "cleared" by anyone higher up controlling the purse strings, and our activities won't cease simply because someone yanks that purse away. Unlike the New Jersey group described earlier, we'll never stand up and say, "Everything stops. There is no more money." when our annual budget dips below $15 million. If one of our groups had an annual budget of $15 *thousand* they'd be in seventh heaven.

Fighting The Endgame

TobakkoNacht – The Antismoking Endgame was designed to be a toolbox, a mini-armory for those fighting against the Antismokers' endgame plans. The opening story itself was meant to give Free Choice advocates the energy to keep fighting by reminding them of how serious the stakes could be in the long run, even if the ending may be a tad unlikely.

Note that I said "may" be unlikely. Remember that the entire basic tale of *TobakkoNacht!* itself, with the exception of the prehistoric introduction, was written in the 1990s. Smoking bans in public plazas of cities, supplies of concentrated nicotine liquid or black market cigarettes posing an immediate threat to children, an international treaty against the tobacco industry enforced by UN sanctions, demonstrations outside "factories of death" mapped and pinpointed by antismoking funded studies, laws against smoking in bordellos, people worrying about protecting children from wisps of smoke outdoors, neurotic souls dodging traffic in the street to avoid sidewalk smokers, suburbanites tagging "safe, smoke-free homes" with blue porch lights,[*][403] or chimpanzees forced to give up smoking cigars for their health ... all of those things were off in the realm of a dystopian psychotic fantasyland that I imagined *could* begin to come about in twenty or thirty years, but which even I, despite being caught up in the midst of the fervor that eventually produced *Dissecting Antismokers' Brains*, never really expected to see becoming reality before the 2020s.

At this point, I really don't know exactly where the antismoking movement will go. Other than continuing to build upon what they've already started, there aren't that many directions left with regard to targeting smokers. They've already begun the initial entries into such areas as job discrimination, medical discrimination, child removals, and housing discrimination – I'd include segregation in public facilities except for the fact that outright banning has become the preferred approach; not just the back of the bus, but no bus at all! I would feel dispirited about my lack of insight into the future directions of the antismoking movement if it weren't for the fact that even as of May, 2013, their main house organ, *Tobacco Control*, wasn't able to clearly define those directions even with the aid of "some of the world's most brilliant tobacco control scholars, strategists and activists, including those who originated the principal endgame concepts."[404] Knowing that I am in such ... good? ... company makes me feel less inadequate.

The most powerful contingent in the antismoking crusade of the twenty-first century is made up of those who control the purse strings and those who are paid out of those purses. That fact results in the odd self-contradiction that those benefitting the most from the arrangement, the professional Antismokers themselves, will never support the out-

---

[*] In 2009 one Winnipeg suburb made the news as antismoking activists handed such lights out to appropriate neighbors. One gets the feeling while reading about the blue light project that the activists behind it would have been thrilled if they could have mandated that symbols be painted on the houses of the smokers, but the blue lights efficiently achieved the same effect in reverse.

right prohibition and criminalization of tobacco use; the disappearance of legal tax funding would land most of them on the unemployment line. Indeed, when proposals have been floated for state prohibitions of all tobacco sales and use, the antismoking organizations have been in the forefront of those who crowd the legislative podiums with testimony as to why it is a bad idea.

There are other areas of attention for those who would seek to limit our carnal pleasures though. At one point in *Dissecting Antismokers' Brains*, I speculated about a campaign against the 900,000 American "Deaths Due To Eating" each year; but the idea that in less than a decade we'd see bans on salty restaurant food, obese children removed from their parents by Child Services, or serious efforts to ban large sodas were far beyond anything I'd have seriously predicted for 2013.

Alcohol may end up being the most vulnerable and acceptable target for our new Prohibitionists. The final Appendix in *Brains* was devoted to the idea of a renewed push in this area, and we're already seeing some of the Smoke-Free Movie nuts redirecting their attention from counting the number of minutes containing inferences about smoking activities to counting the number of minutes with hints of insinuations about alcohol-drinking.[405] When they're done counting the number of on-screen tipplers, the modern temperance crowd might look toward covering 75% of beer cans and bottles with colorful 3-D pictures of battered wives, bloody livers, and post-car-accident dismembered corpses. Throughout 2012 and into 2013, we've seen increasing pressure to raise alcohol prices in the UK and parts of Europe. I wouldn't be the least surprised if we soon begin to see a spate of studies concluding that the European Union's economic collapse must be due to drinking and smoking.

We could very easily see widespread denial of needed medical services to smokers, drinkers, and the weight-challenged under Obamacare-style health plans, both here and abroad; after all, it would simply be seen as "nudging" those unhealthy people into improving their own lives, and who could reasonably argue with that? So far, those who drink an occasional beer or two at home or carry a few extra pounds under their belts haven't felt the touch of the reformer's whip, but that state of affairs may not last for long in the face of healthcare cost budget-crunching.

It's been accepted for a good while now that states without special protective laws can require nicotine testing right along with testing for illegal drugs.* Will smokers become the American untouchable caste, employable only for the dirtiest sorts of jobs that the higher castes refuse to soil their hands with? Will we see large-scale removals of children from smoking parents – even if they only smoke outside† – rather than just the use of smoking as an occasional issue in custody disputes? We're probably already at the point where a parent who knows their sixteen-year-old is smoking and who buys cigarettes for them – to avoid having them seek out alleyways to buy black market smokes – could be targeted by Child Services. And we've actually heard such previously unthinkable proposals as the one recently made in all seriousness by a British Member of Parliament urging that smokers "be driven out of Surrey into concentrated poor areas with low-quality healthcare."[406] Maybe they could give those poor areas special little ghetto names and ask the residents to wear something distinctive if they stepped outside?

The endgame is not only creeping upon us in secret – little slice by little slice – but has been surging forward at full speed as this book was being written. My *Introduction* spoke of the editorial theme of a 2010 issue of the *Tobacco Control*,[407] but as I was writing the first draft of this concluding section in December of 2012, the World Health Organization came out with its own volley, "A Debate On The 'End-Game' Of Tobacco."[408] The orientation was obviously shown in its statement that "The momentum is growing, and more and more governments are taking strong measures to fight tobacco." The authors then followed that up by promoting the concept of forcing smokers to get special, government-approved, "smokers' licenses" that they would have to present when buying tobacco and, presumably, would have to

---

* Outright job discrimination against smokers is actually legal in roughly half of the United States. Efforts to protect individuals from discrimination based on legal, off-the-job lifestyle choices have been heavily attacked by antismoking groups because they know such protections would protect smokers.

† Remember those California articles about such children being "plagued" by smoke? If someone actually pays the membership fee to ASH.org's website they get to see "secret documents" containing hints and tips about using the smoking issue to grab kids in custody disputes.

show if challenged while smoking in public.[409] "Vee vant to zee your papers," could have a whole new meaning in the not-quite-smoke-free future. And as you saw in the *Author's Note* that introduced *The Endgame*, the May, 2013, issue of *Tobacco Control* came out with not just an editorial, but an entire issue devoted to speculations and plans for basically eliminating tobacco use by any substantial number of human beings in the twenty-first century.

The general direction of the antismoking movement is quite clear, though; an extreme stigmatization of smokers, along with a reduction in their numbers that will still allow for a high level of input from their tax funding. The point that the movement is producing significant harm along with any good that may come from reduced levels of smoking is also clear. And finally, the fact that it needs to be fought now, before it gets even worse, is also clear. The question returns to one of how to fight, with essentially no funding base, against an operation that runs on hundreds of millions of dollars every year and is constantly demanding more in its crusade to supposedly protect the children.

Winning The Endgame

The current reality faced by Free Choice activists is simply that we do not have the funding to compete on a level playing field with the antismoking movement. It is truly a David-versus-Goliath conflict, but, as I hope I've communicated in these pages, I believe it's a conflict we can win – with or without funding.

Why do I believe we can win? What could possibly give me hope in such an unbalanced arena? Simply because, as noted throughout this book and *Dissecting Antismokers' Brains*, so much of the foundation of tobacco control activity is based upon lies. If they had stuck with the simple message that smoking was bad for one's health, and with an agenda based upon education rather than control and manipulation, they might have actually been more successful than they have been over the last thirty years in reducing smoking rates and they'd be in a much more secure position today.

As it is, they have wakened the proverbial monster. Many people who would normally have given antismoking activities no real thought at all – people who would, if asked, probably have given them a general thumbs up rather than a thumbs down – have instead grown aware of the danger that their methods may spread to other areas of our lives, threatening our private freedoms and integrity of our families that were previously thought to be sacrosanct. The ever wilder claims made by tobacco control advocates about such things as thirdhand smoke poisoning babies, glimpses of smoking in movies killing tens of thousands of children a year, plain packaging of cigarettes being a do-or-die crisis because children and females are seduced into smoking by pretty colors and font styles, the deadliness of encountering the slightest scent of smoke at beaches and in parks, the little creeping feet of fatal tendrils of smoke searching down hallways and along electrical wiring to hide in fuzzy teddy bears and attack little girls – all these claims are now coming back to haunt them. And unlike the lies of Big Tobacco that Antismokers showcase so often in their materials, the lies of the tobacco controllers aren't from twenty-five or fifty years ago – they're lies that you're likely to have run across in the last fifty weeks, fifty days, fifty hours, or, if you're near a television, even fifty minutes or seconds ago.

The key to winning the endgame in this fight is based upon exposing those lies to a wider public that is predisposed to believe anyone wearing a white coat or a governmental badge of authority. We have to strip off those white coats and badges and show the mechanical engineers, cruise ship dancers, money-grubbing lawyers, and sleazy, populist politicians behind the lies. We have to do that with budgets that are usually nothing more than we can cadge from the corners of our daily pocketbooks, and we have to do it with an energy and a commitment equal to that of those collecting millions of dollars a year in grants, salaries, and fees to fight against us on a full-time, professional basis.

The Internet has to be our main weapon of choice at this point in the battle. While it is not very long, I believe that *Launched In The Trenches* is one of the most important sections of this book. The Internet has grown enormously since the turn of the century, and millions of people who had barely touched a keyboard in their life are today scanning news stories, web pages, and commentaries from around the world as they pound out Tweets on their cell phones. Yes, they'll look at information gleaned from the net with more skepticism than they'll apply to stories read in the papers or seen on TV, but they've also learned that just because something is on the front page of the *Times* or is featured on the evening news, that does *not* make it true. They've seen too many lies from too many presidents; too many nonsense claims by too many supposedly "cognizant authorities"; too many reports that later had to be amended, corrected, or outright withdrawn after unexpected attacks and exposures from outside; and too many official denials followed by clarifications followed by shamefaced admissions. People today are open to listening to alternative voices in a way that they weren't in the past.

Most people are smart enough to dismiss convoluted conspiracy theories, but they're also smart enough to realize that in today's world, there are indeed political and activist movements with enormous amounts of money and power that are seeking to change our attitudes and behaviors in ways far more subtle and invasive than the Madison Avenue ad campaigns of the 1950s and 1960s. But if people see contrary information presented clearly enough, in a literate enough fashion to command respect, and if that information is backed up by facts that can be easily and quickly verified through the click of a key, they will be willing to consider that input and think about how it impacts the beliefs they've formed from mass media consumption. We won't win the endgame with the Antismokers on the money-controlled battlefield of the airwaves where they'd prefer to fight, but we can win it in the trenches of the Internet where their monetary advantage is sapped, their lies can be checked, and they truly do have to meet us on a level playing field.

Exercise some care out in the wilds of Internet, though. As you saw earlier in those four pages of hate-filled postings, there are a lot of unstable people out there who've had their extremism fueled by the antismoking movement. If you take a strong and prominent position in the fight against smokers' persecution, you *will* find yourself attacked in various ways. The name calling can be shrugged off, but there have also been occasional, although rare, reports of people being harassed in their personal lives; instances where bosses or family members were pulled into the fight; and at least one instance where a prominent free choice activist who's been fighting the Antismokers for twenty years, Steve Hartwell of Canada, had an entire, carefully constructed and nurtured website-based organization stolen right out from under him by an Antismoker who then tried to use it as a fake – setting it up so that smokers seeking support in the Free Choice effort would instead be lured in to sermons and postings and ads encouraging them to give up the fight and quit.[410]

If you set up a website or blog, and most particularly if you use the Internet for both open activism and private business affairs, be sure that your activities are compartmentalized and your passwords secure. The antismoking movement, both in the US and abroad, has paid special attention to recruiting and indoctrinating young activists – kids who fit squarely in the demographic of the computer-hacker profile and who'd be strongly motivated to "strike a blow against Big Tobacco" by using their computer skills to attack anyone supporting the Free Choice position.

Don't let yourself be intimidated into anonymity though. Those working on the antismoking side of the aisle often have financial conflicts of interest that they'd prefer to hide. We don't. When we are open about who we are, what our situation is, and why we're doing what we're doing, we command far more respect than the anonymous posters hiding behind nonsense names and squealing about their "right to anonymity" on the Internet. Yes, we can enjoy a measure of anonymity out there in cyberspace, but if we want to be effective in this fight we need to recognize that there are times we need to give that privilege up.

The twenty-first century's revolution in communications technology provided the key that made the "Arab Spring" possible and brought down governments with massive military and economic bases. That same communications technology can build and provide support and communications for ordinary people in other parts of the world who are seeing the danger of having their freedoms gradually stripped away by increasingly intrusive governments. Use it well, use it bravely, but also use it carefully!

Tea Party Pooper

In early 2013, our favorite mechanical engineer, Stanton Glantz, came out with a new study, published, not surprisingly, in what Christopher Snowdon has characterized as "the esteemed, peer-reviewed comic *Tobacco Control*."[411] In that article, Glantz argued that the American equivalent of the Arab Spring, the Tea Party movement, was actually a stealth creation of the tobacco companies, carefully planned and nurtured in secret for over twenty years.[412] As is usual with virtually any claim against Big Tobacco, the fantasy was taken seriously by the media, and the *Huffington Post* weighed in with a story headlined, "Study Confirms Tea Party Was Created by Big Tobacco and Billionaire Koch Brothers."[413]

While the general claim itself is clearly nonsensical, it shows the fear that the antismoking movement has of any activity that encourages the popular questioning of established authority, and their concern over movements that operate outside the more traditional channels they've learned to control over the last thirty years.

As described earlier, tying the enemy to Big Tobacco has always been one of their most powerful weapons. Stanton Glantz was clearly stung badly by the loss of the $800 million tax bonanza in the May 2012 California cigarette tax referendum, and he mistakenly blamed the Tea Party for spoiling his party. In reality, a good bit, quite possibly the decisive bit, of the reason for that loss can be tracked to the outpouring of criticism on the Internet and social media. Antismoking advocates told voters that Proposition 29's taxes would go purely toward cancer research, but Free Choice bloggers and news board posters knew that a substantial chunk of that money would instead be spent to further persecute smokers.

The referendum language sported a nasty Orwellian twist with its specification of "tobacco prevention" as an approved use of the tax. If you examine antismoking literature, you'll find that lobbying to increase cigarette taxes is a prime, perhaps *the* prime weapon in the antismoking arsenal of prevention techniques. What this meant in practical terms was that smokers would be forced to pay a tax increase on themselves that would, itself, be used to promote another, future tax increase on themselves. And another. And another. *Ad infinitum.*

Free Choicers spread this information to tens of thousands of voters who would otherwise have gone to the polls believing they were voting in favor of curing cancer and against the evil plots of Big Tobacco. The spread of that information most assuredly accounted for a swing of at least the fraction of a percent that took $800 million a year out of the hands of Glantz and his friends in the antismoking movement.

But not only was Glantz likely venting his spleen at the wrong target, he was evidently sloppy while doing so. Within a week of *Tobacco Control*'s publication of his hit piece on the Tea Party, the *Huffington Post* came out with a distinctly contrary piece of journalism headlined "National Cancer Institute Funds Tea Party Witch Hunt."[414] The *Post*'s guest piece by the Competitive Enterprise Institute's Bill Frezza exposed that the cancer institute had funded Glantz's research to the tune of $678,952. People who had contributed their hard-earned dollars to the NCI undoubtedly expected their money might be used for something else – perhaps for actually curing cancer. When the Congressional Appropriations Committee asked the NCI how they could have poured that sum of money into political mudslinging, the Institute's representative simply said, "Of course we thought we were funding a different kind of research when these grants were made."[415] Of course. Just a wee misunderstanding. Next time we hand out $678,952 for cancer research, we'll take a minute or two to check on how it will be spent. Meanwhile, please send us another contribution.

Within a month of Glantz's publication, the Cancer Institute issued a formal statement soundly berating him for abusing their generosity by using their funds in this way.[416] Such a public reaction by a funding source is rare in the world of research. The fact that it happened in this instance supports the view that California's Anti-smokers were angry enough about the loss of the tax vote that they were quite willing to cut ethical corners in getting revenge against those they thought responsible. They seemed to completely misunderstand the dynamic behind their defeat: they actually *had* beaten Big Tobacco in the major media, successfully turning the contest into one of "Evil Big Tobacco versus The Children With Cancer" in the eyes of most of the public; but they neglected to fight the Free Choice advocates on the news boards and blogs of the Internet and it was the power of those voices that made the difference between victory and defeat.

Just a few thousand ballots, out of over five million cast, swung the result. While there is no way of really knowing what effect each of many different factors had on the results of that vote, I firmly believe that the Internet efforts of Free Choice activists exposing the abuses of the antismoking movement turned an expected grand antismoking victory into a narrow, but crushing, defeat.

Never one to stop grabbing for money, Glantz has now grabbed the microphone to condemn a new tax proposal that would raise cigarette taxes targeted toward paying for general healthcare. He is urging instead that its treasure be diverted to (Guess what?) California's Tobacco Control Program – where, naturally, it would be used to fund antismoking grants and promote further taxes on smokers.[417] That particular story still has to play out to its conclusion, but hopefully, by this point, Californians have had their full of antismoking grant grubbers.

The importance of this Tea Party tale lies in power of its accomplishment: as noted earlier, the fuel for the antismoking machine is colored green. Take the money away and the entire movement will dissolve back down into ordinary, and decent, efforts at health education… though there'll always be a few nuts nattering on the sidelines. But without the money, their personnel and their muscle will be gone. Antismokers like Stanton Glantz are all too aware of that, and that's why they fight so hard for the almighty buck.

Strategy

The root challenges for the Free Choice movement in the years leading up to 2020 lie in four areas.

(1) Money
(2) Communications
(3) Organization
(4) Victories

We've already discussed money and found no easy answer. Maybe we will find a way to safely fund our movement through Big Tobacco, maybe we'll find slightly less tainted support through casino interests, maybe we'll try a new approach through such Internet crowdfunding vehicles as Kickstarter, or maybe we will just continue to fight more and more effectively without organized and centralized funding. One thing is for sure, however: this is an area where Big Money, although still powerful, can no longer count on automatic victories while hiding its corrupt and ignoble roots behind white cloaks and children.

We've also discussed communications to some extent, examining the importance of the Internet in giving us a means of building a community across distances when we don't have the funding to support regular face-to-face conferences and meetings. The Internet has provided us with the sort of intellectual, emotional, and creative sharing normally reserved for groups whose individuals can interact physically on a regular basis, whether it be on a university campus, within a company boardroom, or even just at international conventions that gather several times a year. We need to build that communications network and make it work more effectively for us in the years to come. Audio and video communication hookups for groups over such Internet avenues as Skype allow for a personal touch and the sharing of ideas in real time that was never available to those without significant funding in the past. Traditional Internet organizing tools such as special interest mailing lists targeted to particular topics or geographic areas will become more powerful as well, encouraging individuals to support each other's local efforts while learning from experiences outside of their own neighborhoods and fights. We need to avoid duplicating each other's efforts, but at the same time we need to recognize that there can be value in similar actions taking place simultaneously in different places and with different emphases.

This last aspect of communications thinking leads us into the question of organization. There have been, and still are, many small Free Choice activist groups that fight smoking bans, taxes, and restrictions. There have been, and still are, a number of larger Free Choice umbrella groups that seek to bring people together across state and national boundaries to share their experiences and learn from each other. And there have been, and still are, Free Choice groups both large and small that are willing to speak to the media when allowed a shot at the microphone. We need to build upon those strengths while also reaching out to new populations, groups like e-cig vapers, college students, casino owners and their patrons, shisha lovers, smokers in the Far East, our heavily beleaguered comrades in Australia and New Zealand, marijuana smokers who are finding their new freedoms under attack, and even gun owners who are finding their old freedoms newly threatened.

In 1994, a resident of Conway, Massachusetts , Samantha Phillipe, saw the writing on the wall in terms of where the antismoking movement was heading and was determined to begin fighting. She founded The Smokers Club and began a weekly online newsletter titled *The United Pro Choice Smokers' Rights Newsletter*. It has been published weekly with virtually no interruptions for almost 20 years with only one significant change. In 2007, recognizing the need to speak not just for smokers, but for any citizen whose rights over their own lives and property were being threatened, she created The Citizens' Freedom Alliance and retitled the online publication as *The Property Rights News-letter*. Today the Alliance and the *Newsletter* support gun owners whose guns are being taken away, bars and restaurants whose livelihoods are threatened by smoking bans, drinkers pressured by "minimum pricing" standards, and private citizens who are finding that even their most fundamental rights to grow their own food or select the foods they want to buy are being taken from them in the name of political correctness.

This widening of spirit is essential to the growth of the Free Choice movement. Some of us may have prejudices against gays, straights, blacks, whites, Catholics, Jews, Muslims, or mainstream blue-blood Protestant gun-owners. Some smokers are angry over the way early e-cig marketers jumped on the bandwagon of smoker vilification in order to sell their products, and some e-cig vapers are still using internalized negative perceptions of smoking to help them stick with their e-cigs.[*]

Anyone with such problems needs to be sure that they stay stowed in the trunk for this ride. Smokers, their families, and their friends are under serious and mounting attack by very powerful interests. Antismoking organizations have had the money to create a large funded class of full-time professional warriors. Those professionals put this battle first in their lives, ignoring any differences or frictions that might weaken their united front. We need to be sure to do

---

[*] My own experiments with vaping indicate that this negative attitude toward smoking may simply be a passing phase among new vapers who'd previously tried and failed to quit smoking. Once they become accustomed to their new enjoyment, I think many who started their vaping experience as "self-hating-smokers," will transform into "self-loving vapers" who'll proselytize smokers more than berate them. The stupidity of Antismokers who've continued to attack these newly nonsmoking vapers has simply worked to Free Choice benefit as we've gained power from a whole new constituency.

the same. And while we're doing so, maybe we'll also discover that some of the differences we thought we had with what we saw as "other" groups of people aren't quite as large as we thought.

Free Choice groups, in comparison to the Antismokers, are guerilla irregulars without clear leaders or even fancy T-shirts. To have any hope of victory at all we need to put irrelevant personal prejudices and divisive issues aside when working together. Smoking issues may not seem like vital buttresses in the foundation of freedom, but it's the corrosion of the little freedoms in life that pave the way for the destruction of the larger ones later on. There are a great number of Antismokers who have made the fight against smoking one of the central themes in their life, some because of career considerations, some because of emotional involvement due to perceived harms and losses, and some simply because of neuroticism. Free Choice supporters cannot dive into this battle half-heartedly and have any hope of winning: the choice is either to surrender or to take the fight as seriously as those you are fighting against.

Finally, we need to make it easier for isolated individuals who are just entering the fight to connect to other like-minded individuals and groups for support and learning so that they can fight more effectively. When someone asks for help fighting a campus ban or an apartment eviction or job discrimination, we may not have all the answers we might wish we had, but we should make certain that we spend the necessary time and energy to provide whatever we do have. We need to make sure such people realize that they've been treated as simply unintended consequences or collateral damage in the larger war waged by those on a crusade for a smoke-free world, and encourage them to become part of that larger fight if they truly want to solve their own local problems. And beyond simply sharing information we need to be sure that newcomers to the fight are included in the exchanges and support networks that are already in place. Yes, it's tempting to safely keep one's little group static. It may seem like it's just the right size to have an effect, and there's always the chance that someone new may not fit in with the group dynamic, but those chances have to be taken if we are to successfully grow in our strength.

That note on more effective organization and fighting brings us to the last challenge that faces us as we head toward the new century's version of the Roaring Twenties. After the fall of Delaware and New

York to strong smoking bans in the early 2000s, Free Choice activists were disheartened to see one major loss after another spreading across North America.[418] Even harder to accept were the spread of copycat bans to Europe, the Mid-East, Asia, and even Australia. While there were a good number of small victories on the levels of institutions, towns, cities, and even a few states and countries, there were many large losses that could not be denied.

However, there is both a negative and a positive way to look at the history of the last ten years. Yes, we've had many losses. But, on the other hand, in addition to a number of victories in the tax arena, we have prevented a far more total victory of the antismoking movement in banning smoking in virtually every workplace or gathering place, indoors or out, in the civilized world if we had not been active.

It's worth remembering that a good number of the victories that Antismokers have claimed have been far less glorious than what they had originally aimed for. In Florida, for instance, bars that do not serve food are still enjoying a Free Choice status, with some of them banning smoking on their own, but most still allow smoking either throughout their establishments or in comfortable smoking areas.

In Nevada, the Antismokers thought they had things pretty well wrapped up outside the casinos, but within a year of the Nevada ban kicking in, the legislature realized what a disaster it had created for the smaller bars and bar/restaurants, particularly those that allowed gambling. Suddenly, hundreds of small businesses had been placed at an enormous competitive disadvantage vis-à-vis the big casinos, and they were dying. Saving them wasn't an easy battle – think in terms of pulling a half-swallowed fish out of the gullet of a large grizzly bear – but Nevada's Free Choice supporters got that particularly important facet of their state ban reversed.

In Pennsylvania, despite years of strident antismoking efforts, the ban on smoking in bars has met almost total failure. At this point in time, there are about 3,000 bars, roughly half the state's total, that officially have exemptions allowing for smoking. Smoking is also still officially allowed in a respectable number of private clubs throughout the state. And finally, while no one has actually done a study of it, it's likely that thousands of other establishments which did not manage to get an official exemption to the law have also worked out ways of accommodating both smokers and non-smokers without throwing people out into the cold. After all, even in California, the heart and soul of the American Antismoking Kingdom itself, remember the earlier-cited study that found half of the state's bars skirting the ban law to various extents. And if my email contacts are trustworthy – and I believe they are – the number of bars that have come to terms with the law by quietly defying it at various levels in the US is incredibly high and that defiance needs to be encouraged. I'll speak a bit more about that in the *Battle of the Bars* a few pages from now.

On the international level, smoking bans have met with great resistance and resentment, despite official governmental lip service and the usual assurances from polls commissioned and designed by anti-smoking organizations. Bans in such countries as Italy, France, Spain, Greece, Bulgaria, Russia, and others may be moderately well-enforced in central city hot spots and tourist areas, but Internet emails and postings on various boards indicate that they're not so well observed in non-tourist neighborhoods and out in the countryside. Paralleling the results of that California study, Germany's antismoking advocates recently chided the government in a 2011 *British Medical Journal* study that claimed roughly 80% of German pubs still allowed smoking – clandestinely or openly.[419]

Finally, the year 2012 saw a world's first in the modern era: the widespread rollback of a ban in an entire nation, The Netherlands. Amazingly, that rollback was even supported by the Dutch Minister of Health, Edith Schippers, a woman now designated as "The Dutch Minister of Death" by Antismokers.[420] Today, small bars and cafés in Holland are once again allowed to have smoking of tobacco (as well as of marijuana of course!) on their premises if they wish. The Dutch Antismokers are obviously extremely unhappy with this, but so far there is little they have been able to do about it. Most of the Dutch people seem to recognize that allowing their citizens Free Choice – between places that choose to allow smoking and places that choose to ban it – is more rational than trying to impose the wishes of a big nanny government upon everyone.

Over the next ten years, we're likely to see a continued slow-down and reversal of antismoking victories in the developed world. The activists have gone too far – they have "jumped the shark" in the public's eye – and they are no longer trusted as much as they were ten or twenty years ago. Their claim of being a little David fighting valiantly against the big Goliath of an all-powerful tobacco industry is laughable today in the face of the draconian limitations placed on that industry over the last few decades. The "deadly threat of secondhand smoke," which garnered a lot of public sympathy when it was first used in arguments to ban smoking in schools, hospitals, public transport, and airplanes has become a joke when it's now applied to smokers at a beach or in a pub with modern ventilation and air-filtration equipment. The trumped up threat of thirdhand smoke is still sucking millions in grant money away from research that might actually cure cancer or benefit people, but it's become a laughingstock in the real world.

In an effort to stay in the headlines and maintain their exorbitant funding levels Antismokers have been pushed into taking ever more extremist positions until today it is clear from the general tone of comments on Internet bulletin boards, and even from the general attitude of news and talk show TV hosts, that their statements are now viewed with the same reservations that were once saved for Big Tobacco. The public doesn't fully realize the extent to which monetary greed drives the antismoking movement – most folks still believe its strongest roots lie in idealism and a desire to see a reduction in smoking-caused illness – but that realization is increasing.

Supporters of Free Choice need to hammer home the message that smoking bans are based on lies aimed at the promotion of behavioral change through social engineering techniques rather than being based on truths derived from actual scientific and medical data. If you want the rat to eat out of the round bowl, you shock it whenever it eats out of the square one. Eventually it learns to eat from the round. People will rightfully object to being treated like rats in an experimental chamber, subjected to little "electric shocks" to guide them into governmentally desired behavioral patterns.

People are *not* rats and they most certainly should not be treated like rats by high and mighty governments that desire to condition them into proper behavior "for their own good" (or, eventually, "for the good of the state"). As the public becomes more and more aware of how such techniques are being used, the general support for smoking bans – and for electric shock equivalents such as outrageous taxation levels and job, housing, and medical discriminations against smokers – will decrease. And as that support decreases, so too will the support for continued funding of groups supporting such things decrease. And, finally, once the money is gone and the gold is no longer beckoning, the great bulk of the neo-prohibitionist movement against tobacco will go down the same disappearing path as the old prohibitionist movement against alcohol.

Yes, we may be better off in terms of our health and as a civilization because of some reasonable and limited restrictions on alcohol use. And we'll likely be better off in terms of our health (though perhaps not so clearly as a civilization) with some reasonable and limited restrictions on smoking. But reasonable and limited restrictions need to be kept to exactly those standards – reasonable and limited – and there are few unbiased observers who would argue that the techniques of the modern antismoking movement have remained within those boundaries.

It's not going to be an easy fight by any means, and victory over the powerful groups that seek to take away our individual choices in the name of the greater good – while also profiting handsomely on the side from their supposedly idealistic efforts – is by no means certain, but it's within our grasp as long as we don't give up. We need to continue improving our coordination, work harder at building our outreach to the various groups being targeted by antismoking campaigns, and be sure that our voice is heard in public debates and discussions on the Internet, in letters to editors, and in the hearing chambers of city halls and state capitals.

Tactics

While accepting the reality that we can never come close to equaling the expenditures of the Antismokers, we need to develop tactics that will afford us a real chance of success in reaching larger numbers of people.

For the most part over the past twenty years, the average mainstream American has thought that smoking bans, in general terms at least, were probably a good thing. After all, they were constantly hearing that there was "no safe level of exposure" to smoke, they were constantly being told that bans didn't hurt business, and they were constantly aware that the two sides of the argument seemed to be mainly between the "good" people (i.e. doctors and government authorities) and the "bad" people (evil money-hungry big tobacco companies and, to a somewhat lesser extent, depraved nicotine addicts). Nonsmokers, a significant majority of the population, were also motivated to support the antismoking position regarding bans and taxes out of normal self-centered convenience: bans meant they didn't have to be concerned about smelling smoke if they happened to find the smell distasteful, and higher taxes on smokers meant lower taxes on them, the nonsmokers.

However, as antismoking claims and demands have become crazier and the economic and social impacts of bans have become more difficult to ignore, more nonsmokers have begun to ask whether things have gone far enough... or possibly even too far. We need to focus our tactics on encouraging such questioning by emphasizing the lies and extremism of our opponents, pointing out their monetary motives, and decrying their use of children and fear in manipulating our behavior. People don't like lies. They don't like extremists. They don't like greed. They don't like people using children as weapons. And they most certainly don't like being manipulated as if they were marionettes on strings or rats in mazes.

Anytime we find ourselves in a public conflict with antismoking advocates, we need to examine how those five things – the lies, the extremism, the greed, the abuse of our love of children, and the callous manipulation through fear – are in play. Whenever we can expose those aspects of an antismoking push clearly enough, we will win support.

There's no danger in the Antismokers knowing that we will do this either: those weapons are absolutely essential to their campaigns at this point, and they're fundamentally incapable of operating without them. If they don't lie, people won't support their pushes for more bans and taxes. If they don't come up with more and more extreme measures and claims, they simply become irrelevant. If they tone down their greed, they'll lose their personnel. (Thirty years ago the anti-smoking movement was largely a volunteer grassroots phenomenon. But as we saw in the New Jersey case cited earlier, today's movement depends on well-paid professionals who'll drop it like a hot potato if the money spigot is turned off.) If they stop using children in their campaign, they have nothing to replace them with that would be nearly as effective as an emotional weapon. And finally, they can't stop trying to frighten people – it's that fear that forms the core of their continued support. Without it, all they'd have left would be a mild sense of dissatisfaction among people who might prefer not to smell smoke too often and who would be quite content to simply select nonsmoking bars, restaurants, and workplaces on their own if they so desired.

To use those weaknesses successfully in a tactical sense, it's of utmost importance that Free Choice proponents make sure we don't open ourselves up to the same weaknesses. When we hear our opponents making stuff up it's certainly tempting to make stuff up right back at them – matching them whopper for whopper. But if we do that, we give away our own strongest weapon: our honesty versus their lies. If they can turn around and point to lies coming from our side, then our advantage is lost. Most people will simply take a quick look, decide neither side can really be trusted, and will then stick with the antismoking side because it at least *seems* to speak with the voice of recognized authority. A reporter won't get fired for reporting something the *Über*-Doctor-General says as a "fact," no matter how crazy that supposed fact might be. But if a journalist reports something odd that the president of Sunny Smilin' Smokers' Sorority says as fact without having double- and triple-checked it … well, their pink slip might not be too far away.

When antismoking organizations resort to extremist language – calling smokers murderers, addicts, or other such things – it gives us an advantage as can we point to the hate that drives them. But if we then turn around and explicitly and repeatedly call them Nazis, thieves, and liars it weakens our position as the great tide of passers-by simply peeks at the fight and concludes it's just two sets of kids calling each other nasty names. Yes, we can point out their explicit lies. Yes, we can point out how their methods and goals represent the antithesis of freedom and embody many of the same ideals pushed by the Nazis in their quest for a perfectly healthy Arian race. And yes, we can point out the misuse of funding, the bloated pay of nonprofit executives, and the criminality of taxes of several hundred percent on a product greatly desired by a population that is largely middle class or below. But while doing all those things, we need to keep in mind that we always want to be better than our opponents – not as offensive, more polite, more reasonable, and more willing to be open to fair discussions with those on the other side. On those occasions when we do make serious charges or use derogatory imagery or parallels – as I'm sure some will argue I have done in the title and opening story of this book – we need to be able to show that there's good reason for doing so and also be open to understanding why people might be disturbed.

While it's often helpful to point to the misuse of funding for illegal lobbying and similar purposes, we should also always emphasize the professional nature of the organizations that push for smoking bans. Yes, they may have a lot of good-hearted volunteers working with them, but the core organizers – the ones doing the planning, running the websites, and paying the expenses for the props, brochures, press events, and conferences – those people are almost always getting paid a tidy sum and it's important that people know that. By contrast, although it may not be by choice, those on our side of the battle can almost always point to our complete innocence with regard to funding. Aside from those who write for or work with institutions that receive more than a few percent of their income from Big Tobacco grants, or the few rare souls who might earn some money from writing articles or books, or the even rarer individuals who actually work with industry-sponsored groups like FOREST, there's virtually no one out there on the Free Choice side who'd have to tick off the "Conflicting Interests" box when describing themselves in this debate.

In terms of manipulation, we need to recognize that everyone seeks, to some degree, to manipulate their opponents and their audience in a debate. No one in a contentious dispute can honestly deny all attempts at manipulation. But in the case of the antismoking movement, they have left themselves vulnerable by the fact that they have so heavily and blatantly sought to manipulate us with something very special: our biologically hardwired love for children. Politicians and dictators have played "the children card" for years, but usually they display at least some level of skill and subtlety in their game. Of course the subtlety doesn't always work, Saddam Hussein's smiling face while patting the head of one of the little American children he was holding hostage was a good example of such a failure.

Antismoking groups are so convinced that they and their motives and methods are beyond questioning that they haven't even bothered with attempts at subtlety. They'll dress kids up in cute little "Disease Costumes" to lobby council members, they'll pile empty children's sneakers up in front of City Hall, they'll talk about 340 children being killed every day by the sight of smoking in movies,[*][421] and they'll repeatedly spout fantastical stories about evil smokers blowing smoke in innocent little faces or smoking up storms in tightly sealed cars with Wee Widdle Wilhelminas choking to death in back-seat bassinets.

People are becoming more aware and resentful of such manipulative techniques. We need to make sure that they are pointed out clearly and emphatically when they're used by antismoking advocates and we need to make sure we don't fall into a similar trap. Our arguments are well-based, and the anger that will arise when the Antismokers' tricks are properly exposed will be more than sufficient. We can bypass the manipulative tricks, particularly any that might use children as a foil.

Finally, there's the more general issue of manipulation, usually through fear and pain, that is practiced by the antismoking movement. As noted earlier, that sort of manipulation, although it has many euphemistic terms – over in the UK they like to refer to it as "nudging" – is basically nothing more nor less than the same sort of behavioral conditioning that's applied to rats with electric shocks.

If you want a smoker not to smoke, you give the smoker something equivalent to an electric shock until they learn to stop smoking. That shock can take the form of a tax increase, it can take the form of making them leave a social situation if they want to smoke, it can take the form of using fear to create pressure from their children, or it can take the form of any sort of unpleasantness at all that can be heaped upon them whenever they move toward the undesired behavior. The next time you see Antismokers grudgingly allowing an "exemption" to a smoking ban, note how they'll try to maintain the unpleasantness of the exemption by such things as limiting the warmth or shelter afforded the smokers or limiting their ability to eat, drink, or even watch television in social groups while smoking. Note how they'll push to make the exemption something that carries its own little

---

[*] Actually, the claim is at the bottom of every fancy, colorful, high-tech page of SceneSmoking.org's web site. It takes the form of a "Surgeon General's Warning" and states, "Viewers Strongly Cautioned – Smoking Kills About 340 Young People A Day – Smoking In Movies Is Not Cool, Healthy, Nor Needed."

electric shock by costing more money or by forcing the smoker to engage in an activity they might well prefer to avoid, whether it be gambling, drinking, standing instead of sitting, or even just trudging through the rain to get to a smoking shelter that's limited to two walls and a partial roof – an enclosure that would be illegal if a farmer used it to shelter pigs.

Choose Your Battlefield

One lesson to be learned from the antismoking playbook is to know what your "make it or break it" conditions are in a fight. If a proposed shelter wouldn't meet the minimum standards to avoid cruelty to pigs it certainly shouldn't be accepted as a "compromise" for smokers. If a 500-acre college campus wants to restrict smoking to 20 square feet on the edge of a far-flung parking lot, that's not a sufficient accommodation to ever be accepted by students.

Battlefield thinking doesn't only apply to physical location or conditions. It also takes into account some of the strategic elements already examined. Areas where the lies of the Antismokers are most easily, convincingly, and quickly illustrated give us enormous advantage in any fight, regardless of how much money the other side has to throw around. Thirdhand smoke, any outdoor bans that don't involve people locked in next to each other, dangers claimed for the levels of smoke normally experienced from neighbors smoking in condos and apartments, and even "save the children" pushes to ban smoking in cars with minors* – these all represent weak points along the battlefront where the Antismokers have left their bare flanks swinging in the breeze just begging for a sling-stone or a sharpened arrow.

I believe most of the above weaknesses have already been explored in previous pages, so there's no need to repeat the analyses. Some are easier to show than others (e.g., the silliness of thirdhand smoke) and some are harder (e.g., explaining peak concentration

---

* Note that these car bans also apply to protecting seventeen-year-olds from smoke in a convertible shooting along at 70 mph with its top down.

trickery and air-exchange rates in cars while your opponent frantically waves gasping children in the air as a distraction). Keep your eyes open for the claims that are easiest to refute, and shy away from being drawn into arguments about ones that are more difficult or where you are not as knowledgeable about the facts, figures, and arguments.

And when you find the right weaknesses, jump on them as fast and heavily as you can. Every time you expose a lie, you'll end up casting real doubt upon the rest of what antismoking advocates say, the equally false claims that are more technical and difficult to expose. We've been criticized at times for supposedly following a tobacco industry strategy: Create Doubt. In a sense, the criticism is true. It's a powerful strategy that was used by the American tobacco industry from the time researchers were first seriously linking smoking to lung cancer in the 1950s. However, we'd be fools to ignore it simply because it's been used by an undesirable entity, particularly since doubting the claims of antismoking advocates is eminently justified. The difference lies in the fact that the industry strove to create doubt in areas where there truly seemed to be very little doubt, whereas we are seeking to create doubt in areas where doubt is clearly the most proper response. We need to create that doubt and then follow it up with attacks on the lies that might be a little harder to demonstrate clearly. An audience that has just grasped a new revelation (i.e., "Antismokers DO Lie!") will be open and eager for more revelations. Free Choice advocates don't have lots of money, but we do have lots of revelations for anyone who listens with open ears!

All Is Not Quiet On The Campus Front

College campuses are likely to be a hot battlefield in this decade, and one on which Free Choice advocates have a solid chance of prevailing. Given the outrageous weakness of outdoor smoke exposure claims,* the

---

* Recall the earlier discussion about campus smoke exposure. Walking through a crowd of smokers at a doorway every single day of the school year might produce something on the order of one extra lung cancer for every 250 million student-years of schooling. I don't want to appear hard-hearted, but if it takes someone that long to get a degree, I think we might be better off if they went into manual labor.

demands of the Antismokers for total campus bans are quite clearly extremist and unreasonable. Even the current situation on many campuses, banning all indoor smoking other than living quarters, is far more extreme than could ever really be justified on health grounds involving secondary smoke exposure.

On some campuses smoking actually has been banned in private living quarters of dormitories, although fear of lawsuits following "unintended consequences" may be hampering the spread of that particular type of ban. It's not easy to track something so vague through Google, but several years ago, after noticing an Internet reference to a past tragedy, I did some searching and turned up two incidents that made the news back in the early 2000s. The first took place in October of 2001 and involved a University of Arkansas freshman who climbed out on a wide window ledge from his fifth floor smoke-banned dorm room for an evening smoke. After he fell to his death a University police lieutenant noted that the practice of getting around the ban by smoking on the ledges was one the school was well aware of. Radio host Neal Boortz passed off the death with the comment "This kid was climbing out on that window ledge to DO DRUGS! He was a drug addict and he died because he was a drug addict."[422] The second took place about a year later, in December of 2002, and involved a Providence College sophomore who slipped off a rain-slicked roof while sneaking a late night smoke with a friend. Again, a little investigation revealed the fact that school authorities knew that students were responding to the ban in this way – and yet the ban remained in place.[423]

While university officials may figure it's not that big a deal if a few smokers miss their graduations, my guess is that the parents involved may not be so blasé about the collateral damages of school policies. I'm generally not a fan of lawsuits, but I hope that in both cases the parents in question considered what responsibility may have lain with the schools. And while I have not been able to find actual hard statistics on the matter, I believe it may well be true that dormitory bans have been soft-peddled in the years since those tragedies despite the pressures from Antismokers.

The fact that the population in question, college students, is generally both fairly well-educated and computer-literate, also works in our favor. The sound bite arguments that play so well on TV don't pack nearly the same punch in a population that knows how to read and analyze a study, or one familiar with the wiles and tricks of debaters. College students are also likely to be more jealous of their freedoms and retain at least some memories of the potential for "the way things used to be" back on the free-spirited campuses of the 1960s and 1970s. The desire for sex and drugs and rock 'n roll may be more constrained than it was back in those days, but it hasn't disappeared – it's merely gone undercover. Whether that's an improvement or not depends upon one's orientation, but it's unquestionably led to a darker atmosphere on today's campuses, one in which students are increasingly expected to inform on each other for rule violations, and one where things that would previously have been treated as only minor infractions are now seen as possible deal-breakers for a student's future advancement and career.

However, nature abhors a vacuum, and the wild left-wing freedom organizations of past decades have now been replaced by more straight-laced, but still freedom-loving, conservative student groups with names like Students For Liberty, Young Republicans, and Young Americans For Freedom. These organizations have become increasingly active in recent years as students have become aware of the slow trickle of their freedoms gradually dripping away. Chapters of such conservative-leaning organizations on hundreds of campuses have become well-known and respected despite many universities' historical leanings toward more liberal political positions.

Antismoking efforts on campuses depend heavily upon keeping the students of a particular campus isolated and making them feel that they are the only ones fighting against "the overwhelming trend toward smoke-free campuses." Smoke Free Campus advocates love to parade the raw number of campuses they claim have banned tobacco (currently somewhere between 700 and 1,000 in the US) while never mentioning that there are actually around 5,000 campuses in the fifty states. Even when they could only brag about 300 campuses under their control, the antismoking organizations tried to make it sound as though it represented an overwhelming majority of campuses.

Seven hundred to a thousand campuses is a respectable number, but it's a bit questionable as to how solid that number actually is. If you visit the campus newspaper websites of some of the schools that have had total campus bans in place for several years, you'll find that the letters columns and comment areas sport an uncomfortable number of complaints from students about how the bans are being ignored. There may well be 750 or more campuses with official full-campus bans… but if a ban is ignored, is it really a ban? At one university, the University of Michigan, the *Michigan Daily* ran a news story on May 8, 2013 touting the success of its ban.[424] The story noted that it was on the list of the thousand "completely smoke free" universities that Americans for Nonsmokers Rights was using to convince other campuses that "Everyone is doing it!". Unfortunately, as you read the story you find out a few things that show the real meaning behind such claims. Two points that stood out for me were that there are "designated ashtrays" all around the campus for smokers, and that over a third of the student body felt smoking on campus hadn't declined *at all* since the start of the ban! Not surprisingly, Antismokers seem to be rather loose in their definitions of "completely smoke-free campuses." That's how they boost the numbers to enhance the bandwagon effect.

Additionally, there have been stories of campuses, both school and hospital, that had so much trouble with total bans that they quietly backtracked without actually changing their official policy. A few secluded and totally unofficial corners where smokers can gather without a hassle are given the blind eye, or, in at least one case I read about in a comments area that I neglected to grab an iCyte image of, a campus went so far as to build a few small and inviting gazebos with comfortable seating and butt bins. According to the commenter, campus authorities just pretended that they, and the smokers using them, simply didn't exist. Meanwhile the corners and gazebos reduce the threat of fire from hidden smoking within the buildings and reduce the number of butts that need to be cleaned up all around the campus when a full ban is officially in place but widely ignored. Meanwhile, by not officially acknowledging such accommodations the schools protect themselves from antismoking groups' threats to warn parents of prospective students that the schools "condone and support student drug use," while also protecting grant funding linked to official campus smoking ban status.

This last element has come into considerable importance recently on the Austin campus of the University of Texas after the Cancer Prevention and Research Institute announced it would tie future research funds to smoking bans. University administrators claimed that continuing to allow students to smoke could cost the University up to $80 million a year in grants.[425] In a similar situation at nearby Texas Tech, where they were looking at a possible loss of just $1 million a year, their vice-president of research said, "I don't know what we want to call it. It's not legislation, it's not a mandate, it's not a federal or state requirement."[426] In a posted comment on that news story, I offered the opinion that it could simply be called "blackmail."

Of particular interest in the case of these universities in Texas is the fact that the actual grant limitations seem to apply only to buildings in which cancer research sponsored by the grants is actually occurring, despite information to the contrary disseminated by campus admin-istrations. Students may be getting falsely told that the campus-wide bans represent the only choice possible if the schools don't want to lose those millions of dollars. In Pennsylvania, just several years earlier, the State University system tried a similar trick with its students while pushing campus-wide bans. Supposedly responsible administrators put on long faces of resignation and regret and sadly informed students that their hands were tied in the decision to impose a total outdoor ban, because their legal consultants had informed them that the state's smoking ban mandated full bans on all state campuses.[427]

In reality, the state ban, titled "The Clean *Indoor* Air Act"(emphasis added), said nothing of the kind. While most of the students at the system's fourteen campuses proceeded to cooperate in ignorance of the actual law, one campus, based in Shippensburg, engaged in some very active protests under the leadership of two determined students, Tom Wing and Allie Bitzer. They mobilized hundreds of others, spread Free Choice flyers and booklets widely, and pretty much completely overturned the efforts for a total ban on that campus.[428] Unfortunately, there weren't clear paths of communication to the other state campuses and I believe most of them simply accepted the ban as being state law rather than child behavior management. That situation turned out to be somewhat temporary though, as you'll see in a few paragraphs.

In both the above situations, Texas and Pennsylvania, college administrations tried to force their smoking bans on the student body by passing the buck – basically using the Nuremburg Defense popu-larized after World War Two when Nazi officers and scientists claimed that they should not be blamed for their actions because they were "simply following orders." That defense didn't hold up for the war criminals and it most certainly shouldn't hold up for university administrators who should be setting good examples for their students – particularly when such "orders" didn't even exist!

What's needed at this point is for a nationwide group like Students For Liberty to take an official stand supporting reasonable campus accommodations for smoking students while also pushing for college course modules to honestly examine the issue of smoking bans from both health and social perspectives. A large national organization that would help establish communication on this issue among groups at different campuses through a well-organized website could go a long way toward balancing the millions of dollars poured into organizations like Smoke Free Campuses. An organized national face of resistance could also go a long way toward countering pressures from funding organizations controlled by advocates who use scientific grants as bludgeons to force conformity rather than as honest tools to support needed research efforts.

A final, and somewhat different, approach to fighting campus bans was demonstrated rather strongly during that battle for Pennsylvania State University. While Shippensburg seemed to be the only campus to successfully organize its students, another force reared its head elsewhere. Unions have collective bargaining agreements that often require renegotiations if a significant change is to be made in workplace requirements or conditions. In absence of a state law, a university requirement that workers have to go off campus to smoke most certainly counts as "a significant change," in their contractual agreements and Pennsylvania's campus workers raised a strong and largely successful objection to the full campus bans. In the fight for the open air at the universities, Free Choice activists should make sure they establish communication with campus workers as well as campus students.

Campuses represent a unique opportunity for Free Choice advocates to reach out to larger numbers of receptive ears. They combine two vital elements mentioned previously. First, the scientific grounds for such outdoor bans is laughably weak, and second, the student population is uniquely designed to be open to and capable of critically examining the facts and arguments on both sides of the debate. That combination, an easily exposed lie, and an educated population open to listening fairly to the arguments on both sides of the issue, is a winning game for Free Choice. Antismoking forces have stretched a lot of truths very thinly in their recent drive to proclaim control over a thousand university campuses by September of 2013. They may very well find that they have rashly overextended themselves.

Battle Of The Bars

As of June 2013, roughly half the US states have extensive or complete bar bans, and a fair number of others have at least partial bar bans of some type. Because smoking is an activity enjoyed by many drinkers and because it's pretty clear that banning smoking in bars reduces bar profits, it's usually easier to find support in fighting bans there than in most other small businesses. It also seems more common to find resistance to bans taking the form of skirting the laws – with owners often allowing smoking during off-hours or late at night or in a back room, even if such back-room smoking is technically illegal. Recall again the California study showing half the state's bars in noncompliance and the complaints from Antismokers in Germany about widespread bar smoking. Also recall two of the more significant Free Choice victories in recent years: the rollbacks of pub bans in Nevada and The Netherlands.

If you're lucky enough to be in a state where the axe hasn't fallen, or if you're in a ban area where there seems to be a fair amount of discontent among bar owners and patrons outside the central city tourist zones, don't overlook the contribution you can make by educating that segment of the public about the technical aspects of the laws and the lies about the health aspects of the bans. If bar owners and their customers realize that a good bit of the push for the bans comes from the desire to "shock the rats" rather than a true desire to protect nonsmokers from smoke, they'll get angry. And that anger is exactly what's needed to energize them to fight the bans both in the active political sense and in the more passive resistance sense.

Technically many state and local ban laws are worded in such a way that bar owners and workers are required merely to post no-smoking signs and inform customers that smoking is not allowed. They are *not* required to physically confront smokers and throw them out to the streets. They are often not even required to cease serving smokers…

particularly not if the smoker has actually, though perhaps only temporarily, ceased smoking before ordering a drink.

There's a reason why the laws have this limitation. The state has no power to impose involuntary servitude upon its citizens. A bar owner or worker cannot be commanded by the government to act as an unpaid, untrained, uninsured, unempowered, and unarmed Citizen Vigilante Enforcer of the law. This is one of the best kept secrets behind the success of smoking bans. Ban supporters universally portray the responsibilities of bar personnel as including actually "stopping" people from smoking – even if there's nothing in the written law itself demanding such action. If bar owners knew and understood this, it's likely that a good number of them would be willing to work with their patrons toward a more livable arrangement than a total ban.

Clearly the bar personnel have to obey the law, but they do not have to go beyond that obedience into volunteering for the role of acting as law enforcers. Some laws are clearly worded in such a way to indicate that owners and employees must make a "good faith effort" to see that the smoking ban is obeyed; however, while giving that message, those owners and employees can also indicate, clearly and honestly, that they personally disagree with the ban and state that if customers smoke they will not be thrown out onto the street, nor punched in their noses, nor have their drinks snatched out of their hands.

If a bar experiences difficulties communicating its policies and expectations clearly to its smoking and nonsmoking patrons, it could make use of a flyer that clearly outlines its policies in print. See Figure 15 for an example of such a flyer that I created for bars in Philadelphia to use.[429] The basic idea behind that flyer is to both provide evidence that the law is being properly observed by a bar's management and workers, and also to take the pressure off workers who might otherwise find themselves constantly caught up in verbal defenses of the bar's policies. If a smoker asks whether smoking is allowed, the bartender or waiter can clearly say it is not and then hand the smoker the flyer by way of explanation. If a nonsmoker complains that someone is smoking, the worker can sympathize while offering the nonsmoker the flyer so that they'll better understand the law and the bar's policy with regard to its enforcement.

Such an approach to the law is by no means flippant. Remember the stories of the actual enforcers, the people who are paid, trained, empowered and insured to enforce the law, and how they refused to carry out those duties without immediate and present armed police backup. There is absolutely no reason in the world for bar workers to be forced into such duties without such pay, training, empowerment, and backup.

While it's likely that using such a resistance technique would create a lot of screaming from antismoking activists, I believe, speaking as a non-lawyer, that resistance based on such an approach would stand up in most courts. While it received very little play in the media and was certainly not publicized to tavern owners in other areas, just two years after its 2002 ban was instituted, New York Supreme Court Justice Paul J. Baisley Jr. ruled that the ban merely required Suffolk bar owners to post "no smoking" signs and "admonish" smokers who light up anyway. He held that the ban did *not* require bars to enforce the law by ordering patrons to stub out their cigarettes or otherwise refuse to serve them, and that to read such a requirement into the law grafted onto it "an onerous, substantive enforcement requirement that the law itself does not impose."[430]

This approach was upheld in 2012 in Ohio. A unanimous Supreme Court ruling was interpreted by the media as upholding the ban, but its actual wording echoed Suffolk's 2004 victory.[431] If Zeno's Bar had used more care in actually obeying the law they might have won their case easily. It's important to note the Court's wording in the decision: "By requiring proprietors of public places and places of employment take reasonable steps to prevent smoking on their premises by posting 'no smoking' signs, removing ashtrays, and *requesting* patrons to stop smoking, the act is rationally related to its stated objective" (my emphasis). The judge did not say proprietors or workers had to put themselves at risk, without proper training or pay, to act as Citizen Vigilante Enforcers of the law. They simply had to request that patrons stop smoking, and could then evidently go on and serve them just as they normally would. They were not required to grab them and throw them out the front door, cease serving them, call the police, or shoot them if they continued to smoke.

Obviously, any bar hoping see its customers happily smoking and drinking again through using such a resistance technique should get proper legal advice first and make sure to follow it exactly, but overall the rulings in these important New York and Ohio cases, as well as hints from other, somewhat less publicized rulings in other states, indicates that this form of resistance could well be what's needed for bars losing the struggle to survive under smoking bans. What's needed however is for bar owners and workers to be made aware of this legal situation. Very few have such awareness, and even fewer have been willing to take the extra steps to act on it despite the economic pressures they're under. Unfortunately, most liquor license holders find themselves in the position of Ayn Rand's *Atlas Shrugged* characters: living in a world of so many laws and regulations that if they annoy the authorities by standing up for their rights regarding one law, those authorities will simply kill them with a different law.[432]

In addition to such a flyer for bars engaged in full resistance, those bars, as well as others that might be a bit more timid, should be encouraged to hang educational/exhortative flyers in places where they commonly post flyers that advertise coming events, drink specials, local bands, and such things. Examples of such flyers can be found at the sites of groups like FORCES.org, and SmokersClub.com (Several wonderfully creative efforts from Club member Bill Brown, based upon World War Two posters, are used as illustrations throughout this book and served as the inspiration for its cover image.) Additionally, some individual pages of The Lies Behind The Smoking Bans are designed specifically to be copied and used as educational flyers for posting or for distribution at events.

The Law Requires Us to Tell You No

SMOKING!

**If you violate this regulation we are required to ask you to stop
We are NOT required to remove ashtrays or stop serving smokers**

Customers should not get involved in disputes over smoking.
If you are disturbed by the sight of a smoker you should make the
staff aware of it. **Note:** Philadelphia Law specifically states:
*"Nothing in this (law) shall be construed to create any
private right of action for enforcement of its provisions..."*

So, Smokers, Nonsmokers, All Of Our Friends ... Enjoy Yourselves And Have Fun!!!

- Please note: The presence of fire-safety devices (ashtrays) on the bar does NOT imply that we allow smoking in here. The devices are given to those who have been asked to stop smoking so they may extinguish their smoking materials safely & without risk to our other customers.

Please also note the main section above regarding enforcement.

Contact Cantiloper@gmail.com to fight the ban!

Figure 15

Smoking Ban Resistance Flyer

The Lies in itself is an excellent tool that should be kept available in every bar fighting a ban before its start or as a way of encouraging its patrons to support efforts for a rollback of a ban in place. A few pages ago I spoke of the need to inspire smokers and their friends to get angry about what is being done to them and their lives. Creating that anger through exposure of the lies about the health and economic effects of bans is exactly what *The Lies* was designed to do. Reading a 500 page book in a dark and noisy bar might be a bit beyond most people, but *The Lies* was specifically designed with bars in mind: it's formed around one-page arguments and analyses, and it's formatted in print big and bold enough to read by candlelight. The booklet is free for downloading and printing from the Internet and I've always left an open invitation to help any group that would like to customize it for their own local battle.[433] The footnoted link to the website hosting the booklet can also be accessed a bit more simply through a "tinyurl" at http://tinyurl.com/SmokingBanLies with the final product being easily printed and bound for durable bar reading.

While *The Lies* is excellent as a static introduction to the justifications for fighting a ban, something a bit more dynamic is needed for a battle that extends over weeks, months, or years. Any local Free Choice group working with bar owners on a resistance project should consider putting out a regular newsletter updating all the bars in the relevant area, or at least the participating bars, with news of what is happening and occasional supplements of new information. Again, I offer a sample of such a newsletter that can be used as a template for local efforts at http://tinyurl.com/KillTheBan for convenience.[434]

A final word of advice for Free Choice advocates hoping to work with bar owners in fighting smoking bans: be ready and willing to deal with some disappointment and irritation while continuing your efforts. Bar owners tend to be fiercely independent and highly competitive. It sometimes seems nearly impossible to get them to work together* and it's often difficult to get any of them to do anything at all unless they can be assured that every other bar owner will be making the same

---

* Bear in mind phrases like "herding kittens" or "nailing Jell-O to a wall" and you'll be well-prepared.

effort. Warn them that efforts may be made to co-opt them at crucial points in the battle: promises that the ban won't be enforced against their establishments, offers of arrangements to delay ban implementation or enforcement for a certain number of months or years, bribes of free advertising or income from smoke-free special events to make up for ban losses, threats to make life more difficult for them as businesses if they don't bend over and cooperate. Any and all of those techniques have been used in the past when antismoking groups felt victory slipping from their grasp, and all of them have had at least some degree of success. Free Choice organizers should certainly aim for and make the best use they can of any aid they can get from bar owners, but it's a mistake to base a campaign too heavily upon dependence on such aid. I saw Audrey Silk learn that lesson in New York City, I learned it myself the hard way in Philadelphia, and I've seen it learned by others in various ban fights around the US. Whether the same situation exists for battles in other countries, I can't say, but I wouldn't bet on things being a lot different.

Parks & Beaches, Cars & Condos

Parks and beaches offer a strong opportunity for resistance because such a small portion of the general public has swallowed antismoking claims about outdoor wisps of smoke posing deadly threats. Cars, condominiums, and private apartments offer a similar opportunity, but have an additional factor going for us: in these areas people have finally begun to recognize the danger of antismoking encroachment upon private lives and rights. All five of these venues are strong from a Free Choice perspective because the lies supporting their ban pushes are so easily shown to be false just by using the fairly simple arguments and information already examined in preceding sections.

Unfortunately, victories in these areas aren't as easy as they should be. Why? Because all of these battlefronts offer ideal ground for the Antismokers to haul out their most powerful weapon: the children. Whether Photo-Shopping pictures of toddlers trapped in smoke-filled cars, painting word images of them gasping their last breaths as tendrils of evil neighbors' smoke sneak out of electric sockets, making up stories about smokers burning innocent children in playgrounds or tales of toddlers rushed to emergency rooms after touching toxic butts lurking in the sand of otherwise idyllic beaches, all of these scenarios just sit there begging for anti-smoking advocates to impale children upon their flagpoles and wave them frantically in the air.

The response to this *always* has to be twofold. First, you must be familiar enough with the technical side of the exposure arguments to present them confidently and convincingly: you need to expose the lies while also having whatever detailed backup references and figures are needed at your fingertips if you are challenged. And second, you must expose the disgraceful technique for what it is and cry shame upon those who abuse our love for our children just to further political goals. Those two responses used together are far more than twice as powerful as either of them separately.

Choose Your Weapons

As noted earlier, the broadcast media is generally not the smokers' friend in this battle. In broadcast interviews we will often find ourselves up against one of three distinct types of opponent:

(1) Ban advocates who are professionals: These can be either professionals within the antismoking movement itself, or professionals in business, the sciences, or education; and in general they tend to be experienced and relaxed when talking to the camera or talking to an audience. They know how to present themselves and their arguments in a powerful and confident way in such situations and, if they are antismoking professionals, they also know the details of their facts, figures, and arguments as well as they know their own names. After all, they're involved with the issue for forty hours or more per week and depend on it for their livelihoods.

(2) Victims or the bereaved: These advocates will believe that they or a loved one has suffered or died because of exposure to secondhand smoke, or that they suffer terrible and life-threatening physical reactions to even sub-microscopic exposures to smoke. They'll often be sincere and may have had their beliefs enforced both by their own doctors (who, of course, will often say almost anything to get people to give up smoking) and by

contact with whatever ban group is pushing a current drive. Because of their suffering they automatically win an edge with listeners or viewers that can be more powerful than any rational argument. Plus, they're almost immune to attack because any but the mildest response to them, no matter how nasty they might have been themselves, will turn an audience against you.

(3) Mothers: Dragging mothers into hearings or before the camera is just one step above dragging children in. The mothers drag the children in with their words, telling tales of choking children and butt-eating babies and addicted adolescents while pleading for whatever antismoking measure, ban, tax, or regulation is up on the docket. Again, just as with facing those who claim victimhood, it's hard to mount a solid response without seeming to be the villain in the eyes of the beholders.

If you have to face any of these opponents in front of the media, it's important to remember to stick with the basics of your arguments, particularly when dealing with the paid professionals, and to avoid seeming to be on the attack if you're facing a victim or a mother. It's a difficult path to walk for Free Choice advocates, particularly since so many of us will have entered the fight only weeks or months earlier and have been engaging in it only after having finished whatever full time jobs we may ordinarily be engaged in. In contrast to the professionals we can sometimes appear amateurish, thereby adding fuel to the fire if the opposition tries to pass us off as simply being cranks not worth serious consideration.

When the style of argument needed to overcome long-held beliefs is thrown into the mix, video appearances become even more questionable as a weapon since the arguments opposing bans and taxes tend to rely far more on detailed analysis and information than the slick and memorably crafted sound bites of their advocates. If they say, "The Surgeon General has said XXX," it might take three or four minutes of explaining to show why the claim is false. But meanwhile, if we say, "So-and-so has said ZZZ," all the ban advocates have to do is say, "Oh, that's an old tobacco industry claim," and 99% of the effectiveness of our statement is wiped out... despite the earlier discussed weakness of such an argument or despite the fact that we can later show that the claim was completely untrue or simply irrelevant to the accuracy of the statement in question.

As a choice of battlefield, the Internet offers a much better playing field than television for a debate in which facts should matter more than style and sound bites. Again as noted earlier, a lot of what we may say in opposing the status quo may now seem counterintuitive to our listeners, an audience who has had contrary opinions and ideas drummed into their minds innumerable times and through many different avenues over the course of decades. Presenting the contrary argument in a medium where those who are skeptical can easily hit a pause button and quickly open a web page to check on a fact works strongly in our favor – and against ban advocates who regularly stretch the truth far beyond its limits.

How long this window will remain open to us is questionable. As noted in the concluding section of *Launched In The Trenches*, anti-smoking advocates may very soon seek to limit our voices by use of censorship justified by the concept that our arguments are ones that should only be accessible to adults. While it's true that it's certainly better not to bombard children with pro-smoking messages, the likely-proposed "fix" for that would be to brand all Free Choice web content with the same domain tags as explicit pornography, tagging them as .xxx domains or some such. This would have the result of hiding our message from the great majority of ordinary adults who might prefer to see it but who earlier excluded (or will have had automatically excluded as an inbuilt preference of their search engines) pornographic material from their general web browsing and searching. The designation of Free Choice sites as deserving of .xxx domain names would effectively render them and their messages invisible, not just to children, but to the great majority of the Internet-browsing public.

Choose Your Targets

In this conflict, it's easy to feel like a general facing a vast invasion force along hundreds of miles of battlefront with only a few battalions of troops. Do you disperse your forces and try to defend everywhere at once? Or do you concentrate the power you have at the enemy's weakest points, make an effort to break through those points, and then force a retreat on other beachheads as your opponent is thrown back on the defensive?

In military terms the second is usually seen as the wiser strategy, and in political terms, facing the enormous resources of the anti-smoking movement, Free Choice advocates must make the same type of difficult choice. Yes, our hearts and our feelings may lie with the businesses in our own town or the people being trampled in our own condo complex, or even our own families, but if it's a lost battle at the moment, if the hopes of a local win are just too remote... it's time to move on and attack at points where the enemy's armor isn't as imposing. This consideration also affects our choice of battle – fighting against bans in casinos and bars is likely to be a much easier battle than fighting against bans in family restaurants, or even in outdoor playgrounds.

The choice goes back again to our earlier discussion: the greatest weakness of antismoking advocacy is its lies. Some of those lies are easier to expose than others. Those are the ones to concentrate on: their lies about the lack of economic effects on the bar and casino industries, their lies about the deadliness of brief encounters with diluted wisps of smoke outdoors or in well-ventilated indoor settings, their lies about such things as thirdhand smoke or the worldwide environmental impact of smoking; all of those lies are relatively easy to expose with simple arguments and using fairly simple numbers. Once people have accepted the idea that they've been lied to about these sorts of easy-to-see things, they're much more likely to be open to the idea that perhaps, just perhaps, they've been lied to about the more arcane research that is falsely claimed to show a real and important threat to personal health from the levels of smoke usually found in most workplace, social, and domestic situations.

Meanwhile, if antismoking forces are compelled to backtrack to fighting for poorly defensible positions they've previously committed to, it will take their energy away from pushing for further territory in even crazier areas. It may even cost them some previously quite "secure" positions such as bans in restaurants or bar/restaurants, forcing them to return to battles they thought they'd won forever. Finally, anything that pulls some of their money and effort from their insatiable pushes for higher taxes will end up weakening them in the long run: their machine runs on money, their lifeblood is money. Again, pull the money plug and the professional Antismokers will disappear back into the swamp from whence they slithered out... just to put it poetically.*

If you're living in a town where bans have been solidly in place for years and where most vulnerable businesses have either closed or learned to adjust, you're unlikely to find a lot of support for changing the status quo. As any successful business owner will tell you, it's usually better to face the devil you know than the devil you don't. A restaurant or bar surviving with at least small margin of profit in a nonsmoking town is going to be slow to want to roll the dice on increasing that margin by changing the law back to the way it was before; and fighting a local bar or restaurant ban without the businesses behind you will be an uphill battle no matter how sound your arguments might be.

So don't waste your energy attacking a target where you can't win. If you want that local ban reversed and things look hopeless, join the battle to fight a ban, regulation, or tax increase someplace else where there's actually a reasonable chance of victory. Every time you help roll back a ban elsewhere or defeat a tobacco tax that would fund the larger antismoking movement, you've weakened the power the Antismokers have in your own area, and you're a step closer to meeting the enemy on a truly level playing field.

Fight where it will count, using whatever skills or resources you can best offer. Don't waste your time, energy, talents, or money fighting losing battles. Point your assets to where they'll make a difference and eventually it will pay off in terms of your own battle and interests.

---

* In terms of target choice, Free Choice activists should *always* be keeping an eye out for the money-grubbers. Tax proposals on tobacco products abound, with the sugar-coating of funding various programs for poor, sick, undereducated children... but often with a tag thrown in to make sure some of the money gets funneled back to the Antismokers pushing them. I'm being repetitious, but it's for a good reason. Money is *absolutely* vital to the Antismoking movement. Remember the threat from the New Jersey group Breathes to stop all its antismoking activities if its budget was reduced to "just" $14 million!

Learn From The Past

Unfortunately, the student champions of individual liberties and rights who fought so hard in the 1960s and 1970s have largely been co-opted by the antismoking movement. The roots of this are fourfold, one of them being a simple quirk of politics and three of them just being unfortunate accidents of timing:

(1) The simple quirk comes from the fact that the hippies of old were rightfully distrustful of corporate shenanigans. Big Tobacco was an easier target than Big Auto, Big Oil, Big Chemical, or Big Food, because it was such a discrete entity unto itself. Its products were used by a disfavored minority, and the revelations from its secret documents cracked open its public façade and showed the corruption beneath that remains hidden in most industries.

(2) The modern movement against smoking coincided with the modern concern about environmental pollution in general. The jump from industrial smokestacks to the little bits of smoke given off by smokers may have been a big one, but the EPA Report of 1992 made that jump possible, even if it was belatedly discredited by the 1998 Osteen court decision. Despite the fact that their "pollution contribution" on a worldwide level is probably on the scale of a single tenth of a single percent of the total, smokers were easily accessible, in-your-face targets for activists concerned about such things as clean air and toxic waste.

(3) As the Vietnam War ended, Americans suffered a guilt complex over how they had treated their soldiers. That guilt complex combined with the fact that quick battlefield rescues combined with modern medicine to save and bring home large numbers of seriously disabled veterans. Most noticeable among them were those who had lost legs or mobility in battle and were finding it difficult for to fit back into life while sitting in a wheelchair. The Americans with Disabilities Act (ADA) of 1990 was passed largely on the

basis of concern about such veterans.[435] Its initial and most visible impact was seen in making facilities wheelchair-accessible, but it has been used by antismoking advocates in more recent years to declare that, since some people may find that high concentrations of smoke can aggravate conditions such as asthma, smoking therefore needs to be banned in all indoor environments where someone with asthma might occasionally want to go. It can be argued that the ADA should not apply to extreme or unusual sensitivities – for example, that outdoor dining should not be prohibited because a few people suffer extreme sensitivity to sunlight and such people would thus be denied jobs waiting tables on a patio rather than safely indoors – but overall, the ADA is simply being used as a weapon by pro-ban groups. Whether such abuse will weaken the ADA's power to help those with real disabilities in the future remains to be seen.

(4) Vietnam-era activists began settling down as parents and voters in the 1980s and 1990s. Most parents of young children tend to forget how wild their own early years were and react in horror to the idea that their own little angels might one day smoke, drink, do drugs, or, heaven forbid, have sex. (Well, ok, most of them will grudgingly admit that the last one will eventually happen… but hopefully not for years and years and years.) So those new parents were prime targets for the Antismokers' favorite propaganda trick, the old "Save The Children" plea. The lifestyle radicals morphed into lifestyle conservatives as they sought restrictions to ensure the perfect safety of their perfect children – while forgetting how much they themselves had chafed against such restrictions twenty or thirty years earlier.

All of this has resulted in the young body-politic, those most often classifying themselves as "Liberals" with a strong distrust in government and corporate politics, moving almost solidly, as a bloc, into the anti-smoking camp. Conversely, the "Conservatives," people who were traditionally in favor of government restrictions on such things as drug use, have found themselves in the position of having to defend alcohol and tobacco use from the encroachments of the Big Government that they had defended so strongly during the Vietnam years.

The reason this concern becomes relevant to us in the Free Choice movement today is that we are in a position quite similar to those in the early US civil rights and peace movements of the 1950s and 1960s. The movers and shakers and thinkers of that era specialized in studying how relatively disempowered and unorganized "radicals" could have significant impacts on larger and more accepted societal positions while maintaining a commitment to nonviolence and a connection to the mainstream. Conservatives today are often horrified at the thought that radical activists such as Saul Alinsky might have left a legacy of valid lessons to be learned, but they shouldn't be. Books such as *Rules For Radicals* should be required reading for anyone seeking social change from a status quo.[436] One of Alinsky's major themes revolved around how groups of people with little organization and no money can fight for their rights against far more powerful and established opponents. The Antismokers have used techniques based on Alinsky's foundations and style quite successfully for years, and to ignore the reality of that success is political suicide.

I recently read David Duke's *My Awakening*.[437] David Duke is generally seen as one of the most highly visible racists and anti-Semites in the United States. Did I like reading his book? No. Do I agree with his views? No. But, just as the quote from Vegetius at the start of this section notes, "If you want peace, prepare for war,"[438] or, as it is sometimes put in slightly different terms, and perhaps more accurately for our own situation, Sun Tzu's military advice: "Know thy enemy."[439] The antismoking movement has a strong left-wing, liberal base as its main political and activist support. Using that base they have been successful in achieving some very far-reaching goals. If you ignore that success you'll never be able to duplicate or defeat it. The co-option of the free-speech-dedicated liberal radicals of the 1960s has been one of the most fundamental successes of today's antismoking movement. The people who fought hard and successfully against a far-better financed and established majority for what they believed was right have experiences and lessons to offer that the Free Choice movement of today needs to study and learn. That's not to say that we should employ all the same methods, since some of those methods involve the very weaknesses we will use to defeat the antismoking industry, but we need to proceed from a base of knowledge, not ignorance.

Conclusion: The True Endgame

In the larger scheme of things, the fight against the Antismokers is a very small part of a much greater war.

We all have beliefs, and some of those beliefs can be very strong... so strong that we're willing to sacrifice our lives for them. That willingness to sacrifice can take the form of how we spend our time and energy throughout our lives, or in actually being willing to risk death in furtherance and support of those beliefs. No one has the right, however, to force their personal beliefs upon others, to make choices about how other people will live their lives or sacrifice them.

This is not to say that we do not have the right to prevent people from harming other people in ways that virtually everyone would agree were clearly harmful, either physically or psychologically. I generally have no right to rob another person of their possessions, to significantly hurt them in their bodies, to make them suffer in their minds, to control their activities, enslave them, or kill them. Some cultures have deviated from those beliefs over time, favoring the concept of "might makes right" and condoning such things as slavery, the infliction of pain for twisted pleasure, or the taking of someone else's work and strife for one's own benefit. But no major belief system in today's world professes that such things are right.*

The members of the antismoking movement seek something that most of them sincerely think is good: the reduction of a practice that they believe results in pain, disease, and early death for many of those

---

* The one major exception to this general rule is the almost universal acceptance of the right of governments to take the fruit of their citizens' labor through taxation. While people may argue about what levels of taxations are fair, or how taxes should be apportioned in their taking or use, few would argue that all taxation should simply be eliminated.

who engage in it, particularly if those who engage in it do so heavily and over most of a lifetime. However in seeking that goal, they've done something evil. They've ignored the fact that many smokers feel some positive benefit from their smoking. If giving their truly honest opinion without the feeling pressured to come out with the "proper" answer, many smokers might state that they prefer to continue smoking and take their chances later in life. Ignoring that fact, deciding that one has the right to make choices for other people as to how those people should live their own lives, is an evil. And the evil is compounded when such people achieve the power to enforce that choice – whether it be overtly by laws and force of arms or covertly through psychological and societal conditioning mechanisms.

As an adult and rational human being, I should have the right to ski, do drugs, get a suntan, live in a cave, sing bawdy songs, jump from airplanes with a parachute or from buildings without a parachute (as long as I've arranged for the cleanup and make sure not to hit anyone), dance to wild music, copulate with another willing adult of either sex, run around nude or wearing a burkha, sit in a lotus position and consume nothing but rainwater for forty days and forty nights, or smoke anything under the sun that burns while drinking anything under the sun that's liquid. Such decisions should clearly be a fundamental part of human rights.

Someone else might think that what I'm doing is wrong. They may not want me to do it because they disapprove of it morally, or because they believe it will hurt me and they want to "save" me from that hurt (with the justification that I have proven myself irrational if I dare to disagree with their analysis); but they have no inherent right to stop me—whether the attempt to stop me is through physical, legal, or psychological force. Once someone has reached whatever a culture decides is the true age of reason,* that person has the right to make his or her own life choices without undue pressures or consequences from others.

The antismoking movement is wrong because – no matter how lofty (or low) its goal might be, no matter how right (or wrong) its

---

* There will always be some disagreement about just what that age is, but there's near universal consensus that it's somewhere between the ages of 12 and 18.

judgment of harm is, no matter how large a majority or tiny a minority it might be – it has no right to push its belief about the balance between the enjoyments and risks of smoking onto other people who believe differently and have made a decision that might place a greater weight upon present enjoyment than future longevity. Is it better to live ten happy years or a hundred miserable years? I don't know the answer to that question, but I do know that I don't have the right to force whatever answer I believe is true onto someone else.

Free Choice in one's way of life is not a relative value. It is an absolute good, and as such, it should never be tampered with unless one's own Free Choice clearly and significantly causes real pain or suffering to others. Those two conditions, the clarity of causality and the degree of significance of a choice's impact on others, will sometimes form a gray area where honest people can debate exactly where lines need to be drawn. But in any such debate, particularly when the force of government and the "men with the guns" are ultimately involved, the benefit of the doubt should always be tilted strongly toward the side of freedom, because freedom to make our own choices is really what life is all about.

That is the only satisfactory finish to the endgame: an ultimate respect for the freedom of the individual to make their own choices in life according to their own beliefs unless those choices clearly and strongly result in significant and inarguable harm to others. The alternative to winning this fight – leaving our children with the legacy of a world where governments condition their populations to unquestioningly do and accept virtually anything a tyranny desires with the justification that individual freedoms need to be limited for vaguely defined benefits to a common good – that alternative is just too frightening. Giving up is simply not an option.

Breathers was written about the same time as *TobakkoNacht!*, but it was only a few years ago that its applicability to the smoking ban battles became overt. As noted earlier in a *Slings And Arrows* entry, Congressman Waxman of California was reported in 2006 as saying about smokers, "They feel they have the right over everyone else to use up the air."[440]

Breathers is dedicated to all the air-users out there, smoking or not. Breathe long and prosper!

Breathers

You're sitting alone, blissfully reading a booklet in a self-serve coffee shop, when suddenly you hear a tinkle. You look up, horrified, as I walk in the door, look in your direction, smile, and exhale a huge cloud of chemical-filled toxic waste.

The cloud expands almost instantly to every corner of the room and you can feel it racing up your nose into your sinuses and down your throat to burrow deep into your lungs as you are forced to breathe, simply in order to stay alive.

You gawk at me, aghast, as I saunter casually over to a table near yours and once again blow out a putrid cloud for you to share. There's a powerful exhaust fan right behind me, but it would need tornadic winds to keep you safe from the deadly mist that pours from my mouth and nose.

I stare at you, puzzled, as I see your eyes widen, hear your dry little choking coughs, and see one hand clutching your throat while the other flutters frantically in the air, as though trying to fan my breath back in my direction.

As you fly out through the door and I see the pamphlet you were perusing at your table, comprehension dawns within me. You are a member of **ASBESTOS: American Solo Breathers Endeared Simply To Ones' Selves**, founded in the second decade of the new millennium. The constant barrage of HoloVision commercials describing the terrifyingly toxic composition of normal human exhalations have filled your mind:

Acetone (used in nail polish remover), toluene (a highly toxic Central Nervous System Depressant), formaldehyde (used to preserve corpses), benzene (a potent carcinogen), and even DiHydrogen Monoxide (DHMO)... a compound so deadly that it strikes down thousands of innocent men, women, and children around the world every year when they accidentally inhale small amounts of this colorless, odorless, but undeniably lethal substance.

As I go to the window and watch you staggering in terror down the street, I see it faintly fog with the DHMO from my breath and I can almost sympathize. Certainly I cannot argue with the reality of the chemicals I forced upon you, nor with the possibility that I may have given you some deadly airborne infection which will end your life in a few short pain- and pus-filled days. Nor can I argue with the fact that before I assaulted your space and filled it with my virulent germs, metabolic toxins, and carcinogens, the air was clearly far cleaner, safer and freer of contagions and filth.

I grimace as you turn the corner and I remember the popular slogan, sung to a cute Pink Floydish jingletune by the CCC (Chorus of Crippled Children) in successive stanzas during the HoloShow public service breaks every morning, afternoon, and evening:

Do not force your Breath on ME, Father!
Do not force your Waste down my THROAT!
Do not force your Breath on ME, Mother!
Your Filthy Breath Will Be STOPPED When WE Get The VOTE!

While the jingle is cute, almost lovable, with its innocent soprano voices and mis-metered structure, I shudder as the last words ring in the ears of my memory. The children's virginal pink lungs will soon outnumber those of us over-50 Breathers, and they will casually and happily vote to slide us under the ocean – where we'll pollute their air no more.

FLASH: Six o'clock HoloNews update: Startling new research indicates that the ASBESTOS political party may be in trouble. It had seemed certain they would soon wield a majority vote in Congress with the reduction of the voting age to twelve, but a new study has shown a surprising lack of sexual reproduction among their membership. ASBESTOS founder, H. Klintoon, could not be reached for comment as she has come to believe that germs can be transmitted through telephones.

Full Holo at Eleven!! Stay tuned!

Recommended Bibliography

Books:

Alinsky, Saul. *Rules For Radicals.* Vintage, 224 pgs. 1989.

      Alinsky provides the guidelines for the powerless to stand up to and defeat the powerful. Alinsky's politics tend toward the extreme liberal, but the tools he provides for the defense of freedom by those without money or power are invaluable for all.

Anonymous. *The Plain Truth About Tobacco.* 150 pgs. 2012.

      This pdf – freely available at http://olivernorvell.com – provides a refreshingly cogent assessment of the real nature of smoking's risks as contrasted with the patently fallacious hype of Tobacco Control's dogma. The myth of the "secondhand smoke peril" is laid bare. *The Plain Truth* incorporates a listing of the findings of secondhand smoke studies relating to lung cancer that, as of June 2013, is the most comprehensive listing of its kind, including qualified studies long overlooked by previous researchers.

Best, Joel. *Damned Lies and Statistics.* U of CA Press, 224 pgs. 2001/2012.

      While this book says almost nothing about smoking, it provides a readable and fascinating introduction to the ways statistics can be misused by those pushing an agenda upon the public. Anyone intimidated by the idea of criticizing "statistical truths" pushed by the media should read this book.

Colby, Lauren A. *In Defense of Smokers.* 1995/2006.

      Larry Colby recently passed away at the age of 81 and was a longtime smokers' rights advocate. His *Defense* tackles not just claims about secondary smoke, but also those connecting smoking itself to cancer and other diseases. Mr. Colby strongly attacks the connection as largely arising from a few early and poorly designed studies and raises some excellent questions. Unfortunately it is no longer available in hardcopy, but for the past several years Mr. Colby has made it freely available on the internet for reading, downloading, and printing at lcolby.com/colbyl.htm.

Di Pierri, Vincent-Riccardo. *Rampant Antismoking Mentality Signifies Grave Danger: Materialism Out of Control*. 578 pgs. 2003.

 Rampant is vast and scholarly. It covers: "the biological (epidemiologic), psychological, social/relational, moral, legal, and metaphysical considerations in indicating that rampant antismoking is not coincidental but symptomatic of dangerous, fully-fledged materialism (rule by superficiality)." Di Pierri's unique Australian perspective on the battle between Free Choicers and Antismokers shines through its pages. Currently out of print, but downloadable at rampant-antismoking.com and prepare for some real reading.

Douglas, William. *The Health Benefits of Tobacco*. Rhino Publishing S.A. 409 pgs. 2004.

 Dr. Douglas, MD, takes a strongly contrarian look at the health effects of tobacco, arguing that smoking is innocent of many charges against it and may even be beneficial overall for those who enjoy it. Taking a holistic and natural approach, Douglas claims health is better measured by the quality of life than simply by a doctor's instruments. Arguing against the negative health effects of primary smoking itself is difficult, but like Rich White, he gives it his best and does it well.

FORCES International. *The ABC's of ETS*. 125 pgs. 2008.

 Freely downloadable analysis of arguments against the shoddy ETS science behind smoking bans. A wonderfully color-coded table shows how few of the studies done on ETS and lung cancer actually pass muster even at the most basic level of simple statistical significance – to say nothing of actual causality. Available online at http://www.data-yard.net/science/abc_ets/abc_ets_2008.pdf.

King, Theodore. *The War On Smokers And The Rise Of The Nanny State*. 206 pgs. 2009.

 King's *War* is the best introduction there is for someone previously unfamiliar with our arguments but who needs something shorter than 300-600 pages. For established Free Choicers, King also offers a fresh perspective as well, as he came to his writing independent of any of the previously established activist groups and developed most of his research and thinking from traditional mainstream news and science sources. An easy, entertaining, and informative read.

Kluger, Richard. *Ashes to Ashes: America's Hundred-Year Cigarette War, the Public Health, and the Unabashed Triumph of Philip Morris.* Vintage. 832 pgs. 1997.

I'd recommend reading Sullum's *For Your Own Good* first, but if you want a similar depth of the legal/commercial history of the war against Big Tobacco but from a somewhat different viewpoint, Kluger will offer it. As with Sullum, it is meticulously researched and presented, but Kluger's sympathies and presentation lean toward the Antismokers – in much the same way that it could be argued that Sullum's lean toward the smokers. Both together are nice… but it's a heckuva lot of reading.

Lair, Felt. *Murdering America.* Publish America. 215 pgs. 2003.

Felt Lair offers an insightful and forward-looking analysis of the attack on American beliefs and principles that lies at the core of antismoking philosophy. He sticks with the essentials and offers a truly unique and valuable analysis of the upcoming battle. Lair also offers a fair part of *Murdering* as a free download, *The Tobacco Dance,* at http://minnesota.smokersclub.com/dance1.htm.

McFadden, Michael J. *Dissecting Antismokers' Brains.* AEthna Press. 384 pgs. 2004.

Brains has three distinct parts: (1) a fifty-page examination of Antismoking "types," describing the current movement as more of a "perfect storm" of differently motivated individuals than an organized conspiracy; (2) an analysis of psychological/linguistic/scientific tricks and misrepresentations used to promote an antismoking world; and (3) ten detailed and varied appendices covering everything from analysis of exposure to the chemicals found in secondhand smoke to the coming health crusade against alcohol. *Brains* pairs well with Snowdon's *Velvet Glove*; a psychological analysis complementing a historical one, with both blasting antismoking pseudoscience along the way.

Nickles, Sara. *Drinking, Smoking, and Screwing.* Chronicle Books. 224 pgs. 1994.

Great set of articles and essays about America's long love of "naughty" pleasures. Not actively political, but wonderfully subversive. More fun than a barrel of Antis!

Oakley, Don. *Slow Burn: The Great American Antismoking Scam (And Why It Will Fail!)* Eyrie Press. 600 pgs. 1999.

This wide-ranging overview freed me up to concentrate on the psychological and scientific aspects of the war in *Brains*. For that I'll be forever grateful. In a very readable style stretching over 600 pages, Oakley outlines the socio-political history of the modern Antismoking Crusade and fires scathing broadsides at those who would attack our freedoms. With over 1,000 carefully footnoted references, it's the best combination of relaxed readability and extensive background information I've seen in our area and can be enjoyed even by those without a specific interest in the subject. It is difficult to find in hard copy, but like some of the others mentioned here, it has been made available for free download by the author. Get it at the FORCES site: http://forces.org/writers/bookcase/oakley/slowburnp.htm and enjoy!*

Snowdon, Christopher. *Velvet Glove, Iron Fist: A History of Antismoking.* 415 pgs. Little Dice. 2009.

Christopher Snowdon treads a path similar to that of *Dissecting Antismokers' Brains*, but he does it from a historical perspective as opposed to a psychological one. Well written, well referenced, and very comprehensive, Snowdon's book covers the history of the antismoking movements in the UK, the US, and elsewhere in depth while also examining the scientific weaknesses at its base.

Sullum, Jacob. *For Your Own Good. The Anti-Smoking Crusade and the Tyranny of Public Health.* Touchstone. 352 pgs. 1999.

Mr. Sullum has been strongly attacked ever since he wrote a critical editorial that was later purchased and reprinted by a tobacco company. That simple purchase opened the door for Antismokers to attack him forever since as being "in the pocket of the tobacco industry." Sullum's book strikes back at them powerfully, honestly

---

* Don't be misled by the free download status of some of these books. Their authors generally wrote them more out of a love of freedom or a respect for science than out of a desire for profit. The longer ones, those such as Oakley's and DiPierri's in particular, will be much more comfortably read if you can find print copies on Amazon or Ebay.

addressing that charge in his introduction and then exposing in great detail their manipulative efforts to control the rest of us "for our own good." His research and presentation is most impressive, and anyone reading this book and Snowdon's *Velvet Glove* will have a solid understanding of the rise of the worldwide antismoking movement.

White, Rich. *Smoke Screens: The Truth About Tobacco.* Lulu.com. 310 pgs. 2009.

Rich White's *Smoke Screens* takes a UK perspective in looking at the antismoking movement and, while still focusing on issues of smoking bans, sloppy science, and unjust taxation, also hits solidly at the claims of dire health effects from primary smoking have been greatly exaggerated. Rich's research nicely complements Douglas's *Health Benefits* and adds a welcome Free Choice activist element.

Free-Choice Web Resources

Most Free Choice groups use the Internet as their main communication tool. Web pages, email lists, and message boards have served to tie together a community that lacks the financial resources for such luxuries as international conferences at fancy resorts. There are dozens, perhaps hundreds, of groups and web pages, far too many to list in any detail here, so I will simply highlight a few of the more comprehensive ones whose links will, in turn, open the door to many more.

Sites And Groups

LesDissidentsDeGeneve.ch

Switzerland's DDG is written largely in French, but is friendly to Google Translate. They fight for citizens' rights to freedom from undue State intrusions such as electronic eavesdropping and, importantly, public smoking bans, and work to educate the Swiss that there is no real harm from "lethal secondhand smoke." Their site is run by a 72-year-old retired multimedia Swiss journalist, Jacques A. Widmer, whose motto is "The Pen is mightier than the Sword." DDG articles also appear on the TICAP site. Jacque's commitment, enthusiasm, and knowledge promise a world of trouble for Swiss Antismokers in the years ahead!

FORCES.org

FORCES provides an incredible resource through its web pages. Years of news, editorials, and archives stretching into the thousands of pages, as well as links to people who believe passionately in what they are fighting for... all that and more can be found here. Forces was in the forefront of noting the importance of Big Pharma's antismoking lobbying power in the 1990s, was known widely for its early and hard-hitting approach to dealing with attacks from Antismokers, and has supported many grassroots Free Choice efforts over the years.

Forestonline.org and Foresteireann.org

The FOREST sites and groups are the only sites listed here that actually have Big Tobacco funding. Since ASH UK has substantial government funding and enjoys almost Ministerial status, no one could ever really point the finger at FOREST as being the bully in the fight there. Simon Clark and John Mallon masterfully fight some of the most vicious antismoking forces east of the Atlantic, and both sites run active blogs. Centuries of British and Irish pub tradition have been laid to waste by the Antismokers, but the FORESTs are determined to reverse that destruction.

Antiprohibition.org

TICAP, The International Coalition Against Prohibition, is an umbrella organization bringing various local and national Free Choice groups into better communication and coordination with each other. They have sponsored three major conferences (including their founding conference) and have regular SKYPE-based meetings for their directors and membership. TICAP offers one of the best multilingual international Free Choice resources available.

TCTactics.org

TCTactics, with TC standing for Tobacco Control, highlights and explains various tactics, dirty tricks, and subterfuges used by anti-smoking organizations and advocates. It was inspired by the creation of a "Tobacco Tactics" website by an antismoking group at the University of Bath, but, despite being unfunded, quickly developed into a far better wiki-style resource than the well-funded professional site it mirrored. Well worth spending some time examining!

NYCClash.com

CLASH (Citizens Lobbying Against Smokers' Harassment) was started by a now-retired Brooklyn police officer named Audrey Silk. It ranks as one of the longest-lasting and most successful local groups in terms of building a membership, engaging in active campaigns, maintaining an active website, and presenting a face and voice for Free Choice to the media. There is almost never a major NY area news story about significant smoking ban or tax issues where a quote from Audrey isn't sought after, and her efforts often extend far beyond New York.

SmokersClub.com

Samantha Phillipe offers the Methuselah of Free Choice Internet resources, with a veritable encyclopedia of information carefully built up over the past twenty years. It also has an archive of over 700 issues of its free weekly newsletter – *very* well worth signing up for! Samantha, as is true of so many smokers' rights activists, puts an enormous amount of energy into her work for freedom, and her newsletter provides an essential weekly summary of Free-Choice news, articles, and commentary. Http://smokersclub.com/mbrlks.htm provides an extensive links page to other groups and sites as well.

Blogs

The Free Choice Blogosphere is ruled by the Brits. Yes, there are good blogs elsewhere, but the UK is clearly the front-runner in having a concentration of long-lasting, dedicated, and talented bloggers who concentrate almost exclusively on articles dealing with the Free Choice battle.

As I was wrestling with succinctly describing the different Free Choice blogs I realized just how difficult it was to characterize them. There's something about blog structures that make them more amorphous than static websites. Yes, I have an inner feel for what I'm likely to be reading if I pop over to Frank's "Banging On About The Smoking Ban" or Phil's "Gasdoc" or Siegel's "Tobacco Analysis," but a good bit of the flavor of each of those blogs is determined by the active audience of posters they attract… and sometimes simply by the state of inebriation or frustration of a blogmeister returning from an antismoking pub or a free-wheeling "smoky-drinky." And a number of those blogs have such a wealth of interesting, well-written, and informative material that they could, individually, provide books purely from their blogmeisters' postings that would rival any of the other Free Choice books on the market.

So what I have decided to do is simply note two sites and one newsletter that offer frequent updates on current blog topics, and then offer links to three blogs with extensive Free Choice relevant "blogrolls" listed and linked in a sidebar. Readers can visit the URLs themselves and see which flavors please or offend their palates.

The Two Sites:
http://encyclopedia.smokersclub.com/38.html (Smokers Club)
http://antiprohibition.org/ (TICAP)

The Newsletter:
http://paper.li/themorrigan1972/1308775990 (Daily Nicotine)

The Blogs:
http://dickpuddlecote.blogspot.com (Dick Puddlecote)
http://velvetgloveironfist.blogspot.com (Velvet Glove, Iron Fist)
http://underdogsbiteupwards.wordpress.com (Leg Iron)

I will make two exceptions to my rule here and point to two more blogs that have some unique material that is particularly well-worth checking and quite valuable for Free Choice advocates. The first is Dr. Michael Siegel's tobaccoanalysis.blogspot.com, a site that has been referenced many times throughout this book. Dr. Siegel remains frustratingly in full support of workplace smoking bans, but I often feel that the only reason he has stuck to that "line in the sand" is because it gives him an aura of "neutrality" that is quite rare in this area of contention and access to a public microphone that would otherwise be denied him. As you follow his blog you will find him consistently ripping apart virtually every antismoking study that comes out... though often with an endnote along the lines of "I can't believe these researchers did this stuff! I've never seen its like before in my life! It's uniquely bad science that brings discredit upon otherwise good and worthy Tobacco Control efforts!" After the 75th repetition or so, that sentiment begins to get a little thin. Nonetheless, he's to be greatly admired for having stood up against bad science at great personal risk, and he's used his exceptionally keen scientific insights to great effect in attacking the bad science that is served up on an almost daily basis by antismoking lobbyists.*

The second blog that stands out is the "Bolton Smokers Club," chaired by blogmeister Junican, (Manchester's James Watson). If you are considering the possibility of growing your own tobacco, his site and archives are well worth checking out. Junican has also performed a worthy service in offering some extraordinarily careful and clear analyses of such things as the McTear lung cancer case in Scotland and in reading, explaining, and evaluating the smoking and lung cancer research of Drs. Doll and Hill, long-cited as being the basic foundation upon which the entire cancer-condemnation of smoking was originally based. His blog address is boltonsmokersclub.com and you will find clear directions there for locating the aforementioned blog items!

THR Sites

While considering the above choice of links, I realized there is still a divide between those who come to the Free Choice movement primarily because they are smokers or friends of smokers, and those who come to it as vapers or as THR (Tobacco Harm Reduction) advocates.† While Jan Johnson's Daily Nicotine Newsletter noted above bridges this gap to some extent, I decided it would be good to add three sites that primarily focus on the e-cig/THR area (they get three because there was really only one section in this book devoted to their

---

* As you saw in *Studies On The Slab* I have a lot of respect for Dr. (Medical Doctor) Siegel's scientific analyses, despite the fact that he has been scathingly attacked by Dr. (Ph.D. in mechanical engineering) Glantz – a former mentor and colleague who has now disowned his unruly intellectual progeny.

† I should not totally ignore a third category of supporters out there, those who come to the fight primarily driven by a respect for the importance of our freedoms and a concern about government encroachments upon those freedoms. In recent years we've begun to see more recognition of the reality of the "slippery slope" that exists for everyone, smokers and nonsmokers alike, once Big Government begins overstepping its bounds.

newly emerging but quite important issue – and I'm feeling guilty about it).

Casaa.org (Consumer Advocates for Smoke-free Alternatives Assn.)
Vapersplace.com (VP Live Network – Check their live podcasts!)
Antithrlies.com (Anti-THR Lies – hosted by Carl Phillips)

The vaping community was originally perceived as being somewhat antismoking since it was largely made up of ex-smokers who were quite eager to disparage and distance themselves from their former habit. However, when Antismokers began heavily attacking vapers purely on the bases that, (1) what they did looked like smoking, and (2) they had the additional chutzpah to enjoy their new practice, vapers realized the true nature of what they were up against and who their natural allies really were.

Some e-cig dealers still blow the horn about giving up the "nasty, dirty habit" while urging smokers to switch to their products, but on the whole the e-cig community is allied with the much larger Free Choice effort and the two groups are working together against a common, intolerant, enemy.

Conclusion

If you familiarize yourself well with the materials contained in even half of these Recommended Bibliography sources, you'll be well-prepared to face down any Antismoker or group of Antismokers you are likely to run into. Very few of them are able to progress beyond the poorly supported sound bites that are simply meant to be repeated and never questioned, and their response to clear challenges usually reveals what they're really made of.

Spoiled Goat Cheese.

Index

Endnotes

(All URLs verified as of May, 2013.)

[1] Wikipedia. "Godwin's law," *Wikipedia.org*. http://en.wikipedia.org/wiki/Godwin's_law.

[2] Editorial. "Imagining things otherwise: new endgame ideas for tobacco control," *Tobacco Control*, October 2010, Volume 19, Issue 5, pp. 349-350. dx.doi.org/10.1136/tc.2010.039727.

[3] Vieira M. "Interview With President Barack Obama," *Today Show*, July 21, 2009.

[4] McFadden MJ. "Obama In Bare-Faced Lie," July 23, 2009. http://pro-choicesmokingdoctor.blogspot.com/2009/07/obama-in-bare-faced-lie.html.

[5] Chen C. "Kristallnacht," *World War II Data* http://ww2db.com/battle_spec.php?battle_id=159.

[6] History.com. "Nazis launch Kristallnacht," *History.com*, http://history.com/this-day-in-history/nazis-launch-kristallnacht.

[7] McFadden MJ. *Dissecting Antismokers' Brains*, Aethna Press, 2004.

[8] James I (King of England). *Counterblaste To Tobacco*, 1604.

[9] http://www.rpa.com/portfolio/ctcp_second-hand-sally/.

[10] American Medical Association. *Report 14 Of The Board Of Trustees (A-01) Annual Tobacco Report (Informational)*, June 2001, p. 123. http://web.archive.org/web/20040421083003/
http://www.ama-assn.org/ama1/upload/mm/interim01/BOT_Reports_1-17.pdf.

[11] Moss L. "Time to Reclaim Our Streets From Persistent Smokers," *The Scotsman*, February 3, 2009. http://www.scotsman.com/news/lyndsay-moss-time-to-reclaim-our-streets-from-persistent-smokers-1-756844.

[12] Ustinova, T. "Boozy ape sent to rehab," *Reuters Life* (Moscow), February 26, 2010. http://uk.reuters.com/article/idUKTRE61P2U220100226.

[13] Hernandez J. "Smoking Ban for Beaches and Parks Is Approved," *New York Times*, February 2, 2011. http://nytimes.com/2011/02/03/nyregion/03smoking.html.

[14] Sebnem A. "Turkish Airstrikes Kill Smugglers Mistaken for Kurdish Separatists," *New York Times*, December 29, 2011. http://nytimes.com/2011/12/30/world/middleeast/turkish-airstrikes-kill-35-along-smuggling-route .html.

[15] Daily Mail Reporter. "Boy, 14, collapses after chewing nicotine gum equivalent to 180 cigarettes ... handed out at school," *UK Daily Mail*, July 21, 2009. http://dailymail.co.uk/news/article-1200918/Boy-14-collapses-overdosing-nicotine-chewing-gum-handed-school.html.

[16] ACLU. "ACLU Victorious In Student Strip Search Case," *ACLU*, April 14, 2009. http://207.170.137.86/pressreleases/2009pr/2009.04.14.asp.

[17] Chacon D. "Medical Marijuana Takes City To New Sales Tax High," *The Gazette* (Colorado). October 7, 2010. http://gazette.com/articles/marijuana-105987-city-medical.html.

[18] Drake B. The Cultivators Handbook of Natural Tobacco, CreateSoace, 2010.

[19] Drake B. *Cultivators Handbook of Marijuana*, Agrarian Reform Company, 1970.

[20] Smith P. "Norway Government Wants to Decriminalize Heroin Smoking," *StopTheDrugWar.org*, March 1, 2013.
http://stopthedrugwar.org/chronicle/2013/mar/01/norway_government_wants_decrimin.

[21] DrugWar.com. "Drug War: Covert Money, Power & Policy," *DrugWar.com*. http://www.drugwar.com/fusarium.shtm.

[22] Gorlick A. "Stanford Researcher's Online Map Pinpoints Cigarette Factories Around The World," *Stanford University News*, October 4, 2010. http://news.stanford.edu/news/ 2010/october/cigarette-citadels-project-100410.html.

[23] Nissen CM. "Stanford Raises The Bar On Useless Tobacco Research," *Tobacco Harm Reduction*, October 10, 2010. http://smokles.wordpress.com/2010/10/05/stanford-raises-the-bar-on-useless-tobacco-research/.

[24] World Health Organization. "Global Progress Report 2012," *World Health Organization*. http://www.who.int/fctc/en/.

[25] Tuffs A. "Full smoking ban is needed in Germany as study finds four out of five pubs flout regulations," *British Medical Journal*, May 14, 2011, Volume 342, Issue 7806. dx.doi.org/10.1136/bmj.d2864.

[26] Nainggolan, L. "Battling Big Tobacco: Physician activism vital on smoking's new frontiers," *TheHeart.org*, November 28, 2011. http://theheart.org/article/1318775.do.

[27] Gallegos A. "International Trade Dispute Threatens to Undo Clove Cigarette Ban," *American Medical News*, April 23, 2012. http://ama-assn.org/amednews/2012/04/23/gvsa0423.htm.

[28] Repace JL. "Air Quality in Marquette Restaurants…," *Michigan Department of Community Health*, April 22, 2011. http://co.marquette.mi.us/departments/health_depart ment/docs/Marquette_Air_Report_Repace_2011.pdf.

[29] Siegel M. "New Study Concludes That Italian Smoking Ban Reduced Heart Attack Rates Within First Year," *The Rest of the Story*, February 14, 2008. http://tobaccoanalysis.blogspot.com/2008/02/new-study-concludes-that-italian.html.

[30] Khamsi R. "Is Obesity Contagious?" *New Scientist Magazine*, July 25, 2007. http://www.newscientist.com/article/dn12343-is-obesity-contagious.html.

[31] McFadden MJ. "BBQ Is Toxic And Will Kill You," *Citizens Freedom Alliance*, http://smokersclubinc.com/modules.php?name=News&file=article&sid=4415.

[32] Russia Today. "Hard to Swallow: Canada junk food to bear ugly warning signs," *Russia Today*, November 4, 2012. http://www.youtube.com/watch?v=lKF6EGF5h8.

[33] Lustig R. "The Toxic Truth About Sugar," *UCSF Public Affairs*, January 31, 2012. http://youtube.com/watch?v=ffoOeW5wZ9s&.

[34] Klepeis N, Ott W, Switzer P. "Real-Time Measurement of Outdoor Tobacco Smoke Particles," *Journal of the Air & Waste Management Association*, 2007, Volume 57, Issue 5, pp. 522-534. dx.doi.org/10.3155/1047-3289.57.5.522.

[35] Picard A. "Outdoor patios as smoky as bars," *The Globe And Mail* (Toronto, Canada), June 7, 2007. http://www.theglobeandmail.com/life/outdoor-patios-as-smoky-as-bars/ article4094934/.

[36] Yates D. "Realtors: Smoke Ban Hurts Sales," *Daily Journal* (San Mateo, CA), October 15, 2007. http://archives.smdailyjournal.com/article_preview.php?type=lnews&title=Realtors: -Smoke-ban-hurts-sales&id=81951.

[37] Wikipedia. "General Motors Streetcar Conspiracy," *Wikipedia.org*. http://en.wikipedia.org/ wiki/General_Motors_streetcar_conspiracy.

[38] Hodgekiss A. "Caffeine is so dangerous that it should be regulated like alcohol and cigarettes, warns leading expert," *Daily Mail* (London, England), February 28, 2013. http://dailymail.co.uk/health/article-2285854/Caffeine-dangerous-regulated-like-alcohol-cigarettes-warns-leading-expert.html.

[39] Garfinkle S. "Smoking Around Kids," *Washington Post*, July 20, 2009. http://voices.washingtonpost.com/parenting/2009/07/smoking_around_kids.html.

[40] Arlinghous C. "Cigarettes and Candy Bars or Smoke Gets in Your Eyes," *Union Leader* (Manchester, NH), May 19, 2005. http://web.archive.org/web/20050525124820/http://www.theunionleader.com/articles_showa.html?article=54927.

[41] Weinstein S. "Raynham voters snuff out smoking ban," *Taunton Gazette* (MA), May 21, 2002. http://archive.tobacco.org/news/94024.html.

[42] "Smoking ban in British waters," *The Times of London*, August 9, 2008. http://archive.tobacco.org/news/269823.html.

[43] Huber GL, *et al.* "Passive Smoking: How Great A Hazard?" *Consumers Research*, Volume 74, Number 7, July 1991, p. 10.

[44] BBC News. "Smokers Get Militant Over Ban," *BBC Online Network News*, September 9, 1998. http://news.bbc.co.uk/2/hi/health/167762.stm.

[45] Asimov I. "The Dead Past," *Astounding Science Fiction*, April 1956.

[46] Matchan, DC. *We Mind If You Smoke*, Pyramid Publications, 1977.

[47] Nichol FD. *Christian Herald*, October 1964. (quoted in Matchan, DC, Ibid.)

[48] Rabin RC. "A New Cigarette Hazard: 'Third-Hand Smoke,'" *New York Times*, January 1, 2009. http://www.nytimes.com/2009/01/03/health/research/03smoke .html.

[49] Levy RA, Marimont RB. "Lies, Damned Lies, & 400,000 Smoking-Related Deaths," *Regulation*, Volume 21, Number 4, 1998, pp. 24-29. http://www.cato.org/sites/cato.org/files/serials/files/regulation/1998/10/lies.pdf.

[50] Dyson C. "Some Teens Hellbent On Smoking To Be 'In'," *Fredricksburg.com*, June 22, 2001. http://fredericksburg.com/News/FLS/2001/062001/06222001/314710.

[51] http://www.saltmonument.org/techinfo/.

[52] Borgerding MF, Bodnar JA, Wingate DE. "The 1999 Massachusetts Benchmark Study: Final Report," July 24, 2000. http://legacy.library.ucsf.edu/documentStore/y/e/k/yek21c00/Syek21c00.pdf.

[53] Parker T. "Arsenic - An Essential Nutrient For Growth," http://articlesnatch.com/Article/Arsenic—An-Essential-Nutrient-For-Growth/1948700.

[54] http://algebra.com/algebra/homework/Exponents-negative-and-fractional/Exponents-negative-and-fractional.faq.question.419964.html.

[55] Monte WC. "Dr. Woodrow C. Monte's Methanol Research – University Of Arizona - Part 10," *SweetPoison.com*. http://www.sweetpoison.com/articles/dr-woodrow-monte10.html.

[56] "Arsenic Toxicity," *ManbirOnline*.com. http://www.manbir-online.com/diseases/arsenic.htm.

[57] Wikipedia. "Cadmium Poisoning," *Wikipedia.org*. http://en.wikipedia.org/wiki/Cadmium_poisoning.

[58] Right Diagnosis. "Symptoms Of Cadmium Poisoning," *RightDiagnosis.com*. http://www.rightdiagnosis.com/c/cadmium_poisoning/causes.htm.

[59] Belgard J. "Air quality in Alexandria businesses has improved since smoking ban enacted, advocate says," *TheTownTalk.com*, February 1, 2013. http://thetowntalk.com/article/20130201/NEWS01/302010325/.

[60] ICdA. "Cadmium exposure and human health," *Cadmium.org*. http://www.cadmium.org/pg_n.php?id_menu=5.

[61] http://www.higheredcenter.org/files/product/community-colleges.txt.

[62] http://www.tobaccofreeu.org/policy/documents/DiamondAwardBooklet.pdf.

[63] Wellnesss Resource Center. "About Gamma," http://wellness.missouri.edu/gamma.html.

[64] http://www.higheredcenter.org/files/product/community-colleges.txt.

[65] "Arizona State University planning campus-wide smoking ban," *Eloy Enterprise*, October 11, 2012. http://www.trivalleycentral.com/eloy_enterprise/education/arizona-state-university-planning-campus-wide-smoking-ban/article_21857904-1330-11e2-93ea-0019bb2963f4.html.

[66] Hall JC, *et al.* "Assessment of exposure to secondhand smoke at outdoor bars and family restaurants in Athens, Georgia, using salivary cotinine," *Journal of Occupational and Environmental Hygiene*, November 2009, Volume 6, Issue 11, pp. 698-704. dx.doi.org/10.1080/15459620903249893.

[67] Gao R. "Smoke-Free Campus Must Become Reality," *DailyTrojan.com*, April 10, 2011. http://dailytrojan.com/2011/04/10/smoke-free-campus-must-become-reality/.

[68] Boffetta P, *et al.* "Multicenter case-control study of exposure to environmental tobacco smoke and lung cancer in Europe," *Journal of the National Cancer Institute*, October 7,1998, Volume 90, Issue 19, pp. 1440-1450. dx.doi.org/10.1093/jnci/90.19.1440.

[69] Macdonald V. "Passive smoking doesn't cause cancer – official," *Sunday Telegraph*, March 8, 1998. http://legacy.library.ucsf.edu/documentStore/m/y/c/myc42c00/Smyc42c00.pdf.

[70] Berlau J. "The Data That Went Up In Smoke," *Investor's Business Daily*, April 8, 1998. http://legacy.library.ucsf.edu/documentStore/c/p/g/cpg81d00/Scpg81d00.pdf.

[71] World Health Organization. "Passive Smoking Does Cause Lung Cancer, Do Not Let Them Fool You," *World Health Organization*, March 9, 1998. http://www.who.int/inf-pr-1998/en/pr98-29.html.

[72] Peck P. "Smoking Ban Saves Lives in Montana Town," *WebMD.com*, April 1, 2003. http://webmd.com/smoking-cessation/news/20030401/smoking-ban-saves-lives-in-montana-town.

[73] Glantz SA. "ANR On Health Benefits Of Local Smokefree Workplace Law," April 1, 2003. https://listsrv.ucsf.edu/OldArchives/hypermail_archives/stanglantz-l/0065.html.

[74] Sargent RP, Shepard RM, Glantz SA. "Reduced incidence of admissions for myocardial infarction associated with public smoking ban: before and after study," *British Medical Journal*, April 22, 2004, Volume 328, Issue 7446, pp. 977-979. dx.doi.org/10.1136/bmj.38055.715683.55.

[75] Pell JP, Haw S, Cobbe S, *et al.* "Smoke-free legislation and hospitalizations for acute coronary syndrome," *New England Journal of Medicine*, July 31, 2008, Volume 359, Issue 5, pp. 482-491. dx.doi.org/10.1056/NEJMsa0706740.

[76] Szabo, L. "Scotland smoking ban credited with fewer heart attacks," *USA Today*, July 30, 2008. http://usatoday30.usatoday.com/news/health/2008-07-30-scotland-smoking-ban_N.htm.

[77] Snowdon C. "Official: Scotland sees large rise in hospital admissions for acute coronary syndrome in second year of smoking ban," *Velvet Glove, Iron Fist*. http://www.velvetgloveironfist.com/index.php?page_id=65.

[78] Siegel M. "Data Released by Scottish National Health Service Show that Conclusions of Earlier Study about Effect of Smoking Ban on Heart Attacks Were Wrong," *The Rest of the Story*, December 1, 2008. http://tobaccoanalysis.blogspot.com/2008/12/data-released-by-scottish-national.html.

[79] Proctor RN. *The Nazi War on Cancer*. Princeton University Press (NJ), 2000.

[80] Fisher RA. "Cigarettes, Cancer, And Statistics," *Centennial Review*, 1958, Volume 2, pp. 151-166. http://www.oxfordreference.com/view/10.1093/acref/9780195122008.001.0001/acref-9780195122008-e-141.

[81] http://www.sourcewatch.org/index.php/Philip_Burch.

[82] Burch P. "Letters," *New Scientist*, April 25, 1974, p. 195.

[83] US Department of Health and Human Services. Smoking and Health: Report of the Advisory Committee of the Surgeon General of the Public Health Service. 1964.

[84] http://www.tobacco.org/resources/history/Tobacco_Historynotes.html.

[85] Lee PN. "Progress in the Prevention of Chest Disease," (One day conference notes), January 27, 1978. http://www.health.gov.bc.ca/guildford/pdf/109/00011035.pdf.

[86] Huber GL, *et al.* "Passive Smoking: How Great A Hazard?" *Consumers Research*, Volume 74, Number 7, July 1991, p. 10. http://tobaccodocuments.org/pm/2046323437-3484.html.

[87] Russell MAH, Cole PV, Brown ET. "Absorption by non-smokers of carbon monoxide from room air polluted by tobacco smoke," *The Lancet*, March 17, 1973, Volume 301, Number 7803, pp. 576-579. dx.doi.org/10.1016/S0140-6736(73)90718-6.

[88] McFadden, MJ. *Dissecting Antismokers' Brains*. Aethna Press, 2004. pp. 64-65.

[89] Matchan D. *We Mind If You Smoke*. Pyramid Press, 1977 p. 124.

[90] Hirayama T. "Non-smoking wives of heavy smokers have a higher risk of lung cancer: a study from Japan," *British Medical Journal*, November 28, 1981, Volume 283, Number 6304. dx.doi.org/10.1136/bmj.283.6304.1466.

[91] Trichopoulos D, Kalandidi A, et al. "Lung cancer and passive smoking," *International Journal of Cancer*, January 1981, Volume 27, Issue 1, pp. 1-4. dx.doi.org/10.1002/ijc.2910270102.

[92] McFadden, MJ. *Dissecting Antismokers' Brains*. Aethna Press, 2004. pp. 231-237.

[93] US Department of Health and Human Services. The Health Consequences of Involuntary Smoking: A Report of the Surgeon General. 1986.

[94] US Environmental Protection Administration. *Respiratory Health Effects of Passive Smoking: Lung Cancer and Other Disorders*. EPA/600/6-90/006F, December 1992.

[95] Thun MJ, Henley SJ, Calle EE. "Tobacco use and cancer: an epidemiologic perspective for geneticists," *Oncogene*, October 21, 2002, Volume 21, Number 48, pp. 7307-7325. dx.doi.org/10.1038/sj.onc.1205807.

[96] Brennan P, Crispo A, *et al.* "High Cumulative Risk of Lung Cancer Death among Smokers and Nonsmokers in Central and Eastern Europe," *American Journal of Epidemiology*, December 15, 2006, Volume 164, Number 12, pp. 1233-1241. dx.doi.org/10.1093/aje/kwj340/.

[97] Congressional Research Service. "Environmental Tobacco Smoke and Lung Cancer Risk," *95-1115 SPR*, November 14, 1995.

[98] http://www.forces.org/evidence/epafraud/files/osteen.htm.

[99] *Osteen Decision* 1998. pp. 88-90. http://archive.tobacco.org/Documents/980717osteen.html.

[100] "Judge Once Worked as Tobacco Lobbyist," *New York Times*, August 23, 1995. http://www.nytimes.com/1995/08/23/business/judge-once-worked-as-tobacco-lobbyist.html.

[101] Cushman JH Jr. "Judge Rules F.D.A. Has Right To Curb Tobacco As Drug," *New York Times*, April 26, 1997. http://www.nytimes.com/1997/04/26/us/judge-rules-fda-has-right-to-curb-tobacco-as-drug.html.

[102] http://pacer.ca4.uscourts.gov/opinion.pdf/982407.P.pdf.

[103] Meier B. "Judge Voids Study Linking Cancer to Secondhand Smoke," *New York Times*, July 20, 1998. http://www.nytimes.com/1998/07/20/us/judge-voids-study-linking-cancer-to-secondhand-smoke.html.

[104] "Delaware's Smoke-Free Public Places Law One of the Toughest in the Nation," *Tobacco Regulation Review*, April 2003, Volume 2, Issue 1, p. 17.

[105] http://archive.tobacco.org/news/121060.html.

[106] http://leg.state.fl.us/statutes/index.cfm?App_mode=Display_Statute&URL=Ch0386/ch0386.htm

[107] American Medical Association. *Report 14 Of The Board Of Trustees (A-01) Annual Tobacco Report (Informational)*, June 2001, p. 123. http://web.archive.org/web/20040421083003/http://www.ama-assn.org/ama1/upload/mm/interim01/BOT_Reports_1-17.pdf.

[108] Goldman LK, Glantz SA. "Evaluation of Antismoking Advertising Campaigns," *Journal of the American Medical Association*, March 11,1998, Volume 279, Number 10, pp. 772-777.

[109] Kolata G. "Stung by Courts, F.D.A. Rethinks Its Rules," *New York Times*, October 15, 2002. http://www.nytimes.com/2002/10/15/health/stung-by-courts-fda-rethinks-its-rules.html.

[110] Associated Press. "Kessler, Koop Refuse To Testify," *Associated Press*, March 5, 1998. http://www.apnewsarchive.com/1998/Kessler-Koop-Refuse-To-Testify/id-969737edf18a146eed0a41448a6072e5.

[111] Bener A. "Colon cancer in rapidly developing countries: review of the lifestyle, dietary, consanguinity and hereditary risk factors," *Oncology Reviews*, April 2012, Volume 5, Number 1, pp. 5-11. dx.doi.org/10.1007/s12156-010-0061-0.

[112] Slaughter A. "Korean vets reveal cold truth about skin cancer," *USA Today*, May 25, 2001. http://usatoday30.usatoday.com/news/health/spotlight/2001-05-25-frostbite.htm.

[113] Enstrom J, Kabat G. "Environmental tobacco smoke and tobacco related mortality in a prospective study of Californians, 1960-98," *British Medical Journal*, May 15, 2003, Volume 326, Issue 7398, p. 1057. dx.doi.org/10.1136/bmj.326.7398.1057.

[114] US Department of Health and Human Services. The Health Consequences of Involuntary Exposure to Tobacco Smoke: A Report of the Surgeon General, 2006.

[115] Klein EG, Forster JL, Erickson DJ, Lytle LA, Schillo B. "Does the type of CIA policy significantly affect bar and restaurant employment in Minnesota cities?" *Prevention Science*, June 2009, Volume 10, Number 2, pp. 168-174. dx.doi.org/10.1007/s11121-009-0122-4.

[116] Heyman D. "Study: Smoking Bans Don't Lead To Layoffs," *Public News Service*, June 24, 2009. http://www.publicnewsservice.org/index.php?/content/article/9420-1.

[117] Caldwell E. "Study: Smoking Bans Do Not Cause Job Losses In Bars And Restaurants," *OSU Research News*. http://researchnews.osu.edu/archive/smokejobs.htm.

[118] University of Minnesota Academic Health Center. "University research finds that smoking bans do not cause job losses in bars and restaurants," May 19, 2009. http://www.health.umn.edu/media/releases/ban051909/index.htm.

[119] Email correspondence with author.

[120] Apollonio D, Bero L. "Creating industry front groups: The tobacco industry and 'Get Government Off Our Back'," *American Journal of Public Health,* March 2007, Volume 97, Issue 3, pp. 419-427. dx.doi.org/10.2105/AJPH.2005.081117.

[121] Scollo M, *et al.* "Review of the quality of studies on the economic effects of smoke-free policies on the hospitality industry" *Tobacco Control,* 2003, Volume 12, Issue 1, pp. 13-20. dx.doi.org/10.1136/tc.12.1.13.

[122] http://clearwaymn.org/wp-content/uploads/2012/12/Unintended-Effects-of-Local-Clean-Indoor-Air-Policies-on-Alcohol-Businesses.doc.

[123] Goetz A. "Conclusions For Sale" *Minnesota Daily,* June 2, 2009. http://www.mndaily.com/2009/06/02/conclusions-sale.

[124] "Unintended Effects of Local Clean Indoor Air Policies on Alcohol Businesses, Alcohol Policies and Alcohol Related Problems," RC-2006-0047. http://clearwaymn.org/wp-content/uploads/2012/12/Unintended-Effects-of-Local-Clean-Indoor-Air-Policies-on-Alcohol-Businesses.doc.

[125] Klein EG, Forster JL, *et al.* "Economic effects of clean indoor air policies on bars and restaurant employment in Minneapolis and St Paul, Minnesota," *Journal of Public Health Management Practice,* July/August 2010, Volume 16, Issue 4, pp. 285-293. dx.doi.org/10.1097/PHH.0b013e3181c60ea9.

[126] Soyfer VN. *Lysenko and the Tragedy of Soviet Science.* Translation by Leo and Rebecca Gruliow, Rutgers University Press, 1994.

[127] Roll-Hansen R. The Lysenko Effect: The Politics Of Science. Humanity Books, 2004.

[128] Personal communications with Dr. Phillips.

[129] Enstrom JE. "Defending legitimate epidemiologic research: combating Lysenko pseudoscience," *Epidemiologic Perspectives & Innovations,* October 2007, Volume 4, Number 11. dx.doi.org/10.1186/1742-5573-4-11.

[130] Enstrom J, Kabat G. "Environmental tobacco smoke and tobacco related mortality in a prospective study of Californians, 1960-98," *British Medical Journal,* May 15, 2003, Volume 326, Issue 7398, p. 1057. dx.doi.org/10.1136/bmj.326.7398.1057.

[131] http://www.bmj.com/content/326/7398/1057?tab=responses.

[132] Kissel A. "Whistleblowing UCLA Professor Now Represented by American Center for Law & Justice," *Fire.org* (Foundation for Individual Rights in Education), June 2, 2011. http://thefire.org/article/13268.html.

[133] Enstrom, JE. "Defending legitimate epidemiologic research: combating Lysenko pseudoscience," *Epidemiologic Perspectives & Innovations,* October 11, 2007. http://archive.biomedcentral.com/1742-5573/content/4/1/11. dx.doi.org/10.1186/1742-5573-4-11.

[134] Morasch C. "CARB critic retained: UCLA extends Enstrom for another year," *Landline Magazine,* September 11, 2011. http://www.landlinemag.com/todays_news/Daily/2011/Sep11/090511/090711-06.shtml.

[135] Kissel A. "Whistleblowing UCLA Professor Wins One More Year," *Fire.org* (Foundation for Individual Rights in Education), September 6, 2011. http://thefire.org/article/13530.html.

[136] Pulse Today. "Retired GP suspended after questioning BMA stance on smoking," *Pulse Today,* July 10, 2012. http://www.pulsetoday.co.uk/retired-gp-suspended-after-questioning-bma-stance-on-smoking/14261198.article.

[137] Kuneman DW, McFadden MJ. "Economic losses due to smoking bans in California and other states," *SmokersClub.com*, 2005. http://kuneman.smokersclub.com/economic.html.

[138] http://web.archive.org/web/20091105045415/http://tobaccoscam.ucsf.edu/fake/hospitality_r esults.cfm?1=1&keyword=kuneman.

[139] Siegel M. "Another Misleading Public Claim: This Time, a TobaccoScam Attempt to Discredit an Individual Who Opposes the Anti-Smoking View," *The Rest of the Story*, October 26, 2005. http://tobaccoanalysis.blogspot.com/2005/10/another-misleading-public-claim-this.html.

[140] Kasprak J. "No-Smoking Laws," September 14, 2000. http://web.archive.org/web/2010053106 1049/http://www.cga.ct.gov/2000/rpt/olr/htm/2000-r-0890.htm.

[141] Noah, T. "Bill Clinton and the Meaning of 'Is'," *Slate.com*, September 13, 1998. http://slate.com/ articles/news_and_politics/chatterbox/1998/09/bill_clinton_and_the_meaning_of_is.html.

[142] Minnesota Gambling Control Board. *Charitable Gambling Impact Study*. March 28, 2008.

[143] Meyer B. "Economic Fears Snuff Out Proposed Smoking Bans," *The Plain Dealer* (OH), February 4, 2009. http://www.cleveland.com/nation/index.ssf/2009/02/economic_fears_snuff _out_propo.html.

[144] Parry W. "Atlantic City moves closer to delaying smoking ban," *USA Today*, October 8, 2008. http://usatoday30.usatoday.com/news/nation/2008-10-08-3003264446_x.htm.

[145] Associated Press. "Atlantic City to have 7-day smoking ban," *USA Today*, October 10, 2008. http://usatoday30.usatoday.com/news/nation/2008-10-10-atlantic-city-smoking_N.htm.

[146] Garrett TA, Pakko MR. "The Revenue Performance of Casinos after a Smoking Ban: The Case of Illinois," *Social Science Research Network*, March 9, 2010. http://papers.ssrn.com/sol3/ papers.cfm?abstract_id=1415034.

[147] Rushton, B. "Casino revenue plummets; gaming lobbyists blame smoking ban," *The State Journal-Register* (Springfield, IL), August 14, 2009. http://www.sj-r.com/news/x1886167264/ Casino-revenue-plummets-gaming-lobbyists-blame-smoking-ban.

[148] Arnott D. "It is a myth that high duties on tobacco lead to increased smuggling," *The Guardian*, February 24, 2011. http://www.guardian.co.uk/commentisfree/2011/feb/24/tobacco-taxes-budget.

[149] Rushton B. "Casino revenue plummets; gaming lobbyists blame smoking ban," *The State Journal-Register* (Springfield, IL), August 14, 2009. http://www.sj-r.com/news/x1886167264/ Casino-revenue-plummets-gaming-lobbyists-blame-smoking-ban.

[150] Laprade T. "Smoking Truths," *The Mississauga News*, May 17, 2006.

[151] California Environmental Protection Agency. "California Identifies Secondhand Smoke as a 'Toxic Air Contaminant'," *California Air Resources Board*, January 26, 2006. http://www.arb.ca.g ov/newsrel/nr012606.htm.

[152] Raven W. "Six-month public smoking ban slashes heart attack rate in community," *UCSF.edu*, April 1, 2003. http://www.ucsf.edu/news/2003/04/4763/six-month-public-smoking-ban-slashes-heart-attack-rate-community.

[153] CBS News. "Did Smoking Ban Cut Heart Attacks?" *CBSNews.com*, April 1, 2003. http://www.cbsnews.com/stories/2003/04/01/health/main547249.shtml.

[154] Associated Press. "Smoking Ban, Heart Attack Drop Tied," *Milwaukee Journal Sentinel*, p. 16G, April 7, 2003. http://news.google.com/newspapers?nid=1683&dat=20030407&id=57saAAAAIBAJ &sjid=gEMEAAAAIBAJ&pg=6336,3890408.

[155] Johnson C. "Three bills seek increased cig taxes," *Independent Record* (Helena, MT), January 28, 2003. http://helenair.com/news/state-and-regional/three-bills-seek-increased-cig-taxes/articl e_7e6cbb50-50e5-5267-9e57-cc07784c7843.html.

[156] http://www.mascotcoalition.org/initiatives/cia/doctors.html.

[157] Burns D, Shanks T, *et al.* "Restrictions on Smoking in the Workplace," *Cancer Control Monograph 12*, pp. 99-128; p. 102. http://cancercontrol.cancer.gov/tcrb/monographs/12/Chapter_3.pdf.

[158] WebMD. "Smoking Ban Saves Lives In Montana Town," *WebMD.com.* http://www.webmd.com/smoking-cessation/news/20030401/smoking-ban-saves-lives-in-montana-town.

[159] American Heart Association. "New Study Links Secondhand Smoke to Heart Attacks," *American Heart Association,* April 5, 2005. http://www.no-smoke.org/pdf/HelenaAHARelease-BMJ.pdf.

[160] ANR Press Release. "British Medical Journal Study Shows Smokefree Law Slashed Heart Attack Rate 40%," *Americans for Nonsmokers Rights,* April 5, 2004. http://no-smoke.org/docume nt.php?id=252.

[161] Neal R. "A Striking Effect," *CBS News,* October 1, 2003. http://www.cbsnews.com/stories/2003/10/01/sunday/main576144.shtml.

[162] Harrington J. "Helena Heart-Attack Study Gets Worldwide Attention," *Independent Record* (Helena, MT), April 6, 2003. http://helenair.com/news/opinion/helena-heart-attack-study-gets-worldwide-attention/article_c9a3b6a2-9279-5fe3-a59d-8097038b9d92.html.

[163] Meikle J. "Fresh Evidence on Passive Smoking" *The Guardian,* April 4, 2004. http://www.guardian.co.uk/society/2004/apr/05/smoking.sciencenews.

[164] Sullum J. "Miracle in Helena," *Reason,* July 2004. http://reason.com/archives/2004/07/01/miracle-in-helena.

[165] Sargent RP, Shepard RM, Glantz SA. "Reduced incidence of admissions for myocardial infarction associated with public smoking ban: before and after study," *British Medical Journal,* April 22, 2004, Volume 328, Issue 7446, pp. 977-979. dx.doi.org/10.1136/bmj.38055.715683.55.

[166] "Tobacco smoke and your health," *National Health Service, UK.* http://www.breathingspace.n hs.uk/your_health.html.

[167] Wikipedia. "Natural Experiment," *Wikipdedia.org.* http://en.wikipedia.org/wiki/Natural_ex periment.

[168] Sargent RP, Shepard RM, Glantz SA. "Reduced incidence of admissions for myocardial infarction associated with public smoking ban: before and after study," *British Medical Journal,* April 22, 2004, Volume 328, Issue 7446, pp. 977-979. dx.doi.org/10.1136/bmj.38055.715683.55.

[169] Ibid.

[170] Email correspondence with author.

[171] http://web.archive.org/web/20030724212153/http://no-smoke.org/HelenaPowerPoint.pdf.

[172] Hitt D. "The Helena Study Chart," *DaveHitt.com,* http://davehitt.com/facts/HelenaPowerPoint.pdf.

[173] National Oceanic and Atmospheric Administration. "Monthly Climatological Summary, 2002." http://www.ncdc.noaa.gov/cdo-web/datasets/GHCNDMS/stations/GHCND:USR0000M HEL/detail.

[174] "The Nation: Trials of the Watergate Jury," *Time Magazine*, January 28, 1974, pp. 13-18. http://www.time.com/time/magazine/article/0,9171,911048,00.html.

[175] Mathews R. "Anti-Smoking Laws and Incidence of Acute Myocardial Infarction: Across 74 US Cities." *Duke Clinical Research Institute*, May 13, 2010. http://my.americanheart.org/idc/groups/ahamah-public/@wcm/@sop/@scon/documents/downloadable/ucm_427365.pdf.

[176] http://tobaccoanalysis.blogspot.com.

[177] http://velvetgloveironfist.blogspot.com.

[178] Snowdon C. Velvet Glove, Iron Fist: A History of Anti-Smoking, Little Dice (England), 2009.

[179] Hillenbrand M. "Study: Smoking Ban Beneficial," *DailyIowan*, January 19, 2010. http://www.dailyiowan.com/2010/01/19/Metro/15051.html.

[180] Siegel M. "Data Show No Effect of Smoking Ban on Heart Attack Admissions in North Carolina," *The Rest of the Story*, November 10, 2011. http://tobaccoanalysis.blogspot.com/2011/02/iowa-researchers-claim-smoking-ban-had.html.

[181] Washington MD, Barnes RL, Glantz SA. "Chipping Away at Tobacco Traditions in Tobacco Country," *Center for Tobacco Control Research and Education*, June 2011. http://escholarship.org/uc/item/7kc398r4.

[182] NC Department of Health and Human Services. "N.C. Heart Attack Rates Down Since Passage of Smoke-Free Law," *NC DHHS*, November 9, 2011. http://www.ncdhhs.gov/pressrel/2011/2011-11-09_heart_attack_down.htm.

[183] Siegel, M. "Data Show No Effect of Smoking Ban on Heart Attack Admissions in North Carolina," *The Rest of the Story*, November 10, 2011. http://tobaccoanalysis.blogspot.com/2011/11/data-show-no-effect-of-smoking-ban-on.html.

[184] Bodden T. "Fewer heart attacks in wake of smoking ban," *Daily Post* (Wales), June 30, 2008. http://www.dailypost.co.uk/news/north-wales-news/2008/06/30/fewer-heart-attacks-in-wake-of-smoking-ban-55578-21170754/.

[185] Siegel M. "Wales Report Claims Smoking Ban Reduced Heart Attacks But Fails to Present Data Which Show an Increase in Expected Number of Heart Attacks," *The Rest of the Story*, December 10, 2009. http://tobaccoanalysis.blogspot.com/2009/12/wales-report-claims-that-smoking-ban.html.

[186] Siegel M. "No Reduction in Heart Attacks in Wales During First Nine Months Following Smoking Ban," *The Rest of the Story*, August 5, 2008. http://tobaccoanalysis.blogspot.com/2008/08/no-reduction-in-heart-attacks-in-wales.html.

[187] Snowdon C. "Pub Closures And The Smoking Ban," *Velvet Glove, Iron Fist*, January 11, 2013. http://velvetgloveironfist.blogspot.com/2013/01/pub-closures-and-smoking-ban.html.

[188] http://4.bp.blogspot.com/_EhRt4AvJLd4/TA9cMnlwIdI/AAAAAAAAA8k/S7xLb-AWQ_0/s1600/England+AMI+-+gilmore.jpg.

[189] Kelland, K. "Smoking Ban Cut Heart Attacks," *Reuters* (UK), June 9, 2010. http://uk.reuters.com/article/2010/06/08/uk-heart-smoking-idUKTRE65764M20100608.

[190] Nursing Times. "Smoking ban 'cuts heart attacks'," *NursingTimes.net*, June 9, 2010. http://www.nursingtimes.net/nursing-practice/clinical-zones/smoking-cessation/smoking-ban-cuts-heart-attacks/5015697.article.

[191] http://www.tobaccotactics.org/index.php/Linda_Bauld.

[192] http://4.bp.blogspot.com/_EhRt4AvJLd4/S5_ZEYU1RsI/AAAAAAAAAz4/xdKvGMnaU9g/s1600-h/closures.JPG.

[193] Bauld L. "The Impact Of Smokefree Legislation In England: Evidence Review," *University of Bath*, March 2011. http://www.dh.gov.uk/prod_consum_dh/groups/dh_digitalassets/documents/digitalasset/dh_124959.pdf.

[194] Kuneman DW, McFadden MJ. "The Impact of State-wide Smoking Bans on Acute Myocardial Infarction Hospital Admissions in California and Other States," 2005. http://www.scribd.com/doc/9679507/bmjmanuscript.

[195] Shetty KD, DeLeire T, White C, Bhattacharya J. "Changes in U.S. hospitalization and mortality rates following smoking bans," *Journal of Policy Analysis and Management*, Winter 2011, Volume 30, Issue 1, pp. 6-28. dx.doi.org/10.1002/pam.20548.

[196] Marlow ML. "Smoking bans and acute myocardial infarction incidence," *Applied Economics Letters*, January 2012, Volume 19, Issue 16, pp. 1577-1581. dx.doi.org/10.1080/13504851.2011.639730.

[197] Mathews R. "Anti-Smoking Laws and Incidence of Acute Myocardial Infarction: Across 74 US Cities." *Duke Clinical Research Institute*, May 13, 2010. http://my.americanheart.org/idc/groups/ahamah-public/@wcm/@sop/@scon/documents/downloadable/ucm_427365.pdf.

[198] Rodu B, *et al*. "Acute Myocardial Infarction Mortality Before and After State-wide Smoking Bans," *Journal of Community Health*, 2012, Volume 37, Number 2, pp. 468-472. dx.doi.org/10.1007/s10900-011-9464-5.

[199] Kuneman DW, McFadden MJ. "The Impact of State-wide Smoking Bans on Acute Myocardial Infarction Hospital Admissions in California and Other States," 2005. http://www.scribd.com/doc/9679507/bmjmanuscript.

[200] Shetty KD, DeLeire T, White C, Bhattacharya J. "Changes in U.S. hospitalization and mortality rates following smoking bans," *Journal of Policy Analysis and Management*, Winter 2011, Volume 30, Issue 1, pp. 6-28. dx.doi.org/10.1002/pam.20548.

[201] Siegel M. "Study of Trends in State Heart Attack Admissions Refutes Conclusions of Helena et al. Studies," *The Rest of the Story*, July 13, 2007. http://tobaccoanalysis.blogspot.com/2007/07/study-of-trends-in-state-heart-attack.html.

[202] Email correspondence with author.

[203] McFadden, MJ. "A Study Delayed…," *Facts And Fears*, American Council on Science And Health, July 12, 2007. http://www.acsh.org/factsfears/newsID.990/news_detail.asp.

[204] http://tidatabase.org.

[205] Ibid.

[206] http://www.panacealink.org/globalink/

[207] Siegel M. "Rest of the Story Author Thrown Off of List-Serve; Censorship Alive in Anti-Smoking Movement, But Little Room for the Truth," *The Rest of the Story*, May 30, 2006. http://tobaccoanalysis.blogspot.com/2006/05/rest-of-story-author-thrown-off-of_30.html.

[208] Shetty KD, DeLeire T, White C, Bhattacharya J. "Changes in U.S. hospitalization and mortality rates following smoking bans," *Journal of Policy Analysis and Management*, Winter 2011, Volume 30, Issue 1, pp. 6-28. dx.doi.org/10.1002/pam.20548.

[209] http://www.no-smoke.org/pdf/SHSBibliography.pdf.

[210] Ellis R. "The Secondhand Smoking Gun," *New York Times*, October 15, 2003. http://www.nytimes.com/2003/10/15/opinion/the-secondhand-smoking-gun.html.

[211] Editors. "Of Smoking Bans and Heart Attacks," *New York Times*, April 27, 2004. http://www.nytimes.com/2004/04/27/opinion/of-smoking-bans-and-heart-attacks.html.

[212] Kaufman M. "Secondhand Smoke Poses Heart Attack Risk, CDC Warns," *Washington Post*, April 23, 2004, p. A01.

[213] Shetty K, DeLeire T, White C, Bhattacharya J. "Changes in U.S. hospitalization and mortality rates following smoking bans," *Journal of Policy Analysis and Management*, Winter 2011, Volume 30, Issue 1, pp. 6-28. dx.doi.org/10.1002/pam.20548.

[214] Marlow ML. "Smoking bans and acute myocardial infarction incidence," *Applied Economics Letters*, January 2012, Volume 19, Issue 16, pp. 1577-1581. dx.doi.org/10.1080/13504851.2011.639730.

[215] Siegel M. "Anti-Smoking Researchers Argue that Mathews Study Shows Significant Effect of Smoking Bans on Acute MI; Lack of Scientific Rigor Apparent," *The Rest of the Story*, August 15, 2011. http://tobaccoanalysis.blogspot.com/2011/08/anti-smoking-researchers-argue-that.html.

[216] Siegel M. "New Study Finds No Significant Decline in Heart Attack Mortality During the First Year in Six States with New Smoking Bans from 1995 to 2003," *The Rest of the Story*, September 14, 2011. http://tobaccoanalysis.blogspot.com/2011/09/new-study-finds-no-significant-decline.html.

[217] Park A. "How Secondhand Cigarette Smoke Changes Your Genes," *Time.com*, August 20, 2010. http://www.time.com/time/health/article/0,8599,2012103,00.html.

[218] Otsuka R, Watanabe H, *et al*. "Acute Effects of Passive Smoking on the Coronary Circulation in Healthy Young Adults," *Journal of the American Medical Association*, July 25, 2001, Volume 286, Number 4. dx.doi.org/10.1001/jama.286.4.436.

[219] Nagda N, Koontz M, *et al*. "Measurement of cabin air quality aboard commercial airliners," *Atmospheric Environment*, Part A, General Topics, August 1992, Volume 26, Issue 12, pp. 2203-2210. dx.doi.org/10.1016/j.bbr.2011.03.031.

[220] Giannini D, Leone A. "The Effects of Acute Passive Smoke Exposure on Endothelium-Dependent Brachial Artery Dilation in Healthy Individuals," *Angiology*, April 2007, Volume 58, Number 2, pp. 211-217. dx.doi.org/10.1177/0003319707300361.

[221] Stacy K. "A Few Whiffs of Smoke May Harm Your Heart ," *WebMD.com*, September 1, 2009. http://www.cbsnews.com/stories/2009/09/01/health/webmd/main5279799.shtml.

[222] Lavi T, Karasik A, *et al*. "The Acute Effect of Various Glycemic Index Dietary Carbohydrates on Endothelial Function in Nondiabetic Overweight and Obese Subjects" *Journal of the American College of Cardiology*, June 16, 2009, Volume 53, Number 24, pp.2283-2287. dx.doi.org/10.1016/j.jacc.2009.03.025.

[223] Siegel M. "Eating Corn Flakes Found to Cause Endothelial Dysfunction," *The Rest of the Story*, July 14, 2009. http://tobaccoanalysis.blogspot.com/2009/07/eating-corn-flakes-causes-endothelial.html.

[224] Siegel M. "Single Bowl of Corn Flakes Causes Heart Damage Similar to That of Habitual Smokers, According to Reasoning of ANR and ASH," *The Rest of the Story*, October 28, 2010. http://tobaccoanalysis.blogspot.com/2010/10/single.html.

[225] McFadden, MJ. *Dissecting Antismokers' Brains*, Aethna Press, 2004, p. 259.

[226] Winickoff JP, *et al*. "Beliefs About the Health Effects of 'Thirdhand' Smoke and Home Smoking Bans," *Pediatrics*, January 1, 2009, Volume 123, Number 1, pp. e74-e79. dx.doi.org/10.1542/peds.2008-2184.

[227] Rabin RC, "A New Cigarette Hazard: 'Third-Hand Smoke'," *New York Times*, January 2, 2009. http://www.nytimes.com/2009/01/03/health/research/03smoke.html.

228 Ballantyne C. "What is third-hand smoke? Is it hazardous? Researchers warn cigarette dangers may be even more far-reaching," *Scientific American,* January 6, 2009. http://www.scientificamerican.com/article.cfm?id=what-is-third-hand-smoke.

229 "Third-Hand Smoke Danger," *Tyrone Times* (N. Ireland), January 6, 2009. http://www.tyronetimes.co.uk/news/health/third-hand-smoke-danger-1-4841349.

230 "Third-Hand Smoke Danger," *Bedford Today* (England), January 6, 2009. http://bedfordtoday.co.uk/news/health/third-hand-smoke-danger-1-4841349.

231 Tumwine J. "Thirdhand Smoke," *Global Health Law,* January 11, 2009. http://globalhealthlaw.wordpress.com/2009/01/11/third-hand-smoke.

232 http://www.nhs.uk/news/2009/01January/Pages/Thirdhandsmoke.aspx.

233 Sullum J. "Thirdhand Smoke Alarm," *Hit And Run,* January 6, 2009. http://reason.com/blog/2009/01/06/thirdhand-smoke-alarm.

234 Siegel M. "Author of Thirdhand Smoke Study Warns Smokers are Contaminated and Emit Toxins; Suggests that Thirdhand Smoke Causes Lead Poisoning," *The Rest of the Story,* January 20, 2009. http://tobaccoanalysis.blogspot.com/2009/01/author-of-thirdhand-smoke-study-warns.html.

235 Kabat, GC. "Is Thirdhand Smoke A Valid Scientific Concept Or A Public Relations Gimmick?" *Columbia University Press Blog,* January 9, 2009. http://www.cupblog.org/?p=493.

236 Feinstein AR. "Justice, Science, and the 'Bad Guys'," *Toxicologic Pathology,* February 1, 1992, Volume 20, Number 2, pp. 303-305. dx.doi.org/10.1177/019262339202000217.

237 Beigs. "Propaganda in the New York Times: The Sinking of the Lusitana," *Everything2.com,* August 3, 2002. http://everything2.com/title/Propaganda+in+the+New+York+Times%253A+The+Sinking+of+the+Lusitania.

238 New Jersey Group Against Smokers Pollution. (2013: Headlines reduced to body text.) http://www.njgasp.org/ths.htm.

239 Rehan V. "Thirdhand smoke: a new dimension to the effects of cigarette smoke on the developing lung," *American Journal of Physiology, Lung Cellular and Molecular Physiology,* July 2011, Volume 301, Issue 1, pp. L1-L8. dx.doi.org/10.1152/ajplung.00393.2010.

240 Winickoff JP, et al. "Beliefs About the Health Effects of 'Thirdhand' Smoke and Home Smoking Bans," *Pediatrics,* January 1, 2009, Volume 123, Number 1, pp. e74-e79. dx.doi.org/10.1542/peds.2008-2184.

241 Matt GE, Quintana PJ, Hovell MF, et al. "Households contaminated by environmental tobacco smoke: sources of infant exposures," *Tobacco Control,* March 2004, Volume 13, Issue 1, pp. 29-37. dx.doi.org/10.1136/tc.2003.003889.

242 "Babies 'suffer third hand smoke'," *Irish Examiner,* August 7, 2006. http://irishexaminer.com/breakingnews/world/babies-suffer-third-hand-smoke-271088.html.

243 Snowdon C. "Beyond Belief," *Velvet Glove, Iron Fist,* 2009. http://www.velvetgloveironfist.com/thirdhandsmoke.php.

244 Sleiman L. *Proceedings of the National Academy of Sciences of the United States,* April 13, 2010, Volume 107, Number 15, pp. 6576-6581. dx.doi.org/10.1073/pnas.0912820107.

245 Sebelius K. *Report on Carcinogens,* U.S. Department of Health and Human Services, Public Health Service National Toxicology Program, 12th Edition, 2011. http://ntp.niehs.nih.gov/ntp/roc/twelfth/profiles/TobaccoRelatedExposures.pdf.

[246] Jarvis D, Leaderer B, Chinn S, Burney P. "Indoor nitrous acid and respiratory symptoms and lung function in adults," *Thorax*, June 2005, Volume 60, Issue 6, pp. 474-479. dx.doi.org/10.1136/thx.2004.032177.

[247] Lee K, Xue J, *et al.* "Nitrous acid, nitrogen dioxide, and ozone concentrations in residential environments," *Environmental Health Perspectives*, February 2002, Volume 110, Issue 2, pp. 145-150. http://www.ncbi.nlm.nih.gov/pubmed/11836142.

[248] Hohenstein Institute. "When Baby Smokes Too!" September 22, 2010. http://www.njgasp.org/hohenstein_when_baby_smokes_too-9-2010.pdf.

[249] Banzhaf, JFIII. "Tobacco Smoke Residue Causes 'Massive Damage' in Babies' Skin - New Study," October 5, 2010. *ASH.org*. http://www.prlog.org/10976443-tobacco-smoke-residue-causes-massive-damage-in-babies-skin-new-study.html.

[250] Siegel M. "Stop and Think About This: Anti-Smoking Groups are Telling the Public that Touching a Smoker's Clothes Can Cause Massive Skin and Neurological Damage," *The Rest of the Story*, October 14, 2010. http://tobaccoanalysis.blogspot.com/2010/10/stop-and-think-about-this-anti-smoking.html.

[251] Vickij592000@yahoo.com. Posted board comment, 02:49 AM, April 2, 2009. http://www.sciam.com/article.cfm?id=what-is-third-hand-smoke#comments.

[252] Hadro M. "CNN Anchor Shamelessly Lauds Mayor Bloomberg's Newest Smoking Ban," May 24, 2011. http://newsbusters.org/blogs/matt-hadro/2011/05/24/cnn-anchor-shamelessly-lauds-mayor-bloombergs-newest-smoking-ban.

[253] http://www.kssmokefree.org/download/Law_HB2221.pdf.

[254] http://cosweb.cityofshawnee.org/web/book.nsf/0fec651ce7cb4f5286256fc8007 57582/6a14334 b37d56abe8625769d0052f80b?OpenDocument.

[255] http://ci.somerville.ma.us/sites/default/files/SomervilleSmokingRegs03-11-13.pdf.

[256] http://www.ncga.state.nc.us/enactedlegislation/statutes/html/bysection/chapter _130a/gs_130a-492.html.

[257] http://www.thompsonhine.com/publications/publication946.html.

[258] http://web.archive.org/web/20101125042116/http://montgomerycountync.com/online_forms/Ordinances/nonsmoking_draft.pdf.

[259] Cheek J. "Don't Buy The Ventilation Lie," *Burning Issues*, July 2001, Volume 4, Number 1, p. 9. http://www.trdrp.org/newsletter/2001/Nslttr701.pdf.

[260] Repace JL. "A Killer On The Loose," http://repace.com/pdf/killer1.pdf.

[261] Boucher P. "Rendez-vous with James Repace," *Tobacco.org*, April 26, 2000. http://www.tobacco.org/News/rendezvous/repace.html.

[262] http://www.repace.com/Repace-CV.pdf.

[263] Repace JL. "Indoor and Outdoor Pollution Carcinogen Pollution on a Cruise Ship," Address at the 14th Annual Conference of the International Society of Exposure Analysis, *Repace.com*, October 2004. http://www.repace.com/pdf/ISEA_2004_W1B07_Repace.pdf.

[264] Klepeis NE, Ott W, Switzer P. "Real-Time Measurement of Outdoor Tobacco Smoke Particles," *Journal of the Air & Waste Management Association*, February 29, 2007, Volume 57, Issue 5, pp. 522-534. dx.doi.org/10.3155/1047-3289.57.5.522.

[265] Ibid.

[266] Hall JC, Bernert JT, Hall DB, *et al.* "Assessment of exposure to secondhand smoke at outdoor bars and family restaurants in Athens, Georgia, using salivary cotinine," *Journal of Occupational and Environmental Hygiene,* November 2009, Volume 6, Issue 11, pp. 698-704. dx.doi.org/10.1080/15459620903249893.

[267] Benowitz NL, Jacob P. "Metabolism of nicotine to cotinine studied by a dual stable isotope method," *Clinical Pharmacology and Therapeutics,* November 1994, Volume 56, Issue 5, pp. 483-493. dx.doi.org/10.1038/clpt.1994.169.

[268] Hall JC, Bernert JT, Hall DB, *et al.* "Assessment of exposure to secondhand smoke at outdoor bars and family restaurants in Athens, Georgia, using salivary cotinine," *Journal of Occupational and Environmental Hygiene,* November 2009, Volume 6, Issue 11, pp. 698-704. dx.doi.org/10.1080/15459620903249893.

[269] AllExperts. "Paranormal Phenomena," *AllExperts.com,* January 22, 2009. http://en.allexperts.com/q/Paranormal-Phenomena-3278/2009/1/Smelling-Cigarette-Smoke.htm.

[270] Tominey C. "Mystery of India's 'cigarette smoking' saint," *Daily and Sunday Express* (England), March 17, 2013. http://www.express.co.uk/news/uk/384855/Mystery-of-India-s-cigarette-smoking-saint.

[271] Carter G. "A Man Who Knows Everything," *Austin Chronicle* (Letter), June 3, 2003. http://www.austinchronicle.com/columns/2003-06-13/163631/.

[272] Chapman K. "Hancock Is On The Right Track," January 16, 2007. http://ken-chapman.blogspot.com/2007/01/hancock-is-on-right-track.html.

[273] Signore JD. "Full Smoking Ban in Parks Stubbed Out by Bloomberg," *Gothamist,* September 16, 2009. http://gothamist.com/2009/09/16/full_smoking_ban_in_parks_stubbed_o.php.

[274] Stutts J, Reinfurt D, *et al.* "The Role Of Driver Distraction In Traffic Crashes," *AAA Foundation for Traffic Safety,* May 2001, p. 4. http://safedriver.gr/data/84/distraction_aaa.pdf.

[275] http://www.nsc.org/news_resources/Resources/Documents/The role of driver distraction in traffic crashes.pdf.

[276] Offermann FJ, Colfer R, *et al.* "Exposure To Environmental Tobacco Smoke In An Automobile," *Proceedings of the 9th International Conference on Indoor Air Quality and Climate,* Monterey, CA, June 30 - July 5, 2002, Paper Number 2C3p1, p. 2002, 506. http://tobacco.cleartheair.org.hk/wp-content/uploads/2008/03/ets-exposure-automobile.pdf.

[277] Ott W, Klepeis NE, Switzer P. "Air change rates of motor vehicles and in-vehicle pollutant concentrations from secondhand smoke," *Journal of Exposure Science and Environmental Epidemiology,* May 2008, Volume 18, Issue 3, pp. 312-325. dx.doi.org/10.1038/sj.jes.7500601.

[278] MacKensie R, Freeman B. "Second-hand smoke in cars: How did the '23 times more toxic' myth turn into fact?" *Canadian Medical Association Journal,* May 18, 2010, Volume 182, Number 8, pp. 796-799. dx.doi.org/10.1503/cmaj.090993.

[279] http://www.epa.gov/ttn/caaa/t1/memoranda/pmfinal.pdf, p. 42/47.

[280] Jones MR, Navas-Acien A, *et al.* "Secondhand tobacco smoke concentrations in motor vehicles: a pilot study," *Tobacco Control,* October 2009, Volume 18, Issue 5, pp. 399-404. dx.doi.org/10.1136/tc.2009.029942.

[281] Johns Hopkins. "Secondhand Smoke Levels Higher in Cars than in Bars or Restaurants," *Johns Hopkins Bloomberg School of Public Health,* August 25, 2009. http://www.jhsph.edu/news/news-releases/2009/navas-acien-car-smoke.html.

[282]   Jones MR, Navas-Acien A, *et al.* "Secondhand tobacco smoke concentrations in motor vehicles: a pilot study," *Tobacco Control,* October 2009, Volume 18, Issue 5, pp. 399-404. dx.doi.org/10.1136/tc.2009.029942.

[283]   Nebot M, Lopez MJ, Gorini G, *et al.* "Environmental tobacco smoke exposure in public places of European cities," *Tobacco Control,* February 2005, Volume 14, Issue 1, pp. 60-63. dx.doi.org/10.1136/tc.2004.008581.

[284]   http://www.famri.org/researchers/resources/CIA_RFA_2009.pdf.

[285]   Wilson F, Stimpson J. "Trends in Fatalities From Distracted Driving in the United States, 1999 to 2008," *American Journal of Public Health,* November 2010, Volume 100, Number 11, pp. 2213-2219. dx.doi.org/10.2105/AJPH.2009.187179.

[286]   Science Daily. "Secondhand Smoke Linked To Risk Of Tooth Loss," *ScienceDaily.com,* April 4, 2007. http://www.sciencedaily.com/releases/2007/04/070403153859.htm.

[287]   Mogell KA. "Periodontal Disease and secondhand smoke," *DrMogell.com,* June 5, 2007. http://www.drmogell.com/blog.htm.

[288]   Sims, J. "Secondhand Smoke Harms Children's Health," *DeltaDental.com,* December 13, 2010. http://oralhealth.deltadental.com/Search/22,21375.

[289]   MedlinePlus. "Secondhand Smoke," *National Institutes of Health,* http://www.nlm.nih.gov/medlineplus/secondhandsmoke.html.

[290]   Nogueira-Filho G, Rosa BT, César-Neto JB, *et al.* "Low- and High-Yield Cigarette Smoke Inhalation Potentiates Bone Loss During Ligature-Induced Periodontitis," *Journal of Periodontology,* April 2007, Volume 78, Number 4, pp. 730-735. dx.doi.org/10.1902/jop.2007.060323.

[291]   Rehan V. "Thirdhand smoke: a new dimension to the effects of cigarette smoke on the developing lung," *American Journal of Physiology, Lung Cellular and Molecular Physiology,* July 2011, Volume 301, Issue 1, pp. L1-L8. dx.doi.org/10.1152/ajplung.00393.2010.

[292]   UPI. "Thirdhand Smoke Hurts Infant Lungs," *UPI.com,* April 19, 2011. http://upi.com/Health_News/2011/04/19/Thirdhand-smoke-hurts-infant-lungs/UPI-49871303262530.

[293]   Mandel H. "Unborn babies at risk from third-hand smoke," *Examiner.com,* April 20, 2011. http://www.examiner.com/article/unborn-babies-at-risk-from-third-hand-smoke.

[294]   California State News. "Thirdhand Smoke Dangerous to Unborn Babies' Lungs," *california.statenews.net,* April 20, 2011. http://california.statenews.net/story/771641.

[295]   India Times. "Thirdhand Smoke Affects Infant's Lungs," *Indiatimes.com,* May 12, 2011. http://articles.timesofindia.indiatimes.com/2011-05-12/health/29450221_1_smoke-lung-childhood-exposure.

[296]   ModernPregnancyTips. "Thirdhand Smoke Can Damage Unborn Babys Lungs," *modernpregnancytips.com.* http://www.modernpregnancytips.com/pregnancy-health/third-hand-smoke-can-damage-unborn-babys-lungs.

[297]   ScienceDaily. "'Thirdhand Smoke' Poses Danger to Unborn Babies' Lungs, Study Finds," *ScienceDaily.com,* April 19th, 2011. http://sciencedaily.com/releases/2011/04/110419101231.htm.

[298]   Radowitz JV. "Passive Smoking Blood Pressure Risk," *The Independent* (England), May 2, 2011. http://www.independent.co.uk/life-style/health-and-families/health-news/passive-smoking-blood-pressure-risk-2277692.html.

[299]   Jha A. "Passive smoking raises blood pressure in boys, study reveals," *The Guardian,* May 1, 2011. http://www.guardian.co.uk/science/2011/may/01/passive-smoking-blood-pressure-boys.

[300] Tweed C. "Second Hand Smoke Could Lead to Hypertension among Boys," *TopNews.us*, May 3, 2011. http://topnews.us/content/239492-second-hand-smoke-could-lead-hypertension-among-boys.

[301] Mirror. "Second-hand smoke can raise boys' blood pressure and cause heart disease," *Mirror.co.uk*, May 2, 2011. http://www.mirror.co.uk/news/health-news/2011/05/02/second-hand-smoke-can-raise-boys-blood-pressure-and-cause-heart-disease-115875-23101325/.

[302] Baumgartner J, Witt W, et al. "Environmental Tobacco Smoke Exposure and Blood Pressure in Children and Adolescents: Results from the 1999-2006 National Health and Nutrition Examination Survey (NHANES)," *[2805.5] Platform Session: Environmental /International Epidemiology*, May 1, 2011.

[303] King J. "Second hand smoke higher risk to boys," *imperfectparent.com*, May 2, 2011. http://www.imperfectparent.com/topics/2011/05/02/second-hand-smoke-higher-risk-to-boys/.

[304] Siegel M. "Study Finds Secondhand Smoke Associated With Higher Blood Pressure in Boys, But Lower Blood Pressure in Girls; Concludes Effect is Real Only for Boys," *The Rest of the Story*, May 2, 2011. http://tobaccooanalysis.blogspot.com/2011/05/study-finds-secondhand-smoke-associated.html.

[305] Simonetti GD, Schwertz, R, *et al.* "Determinants of Blood Pressure in Pre-school Children The Role of Parental Smoking," *Circulation*, January 25, 2001, Volume 123, Issue 3. dx.doi.org/10.1161/CIRCULATIONAHA.110.958769.

[306] Biliuti S. "Smoking Exposure Early in Life Causes High Blood Pressure," *Softpedia.com*, January 11, 2011. http://news.softpedia.com/news/Smoking-Exposure-Early-in-Life-Causes-High-Blood-Pressure-177464.shtml.

[307] http://www.trdrp.org/fundedresearch/grant_page.php?grant_id=7983.

[308] Henriksen L, Schleicher NC. "Targeted Advertising, Promotion, and Price For Menthol Cigarettes in California High School Neighborhoods," *Nicotine And Tobacco Research*, June 2011, Volume 14, Issue 1, pp. 116-121. http://ntr.oxfordjournals.org/content/14/1/116. dx.doi.org/10.1093/ntr/ntr122.

[309] Henderson P. "Newport ads target black youth - Stanford study," *Reuters*, June 24, 2011. http://www.reuters.com/article/2011/06/24/tobacco-lorillard-idUSN1E75M1XL20110624.

[310] Strulovici-Barel Y, Omberg L, *et al.* "Threshold of Biologic Responses of the Small Airway Epithelium to Low Levels of Tobacco Smoke," *American Journal of Respiratory and Critical Care Medicine*, December 2010, Volume 182, Issue 12, pp. 1524-1532. http://ajrccm.atsjournals.org/content/182/12/1524.full.

[311] Park A. "How Secondhand Cigarette Smoke Changes Your Genes," *Time.com*, August 20, 2010. http://www.time.com/time/health/article/0,8599,2012103,00.html.

[312] Phys.org. "Cigarette smoke causes harmful changes in the lungs even at the lowest levels," *Phys.org*, August 20, 2010. http://phys.org/news201461912.html.

[313] Lalwani A, Ying-Hua L, Weitzman M. "Secondhand Smoke and Sensorineural Hearing Loss in Adolescents" *Archives of Otolaryngology – Head and Neck Surgery*, July 2011, Volume 137, Issue 7, pp. 655-662. dx.doi.org/10.1001/archoto.2011.109.

[314] Hart A. "Newspapers aren't warning young people about possibly going deaf from smoking," *Examiner.com*, September 22, 2011. http://examiner.com/article/newspapers-aren-t-warning-young-people-about-possibly-going-deaf-from-smoking.

315 "Daddy Don't Smoke" *Deccan Chronicle* (India), Aug. 4, 2010. http://web.archive.org/web/20100804094549/http://deccanchronicle.com/chennai/daddy-don%E2%80%99t-smoke-439.

316 "Daddy Don't Smoke," *Health Life*, August 6, 2010. http://smoking-quit.info/daddy-don't-smoke.

317 University of California, Riverside. "UC Riverside Receives Six Grants for Tobacco-related Research," *Press Release*, August 4, 2010. http://newsroom.ucr.edu/news_item.html?action=page&id=2403.

318 Siegel M. "California Taxpayers to Spend a Quarter Million Dollars Studying Effects of Thirdhand Smoke Dust on Skin," *The Rest of the Story*, August 9, 2010. http://tobaccoanalysis.blogspot.com/2010/08/california-taxpayers-to-spend-quarter.html.

319 University of California. "University of California, Riverside Receives Six Grants for Tobacco-related Research," *Press Release*, August 4, 2010. http://newsroom.ucr.edu/news_item.html?action=page&id=2403.

320 Ebright O. "2.5M California Children Plagued by Secondhand Smoke," *NBC 4* (Southern California), October 28, 2011. http://www.nbclosangeles.com/news/health/Secondhand-Smoke-132808978.html.

321 Driscoll G. "Nearly one million California children still at risk of secondhand smoke exposure," *HealthPolicy.ucla.edu.* http://web.archive.org/web/20111230163558/http://healthpolicy.ucla.edu/NewsReleaseDetails.aspx?id=88.

322 Ebright O. "Secondhand Smoke Study Flawed; New Data Posted," *NBC 4* (Southern California) November 2, 2011. http://www.nbclosangeles.com/news/health/California-Children-Secondhand-Smoke-133118288.html.

323 Wikipedia. "Four Horsemen of the Apocalypse," *Wikipedia.org.* http://en.wikipedia.org/wiki/Four_Horsemen_of_the_Apocalypse#Pestilence.2C_War.2C_Famine.2C_and_Death.

324 Lerner M. "New Study: Minneapolis Smoking ban cut air pollution by 99%," *ASH/Star-Tribune* (Minneapolis, MN), September 16, 2005. http://no-smoking.org/sept05/09-16-05-3.html.

325 Lerner M. "New Study: Minneapolis Smoking ban cut air pollution by 99%," *DCS/Star-Tribune* (Minneapolis, MN), September 16, 2005. http://www.discount-cigars-store.com/news/study_smoking_ban_cut_air_pollution_by_99_.html.

326 http://www.nycclash.com/OSHAaction.html.

327 Gori GB. *Virtually Safe Cigarettes: Reviving an Opportunity Once Tragically Rejected,* Ios Pr Inc, 2000.

328 http://www.pbs.org/wgbh/nova/body/safer-cigarettes-history.html.

329 http://casaa.org/E-cigarette_History.html.

330 Lipton E, Barboza D. "As More Toys Are Recalled, Trail Ends In China," *New York Times*, June 19, 2007. http://nytimes.com/2007/06/19/business/worldbusiness/19toys.html?pagewanted=all.

331 Lee M. "Cadbury pulls melamine-laced chocolate made in China," *USA Today*, September 29, 2008. http://usatoday30.usatoday.com/money/industries/food/2008-09-29-cadbury-chocolate-recall_N.htm.

332 Dautzenberg B, Birkui P, Noël M, *et al.* "E-Cigarette: A New Tobacco Product for Schoolchildren in Paris," *Open Journal of Respiratory Diseases*, February 2013, Volume. 3, Number 1, pp. 21-24. dx.doi.org/10.4236/ojrd.2013.31004.

333 Etter JF. "Electronic cigarettes: a survey of users," *BMC Public Health*, May 4, 2010, Volume 10, Article 231. dx.doi.org/10.1186/1471-2458-10-231.

[334] Stepanov I, Jensen J, Hatsukami D, Hecht SS. "Tobacco-specific nitrosamines in new tobacco products," *Nicotine & Tobacco Research*, April 2006, Volume 8, Number 2, pp. 309-313. http://www.ncbi.nlm.nih.gov/pubmed/16766423#.

[335] http://www.fda.gov/downloads/Drugs/ScienceResearch/UCM173250.pdf.

[336] Material Safety Data Sheet – Acetylsalicylic acid. http://fscimage.fishersci.com/msds/00300.htm.

[337] Schripp T, Markewitz D, *et al.* "Does e-cigarette consumption cause passive vaping?" *Indoor Air*, Volume 23, Issue 1, pp. 25-31. dx.doi.org/10.1111/j.1600-0668.2012.00792.x.

[338] Siegel M, Cahn Z. "Electronic cigarettes as a harm reduction strategy for tobacco control: A step forward or a repeat of past mistakes?" *Journal of Public Health Policy*, January 2011, Volume 32, Issue 1, pp. 16-31. dx.doi.org/10.1057/jphp.2010.41.

[339] McNeill C. "Enjoy indoor fun on National Stay Out of the Sun Day," *Examiner.com*, July 2, 2013. http://www.tampabay.com/news/humaninterest/enjoy-indoor-fun-on-national-stay-out-of-the-sun-day/2129597.

[340] http://www.njgasp.org/E-Cigs_White_Paper.pdf.

[341] Atkinson J. "Safer Alternative To Cigarettes To Be Banned by EU," *Huffington Post*, March 7, 2013. http://www.huffingtonpost.co.uk/janice-atkinson/safer-alternative-to-cigarettes-banned-by-eu_b_2827043.html.

[342] Gagnon D. "Man With Electronic Cigarette Not Charged In Flight Diversion," *Bangor Daily News* (Bangor, ME), January 27, 2011. https://bangordailynews.com/2011/01/27/news/bangor/man-with-electronic-cigarette-not-charged-in-flight-diversion/.

[343] http://www.icyte.com/saved/www.trdrp.org/494553.

[344] Evangelista A. "Thirdhand Smoke Health Threat Suspected," *Your University of California*, October 2010. http://www.universityofcalifornia.edu/youruniversity/archive/2010/october/thirdhand-smoke-health-threat-suspected.html.

[345] Register K. "Cigarette Butts As Litter—Toxic As Well As Ugly," *Underwater Naturalist—Bulletin of the American Littoral Society*, Volume 25, Number 2, August 2000.

[346] Slaughter E, *et al.* "Toxicity of cigarette butts, and their chemical components, to marine and freshwater fish," *Tobacco Control*, 2011, Volume 20, Issue Supplement 1, pp. i25-i29. dx.doi.org/10.1136/tc.2010.040170.

[347] Angell M. "Drug Companies & Doctors: A Story of Corruption," *The New York Review of Books*, January 15, 2009. http://www.nybooks.com/articles/archives/2009/jan/15/drug-companies-doctorsa-story-of-corruption/?page=2

[348] Kuntz T. "Word for Word/The National Smokers Alliance; Got a Light? How About The Flame of Freedom?" *New York Times*, Sept. 21, 1997. http://nytimes.com/1997/09/21/weekinreview/word-for-word-national-smokers-alliance-got-light-about-flame-freedom.html.

[349] Wikipedia. "National Smokers Alliance," *Wikipedia.org.* http://en.wikipedia.org/wiki/National_Smokers_Alliance.

[350] American Medical Association. *Report 14 Of The Board Of Trustees (A-01) Annual Tobacco Report (Informational)*, June 2001, p. 123. http://web.archive.org/web/20040421083003/http://www.ama-assn.org/ama1/upload/mm/interim01/BOT_Reports_1-17.pdf.

[351] Siegel R. "Activists protest plan to cut off anti-smoking programs," *Asbury Park Press* (Asbury Park, NJ), January 8, 2002.

[352] Zwillich T. "Congressional Overhaul Of National Cancer Effort Recommended," *Reuters Health* (Washington), October 11, 2001.

[353] American Medical Association. *Report 14 Of The Board Of Trustees (A-01) Annual Tobacco Report (Informational)*, June 2001, p. 123. http://web.archive.org/web/20040421083003/http://www.ama-assn.org/ama1/upload/mm/interim01/BOT_Reports_1-17.pdf.

[354] Peterson M. "Tobacco Policies 'Abysmal' in US States, Lung Group Says" *Bloomberg Business Week*, January 23, 2012. http://www.bloomberg.com/news/2012-01-19/tobacco-policies-abysmal-in-u-s-states-lung-association-says.html.

[355] Proctor R. *The Nazi War On Cancer*, Princeton University Press, 2000.

[356] Proctor R. "Puffing On Polonium," *New York Times*, December 1, 2006. http://www.nytimes.com/2006/12/01/opinion/01proctor.html.

[357] Pianezza ML, Sellers E, Tyndale R. "Nicotine metabolism defect reduces smoking," *Nature*, June 25, 1998, Volume 393, Number 750. dx.doi.org/10.1038/31623.

[358] Fletcher JM. "Why Have Tobacco Control Policies Stalled? Using Genetic Moderation to Examine Policy Impacts," *PLOS One*. dx.doi.org/10.1371/journal.pone.0050576.s005.

[359] Schrand JR. "Does insular stroke disrupt the self-medication effects of nicotine?" *Medical Hypotheses*, September 20, 2010, Volume 75, Issue 3, pp. 302-304. dx.doi.org/10.1016/j.mehy.2010.03.009.

[360] Naqvi N, Rudrauf D, *et al.* "Damage to the Insula Disrupts Addiction to Cigarette Smoking," *Science*, January 26, 2007, Volume 315, Number 5811, pp. 531-534. dx.doi.org/10.1126/science.1135926.

[361] US Department of Health and Human Services. The Health Consequences of Involuntary Exposure to Tobacco Smoke, Report of the Surgeon General, 2006. p. 11.

[362] Ellin A. "Cleaning Up Baby Products," *New York Times*, May 5, 2009. http://www.nytimes.com/2009/05/28/fashion/28skinside.html?_r=0.

[363] McFadden MJ. "Secondary Smoke, Alcohol, and Deaths," *BMJ.com*, April 28, 2005. http://www.bmj.com/rapid-response/2011/10/30/secondary-smoke-alcohol-and-deaths.

[364] Turney E. "Pub crisis: 27 close each week," *MorningAdvertiser.co.uk*, March 5, 2008. http://www.morningadvertiser.co.uk/General-News/Pub-crisis-27-close-each-week.

[365] Ibid.

[366] Modern Brewery Age. "British pubs closing at rate of 39 a week," *TheFreeLibrary.com*, January 19, 2009. http://www.thefreelibrary.com/British+pubs+closing+at+rate+of+39+a+week.-a0209903347.

[367] British Beer And Pub Association. "Pub closures rise to record 52 a week," *BB&PA*, July 21, 2009. http://www.beerandpub.com/news/pub-closures-rise-to-record-52-a-week.

[368] Sunday Mirror. "Home truths about pubs," *Sunday Mirror* (England), January 1, 2008. http://www.highbeam.com/doc/1G1-173616344.html.

[369] Springen K. "States: Time to Stub Out Smoking," *Newsweek Magazine*, February 26, 2006. http://www.thedailybeast.com/newsweek/2006/02/26/states-time-to-stub-out-smoking.html.

[370] Action on Smoking and Health. "Cigarette Tax Facts, Including Intl, Historical, and State Tables," *ash.org.* http://ash.org/cigtaxfacts.html.

[371] Doyle K. "Public Transport Smokers Targeted," *Herald ie.* (Dublin, Ireland) June 3, 2008. http://www.herald.ie/news/public-transport-smokers-targeted-27874396.html.

[372] Cardwell D. "City Tries to Shut Club It Says Flouts Smoking Ban," *New York Times*, March 14, 2010. http://nytimes.com/2010/03/15/nyregion/15smoke.html.

[373] Lee JP, *et al*. "Unobtrusive Observations Of Smoking In Urban California Bars," *Journal of Drug Issues*, October 2003, Volume 33, Issue 4, pp. 983-999. dx.doi.org/10.1177/002204260303300410.

[374] Email correspondence with author.

[375] Tilkin D. "Smoking: The new child abuse?" *Katu.com*, April 27, 2006. http://archive.tobacco.org/news/222816.html.

[376] Kinsley M. "Let The Guy Smoke," *Washington Post* (Washington, D.C.), November 20, 2008, p. A23. http://www.washingtonpost.com/wp-dyn/content/story/2008/11/20/ST2008112001553.html.

[377] McFadden MJ. *Dissecting Antismokers' Brains*, Aethna Press, 2004, p. 333.

[378] Forsythe J. *Smoke-Free Outdoor Public Spaces: A Community Advocacy Toolkit*, Physicians for a Smoke-Free Canada, Ottawa, Ontario, September 2010, p. 33.

[379] http://greenbelt.patch.com/articles/witness-testimony-ends-in-secondhand-smoke-trial.

[380] http://icyte.com/saved/greenbelt.patch.com/544357 (Original page deleted September, 2012).

[381] http://icyte.com/saved/www.smokefreedc.org/538500 (Original page deleted September, 2012).

[382] http://icyte.com/saved/kansascity.com/616808 (Original page deleted September, 2012).

[383] http://icyte.com/saved/counselheal.com/611378.

[384] http://icyte.com/saved/galvestondailynews.com/616806 (Original page deleted September, 2012).

[385] http://icyte.com/saved/columbian.com/605920.

[386] Hitt D. "Name Three," *DaveHitt.com*, April 2004. http://davehitt.com/2004/name_three.html.

[387] Blakeslee S. "Nicotine: Harder To Kick ... Than Heroin," *New York Times*, April 5, 1987. http://www.nytimes.com/1987/03/29/magazine/nicotine-harder-to-kickthan-heroin.html.

[388] US Department of Health and Human Services. *Nicotine Addiction – A Report of the Surgeon General*, 1988.

[389] http://www.icyte.com/saved/www.theglobeandmail.com/591390

[390] Orwell G. *1984*, Signet Books, 1961.

[391] Email communication with author.

[392] BBC News. "Smokers Get Militant Over Ban," *BBC Online Network*, September 9, 1998. http://news.bbc.co.uk/2/hi/health/167762.stm.

[393] Assorted authors. "The Tobacco Endgame," *Tobacco Control*, May 2013, Volume 22, Issue Sup 1. http://tobaccocontrol.bmj.com/content/22/suppl_1.toc.

[394] http://celebratesmokefreecasinos.wordpress.com/join-the-celebration/.

[395] Harris JK, Carothers BJ, Luke DA, et al. "Exempting casinos from the Smoke-free Illinois Act will not bring patrons back: they never left," *Tobacco Control*, June 2011. dx.doi.org/10.1136/tc.2010.042127.

[396] Graves S. "Sara Summers Stein," *CapitolWord*, Volume 158, Number 92, p. e1057. http://capitolwords.org/date/2012/06/18/E1057-3_sara-summers-stein/.

[397] Clark M. "Investigation shows partial smoking ban on Atlantic City casino floors has little effect," *pressofAtlanticCity.com*, April 25, 2011. http://pressofatlanticcity.com/communities/atlantic-city_pleasantville_brigantine/investigation-shows-partial-smoking-ban-on-atlantic-city-casino-floors/article_5a00756e-6de0-11e0-afd1-001cc4c03286.html.

[398] http://Kickstarter.com.

[399] http://Indiegogo.com.

[400] FoodMattersMovie. "Pharmaceutical Drug Commercial Spoof," *FoodMattersMovie.com*, February 12, 2009. http://youtube.com/watch?v=yLR2OKesTw0.

[401] Olsen G. Confessions of an Rx Drug Pusher, iUniverse, 2009.

[402] Olsen G. "Inside Chemical Drug Corporations," *gwenolsen.com*, December 16, 2011. http://youtube.com/watch?v=Re12KzdZ4Wk.

[403] Rollason K. "Blue lights combat smoking in Cross Lake," *Winnipeg Free Press*, January 23, 2009. http://winnipegfreepress.com/local/-Blue-lights-help-combat-smoking-in-Cross-Lake38215029.html.

[404] Warner K. "An Endgame For Tobacco?" *Tobacco Control*, May 2013, Volume 22, Issue Supplement 1. dx.doi.org/10.1136/tobaccocontrol-2013-050989.

[405] Lyons A, McNeill A, Gilmore I, Britton J. "Alcohol imagery and branding, and age classification of films popular in the UK," *International Journal of Epidemiology*, Volume 40, Issue 5, pp. 1411-1419. dx.doi.org/10.1093/ije/dyr126.

[406] Surrey Today. "People with self-inflicted illnesses should move away from Surrey," *Surrey Today*, June 1, 2012. http://www.thisissurreytoday.co.uk/People-self-inflicted-illnesses-away-Surrey/story-16250767-detail/story.html.

[407] Editorial. "Imagining things otherwise: new endgame ideas for tobacco control," *Tobacco Control*, October 2010, Volume 19, Issue 5, pp. 349-350. dx.doi.org/10.1136/tc.2010.039727.

[408] World Health Organization Regional Office For Europe. "A debate on the 'end-game' of tobacco," December 12, 2012. http://www.euro.who.int/en/what-we-do/health-topics/disease-prevention/tobacco/news/news/2012/12/a-debate-on-the-end-game-of-tobacco.

[409] Chapman S. "The Case for a Smoker's License," *PLoS Medicine*, November 2012, Volume 9, Issue 11, e1001342. dx.doi.org/10.1371/journal.pmed.1001342.

[410] Email correspondence with author.

[411] Snowdon C. "Stanton's Spanking," Velvet Glove, Iron Fist, March 10, 2013. http://velvetgloveironfist.blogspot.co.uk/2013/03/stantons-spanking.html.

[412] Fallin A, Grana R, Glantz SA. "'To quarterback behind the scenes, third-party efforts': the tobacco industry and the Tea Party," *Tobacco Control*, February 8, 2013. dx.doi.org/10.1136/tobaccocontrol-2012-050815.

[413] DeMelle B. "Study Confirms Tea Party Was Created by Big Tobacco and Billionaire Koch Brothers," *Huffington Post*, February 11, 2013. http://huffingtonpost.com/brendan-demelle/study-confirms-tea-party-_b_2663125.html.

[414] Frezza B. "National Cancer Institute Funds Tea Party Witch Hunt," *Huffington Post*, February 19, 2013. http://huffingtonpost.com/bill-frezza/national-cancer-institute_1_b_2703072.html.

[415] http://www.ustream.tv/recorded/29747868.

[416] Snowdon C. "Stanton's Spanking," *Velvet Glove, Iron Fist*, March 10, 2013. http://velvetgloveironfist.blogspot.co.uk/2013/03/stantons-spanking.html.

[417] Glantz SA. "Another ill-conceived tobacco tax initiative emerging in California," *UCSF.edu*, March 10, 2013. http://tobacco.ucsf.edu/another-ill-conceived-tobacco-tax-initiative-emerging-california.

[418] Wikipedia. "List of smoking bans in the United States," *Wikipedia.org*. http://en.wikipedia.org/wiki/Smoking_bans_in_the_united_states.

[419] Tuffs A. "Full smoking ban is needed in Germany as study finds four out of five pubs flout regulations," *British Medical Journal*, May 14, 2011, Volume 342, Issue 7806. dx.doi.org/10.1136/bmj.d2864.

[420] Nainggolan, L. "Battling Big Tobacco: Physician activism vital on smoking's new frontiers," *TheHeart*, November 28, 2011. http://www.theheart.org/article/1318775.do.

[421] Breathe California. http://scenesmoking.org.

[422] Boortz N. "Cowardly Islamic Warriors," *Newsmax.com*, October 24, 2001.

[423] http://www.traditionalroofing.com/TR8_bits.html.

[424] http://michigandaily.com/article/smoking-ban-shows-success.

[425] Hamilton R. "UT Austin Bans Smoking On Campus," *kvia.com*, April 11, 2012. http://www.kvia.com/news/UT-Austin-Bans-Smoking-On-Campus/-/391068/15242966/-/mk718 mz/-/index.html.

[426] Hamilton R. "With Billions in Grant Money, Leverage to Curb Smoking," *New York Times*, February 18, 2012. http://www.nytimes.com/2012/02/19/health/texas-cancer-institute-uses-might-to-curb-campus-smoking.html.

[427] Luff R. "Smoking banned on campus of Shippensburg University—indoors and out," *Public Opinion*. http://www.publicopiniononline.com/ci_10474147.

[428] Heberlig D. "Shippensburg University students react to smoking ban," *The Sentinel* (Shippensburg, PA), September 19, 2008. http://cumberlink.com/news/local/shippensburg-university-students-react-to-smoking-ban/article_10a39454-444f-5acd-a255-020cbc02ca69.html.

[429] McFadden MJ. "The Lies Behind The Smoking Bans!" *Citizens Freedom Alliance*, 2013. http://kuneman.smokersclub.com/PASAN/StilettoGenv5h.pdf.

[430] New York Lawyer. "NY Bars Defeat Smoking Ban - Sort Of," *nylawyer.com*, October 26, 2004. http://web.archive.org/web/20041027203442/http://nylawyer.com/news/04/10/102604f.html.

[431] Rowland D. "Ohio SmokeFree Workplace Act Is Constitutional," *The Columbus Dispatch*,(Columbus, OH) May 23, 2012. http://dispatch.com/content/stories/local/2012/05/23/smoke-free-workplace-act-is-constitutional.html.

[432] Rand, A. *Atlas Shrugged*. Signet Books, NY, 1957, p. 411.

[433] McFadden MJ. "The Lies Behind The Smoking Bans!" *Citizens Freedom Alliance*, 2013. http://kuneman.smokersclub.com/PASAN/StilettoGenv5h.pdf.

[434] McFadden MJ. "Kill The Ban!" *Pennsylvania Smokers Action Network*, August 2007. http://kuneman.smokersclub.com/PASAN/KillTheBan_4c.pdf.

[435] Wikipedia. "Americans With Disabilities Act of 1990," *Wikipedia.org*. http://en.wikipedia.or g/wiki/Americans_with_Disabilities_Act_of_1990.

[436] Alinsky S. *Rules For Radicals*, Vintage Books, 1972.

[437] Duke D. *My Awakening*, Free Speech Press, 1998.

[438] Wikipedia. "*Si vis pacem, para bellum*," *Wikipedia.org*. http://wikipedia.org/wiki/Si_vis_pac em,_para_bellum.

[439] Wikipedia. "Sun Tzu," *Wikipedia.org*. http://wikiquote.org/wiki/Sun_Tzu.

[440] Friedman L. "Smoke blows other way in Capitol," *Daily News* (Los Angeles, CA), November 25, 2006. http://www.dailynews.com/ci_4719265.

In Memoriam

According to antismoking folklore, the author should have been dead several years before this book was written. So therefore I think it only proper to have a memorial page for myself. In the event that I am no longer extant, please take a moment to remember me and my fight throughout life for freedom, tolerance, and fairness. Thank you.

If I have passed away due to cancer, the statistical cause may be somewhat indeterminate. It may have been caused by drinking and smoking (perhaps a 20% chance), by my excessive consumption of chocolate milk (maybe a 15% chance), by my bicycling in traffic fumes (plus 15% or so), by my playing Ping-Pong in an asbestos-rich Brooklyn basement years ago or my working on extensive home renovations in a 1920s-era Philadelphia row-home for 30+ years (another 15%) or just from plain old nasty little cancer cells sprouting as part of normal cell-duplication randomness and/or nuclear testing radiation, fluoridations, little green aliens, and the NSA's micro-ultra-infra-waving surveillance of my comfortably subversive domicile (the last 30% of the odds). Oh. Wait. There's 5% left. Hmmm... might have come from living with a parrot for several years, or listening to Kenn Kweder play too much live music, or maybe the quiet squirrel in my kitchen.

If I have passed onward due to heart or circulatory problems, the root cause is probably the thousands of hours I've spent sitting here fighting for Free Choice through my keyboard. If I've been flattened by a car, the fault is 100% the car's. Vote for a $10/gallon tax increase.

Antismokers will naturally disagree with those figures and they are welcome to resume the stratistical argument with me when they pass through the pearly gates or (more likely) when they call me on a cell phone from Heaven's sub-basement and furnace room.

However, at the moment, as I'm typing this, I seem to still be here and feel fine. Given such evidence, please rest assured the news of my demise may be premature. Check your local listings for updates.

And keep on fighting!

Michael J. McFadden
July 4th, 2013

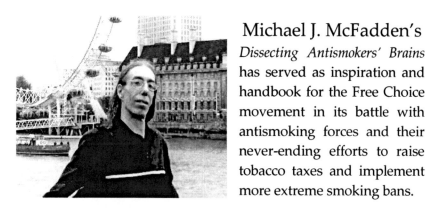

Michael J. McFadden's

Dissecting Antismokers' Brains has served as inspiration and handbook for the Free Choice movement in its battle with antismoking forces and their never-ending efforts to raise tobacco taxes and implement more extreme smoking bans.

TobakkoNacht – The Antismoking Endgame complements his earlier work by providing a toolbox, armory, and map for carrying the fight into the second decade of the 2000s.

Mr. McFadden grew up in Brooklyn, graduated *PBK* with honors from Manhattan College's Peace Studies and Psychology programs, and won a full fellowship to the Wharton School to study statistics and propaganda analysis.

He left Penn after two years to host a Quaker training center in non-violent activism and has worked at various levels in the areas of peace activism and social change ever since. His activities have ranged from ecologically-friendly transport advocacy to being formally commissioned by Queen Elizabeth II to plan and conduct nonviolence training workshops in Canada. He has worked as a lab tech, peace canvasser, online conference coordinator with CNN, and a free-lance book editor. He serves without compensation on the Boards of Directors of both The International Coalition Against Prohibition and FORCES International, and works with many grassroots Free Choice groups.

He currently lives and bicycles in West Philadelphia where he shares a small row house with a psychotic cat, a quiet squirrel, and assorted other creatures.

Brainy

As you turn the last page here, you'll see the cute little cartoon that greeted you when you first opened this book. You might wonder why it is, in such a serious book, that this little cartoon is given such prominence and also just where it came from.

I wrote *Dissecting Antismokers' Brains* almost ten years ago. It was an important book at the time for several reasons. First of all, it offered some serious questioning and criticism of the blatant misuse of science in the promotion of smoking bans, but secondly, and more importantly in my view, was its focus on the psychological aspect of the battle building between smokers and Antismokers.

Antismokers themselves were trying to deny the existence of the term as a self-description. They wanted the public to perceive them as simply being representative of ordinary day-to-day nonsmokers. It was fairly clear that it was a false claim at that point in history, but they knew that if they threw enough money and media into building the perception that they had a shot at making it stick. I saw the danger in that, and resented the almost Orwellian manipulation that it represented, and thus decided to not only use the antismoking term prominently in the title of my book but also to capitalize the noun itself whenever I used it.

I had the good fortune of connecting to an excellent and talented cover artist at the time, a man named Sam Ryskind. I had a vision for the cover of my book, one designed to grab attention in the dim light of bars under ban attacks, and one that would vault over the preconceptions that any book about smoking must naturally be against smoking, and that vision rested upon a pure white cover with just the three big black words of the title on it while my name would be in much smaller type at the bottom.

Sam's artistic mind rebelled against such sterility and he created Brainy, as I grew to think of his graphic, and urged that I use it on my cover. It was indeed an excellent conception for the book, both because of the psychological aspects already mentioned and because the first fifty pages of Brains was devoted purely to an examination of the psychological drives behind all the various types of Anti-smokers. Eventually, several years after first publication, I relented and put Brainy on my cover.

He continues to be of prime importance today. The brain games played through the media by antismoking advocates in the presentation of every study and news story about smoking you see or read are manipulative, dangerous, and as rotten as anything Big Tobacco ever dreamed up with their white-coated "doctors" telling us on TV how the lily-pure smoke from their cigarettes would soothe our throats and make us into athletes.

Today they've been replaced by lavishly-funded Government and quasi-government agencies run by people and groups with various mixtures of selfish or pseudo-idealistic motivations, and I believe these agencies and organizations represent as dangerous an attack on our freedoms as any military force in the world. Guns can be fought with guns, or even with rocks if necessary; but if our minds are perverted to the point where we don't realize we're losing our freedoms, then we are truly lost. If you haven't read *Brains* yet, do so. It's the foundation that *TobakkoNacht* was built on and the insights and tools it provides are essential for anyone hoping to take on antismoking professionals face-to-face. Together they make for a powerful weapon to fight the money machine of the Antismokers.

Sam Ryskind's Brainy is at the beginning and at the end of *TobakkoNacht* to remind all of us to resist those who would try to control our behaviors through manipulating our minds and emotions. It couples with the quote from Supreme Court Justice Douglas as both the closing of *Brains* and the opening of *TobakkoNacht* in recognizing how insidious and subtle that control can be, and the necessity of fighting its grasp before it's too late.

A democratic republic that allows its policies to be built on the basis of lies, and a citizenry that accepts those lies as being the norm, is a republic and a citizenry in very deep and serious trouble.

- Michael J. McFadden